MARY AND JOHN GRAY LIBRARY
LAMAR UNIVERSITY

Purchased
with the
Student Library Use Fee

Antenna-Based Signal Processing Techniques for Radar Systems

The Artech House Antenna Library

Helmut E. Schrank, *Series Editor*

Bhartia, Prakash, *Millimeter-Wave Microstrip and Printed Circuit Antennas*

Brookner, Eli, et al., *Practical Phased-Array Antenna Systems*

Corzine, Robert G., and Joseph A. Mosko, *Four-Armed Spiral Antennas*

Djordjevic, A.R., M.B. Bazdar, G.M. Bazdar, G.M. Vitosevic, T.K. Sarkar, and R.F. Harrington, *Analysis of Wire Antennas and Scatterers: Software and User's Manual*

Evans, Gary E., *Antenna Measurement Techniques*

Hansen, Robert C., ed., *Moment Methods in Antennas and Scattering*

Hirasawa, K., and M. Haneishi, *Analysis, Design, and Measurement of Small and Low-Profile Antennas*

Hirsch, Herbert L., and Douglas C. Grove, *Practical Simulation of Radar Antennas and Radomes*

Johnson, Richard C., *Designer Notes for Microwave Antennas*

Kitsuregawa, Takashi, *Advanced Technology in Satellite Communication Antennas: Electrical and Mechanical Design*

Kumar, A., *Fixed and Mobile Terminal Antennas*

Kumar, A., and H.D. Hristov, *Microwave Cavity Antennas*

Rohan, Paul, *Introduction to Electromagnetic Wave Propagation*

Scott, Craig, *Modern Methods of Reflector Antenna Analysis and Design*

Slater, D., *Near-Field Antenna Measurements*

Sletten, Carlye J., ed., *Reflector and Lens Antennas, Version 2.0: Analysis and Design Using Personal Computers*

Sletten, Carlyle J., ed., *Reflector and Lens Antennas: Software User's Manual and Example Book*

Weiner, M.M., et al., *Monopole Elements on Circular Ground Planes*

Wolff, Edward A., *Antenna Analysis, Second Edition*

Yamashita, E., ed., *Analysis Methods for Electromagnetic Wave Problems*

Antenna-Based Signal Processing Techniques for Radar Systems

Alfonso Farina
of
Radar and C^2 Division
of the Defense Systems Group
Alenia (Aeritalia & Selenia) SpA

Artech House
Boston • London

Library of Congress Cataloging-in-Publication Data

Farina, A. (Alfonso), 1948-
    Antenna-based ECCM techniques for radar systems / Alfonso Farina.
      p. cm.
    Includes bibliographical references and index.
    ISBN 0-89006-396-6
    1. Electronic counter-countermeasures. 2. Radar–Antennas.. I. Title.
UG485.F35   1991   91-30218
623'.7348–dc20   CIP

© 1992   ARTECH HOUSE, INC.
685 Canton Street
Norwood, MA 02062

All rights reserved. Printed and bound in the United States of America. No part of this book may be reproduced or utilized in any form or by any means, electronic or mechanical, including photocopying, recording, or by any information storage and retrieval system, without permission in writing from the publisher.

International Standard Book Number: 0-89006-396-6
Library of Congress Catalog Card Number: 91-30218

10 9 8 7 6 5 4 3 2 1

To Franca
with love

Contents

Foreword		xi
Preface		xiii
Chapter 1	Introduction to ECM and ECCM Techniques for Radar Systems	1
1.1	Electronic Warfare Taxonomy and Terminology	1
1.2	Electronic Countermeasures (ECM)	2
1.3	An Overview of ECCM Techniques	5
	1.3.1 Antenna-Related ECCM	6
	1.3.2 Transmitter-Related ECCM	8
	1.3.3 Receiver and Signal Processing-Related ECCM	9
1.4	Focus of the Book	10
Chapter 2	Low Sidelobe Antennas	13
2.1	Introduction	13
2.2	Relevance of Low and Ultra-Low Sidelobe Antennas in Radar Systems	14
2.3	Basic Principles to Achieve Low and Ultra-Low Sidelobe Antennas	19
	2.3.1 General Guidelines	19
	2.3.2 Aperture Tapering Functions for Low Sidelobes	21
2.4	Reflector Antennas with Low and Ultra-Low Sidelobes	22
2.5	Array Antennas with Low and Ultra-Low Sidelobes	25
2.6	Analysis of Random Error Effects on Average Sidelobe Level of Arrays	28
2.7	Digital Beam-Forming (DBF) Technique	34
	2.7.1 Basic Concepts	34
	2.7.2 Beam-Forming Techniques	38
	2.7.3 Analog-to-Digital Conversion Options and Related Processing Schemes	43

	2.7.4	Performance Limitations in a Digital Beam-Forming Network	48
	2.7.5	Calibration Procedures	55
	2.7.6	Implementation of DBF	56
	2.7.7	Existing DBF Systems	57
2.8	Concluding Remarks		58

Chapter 3   Sidelobe Blanking System — 59
- 3.1 Introduction — 59
- 3.2 Working Principle and Processing Schemes — 65
- 3.3 Performance Evaluation — 73
- 3.4 Design Methodology — 91
- 3.5 Concluding Remarks — 92

Chapter 4   Sidelobe Canceler (SLC) System — 95
- 4.1 Introduction and Working Principle of SLC — 95
- 4.2 Implementation Schemes — 103
  - 4.2.1 Closed-Loop Methods — 104
  - 4.2.2 Direct Solution Methods — 120
  - 4.2.3 Data-Domain Methods — 138
- 4.3 Performance Evaluation — 164
  - 4.3.1 Practical Limitations of SLC Nulling Capabilities — 164
  - 4.3.2 Calculation of Cancellation Ratio (CR) — 181
- 4.4 Concluding Remarks — 213

Chapter 5   Adaptive Arrays — 217
- 5.1 Introduction — 217
- 5.2 Model of the Environment — 220
- 5.3 Jammer Cancellation and Enhancement of Target Signal — 224
  - 5.3.1 Basic Principles — 225
  - 5.3.2 Eigenanalysis Applied to Adaptive Arrays for Jammer Cancellation — 239
  - 5.3.3 Performance Limitations due to Mismatching of Receiving Channels — 259
  - 5.3.4 Effects of Correlated Sources — 260
- 5.4 Suppression of Jammers by Multiple-Beam Signal Processing — 273
- 5.5 Deterministic Spatial Filtering — 277
  - 5.5.1 Principles — 277
  - 5.5.2 Applications — 279
  - 5.5.3 Examples — 281
- 5.6 Phase-Only Nulling — 283
  - 5.6.1 Null Placement in *a priori*-Known Directions — 286
  - 5.6.2 Adaptive Methods — 291
- 5.7 Adaptive Arrays with Angular Superresolution — 292

	5.7.1	Classical Beam-Forming Methods	295
	5.7.2	Linear Prediction Applied to DOA Estimate	298
	5.7.3	Minimum Variance Applied to DOA Estimate	306
	5.7.4	DOA Estimate Based on Eigenanalysis	313
	5.7.5	Application of Singular Value Decomposition	324
	5.7.6	Superresolution of Coherent Sources by Adaptive Array Techniques: Problems and Remedies	332
5.8	Experimental Systems and Realizations with Adaptive Arrays		336
5.9	Concluding Remarks		339
Bibliography			347
Index			367

Foreword

This book is one of several that Alenia (Aeritalia & Selenia) SpA, the leading aerospace company of the IRI-Finmeccanica Group, is promoting to illustrate the wide spectrum of its research and development activities.

Signal processing for radar systems is a theme that Selenia SpA (now a part of Alenia SpA) has mastered since the beginning of the 1970s. Antenna technology is another major strength in Alenia radar experience. Both technologies constitute the backbone of today radar systems for both defense and civilian applications. Alenia radar systems, in operation worldwide, are the result of a continuous effort to find effective and often novel technical solutions.

Dr. Farina's book focuses on the intimate connection between signal processing and antennas. It describes four major techniques aimed at efficiently suppressing interference, thus giving the radar superior performance in target detection and location. These techniques are: (1) antenna design for low and ultralow sidelobes, (2) sidelobe blanking against pulsed interferences, (3) sidelobe cancellation to contrast high-duty-cycle interferences, and (4) the full fledged adaptive array system concept.

Books written by practicing engineers are in my opinion especially valuable because they quite often carry that special flavor of theory applied to real-life situations. This volume, written by an engineer who has been very much involved in advanced radar activities, is an excellent example of such a book and is presented to the international radar community as a timely synthesis of the somewhat difficult but nevertheless fascinating research and development topics that have now reached a solid level of maturity.

Raffaele Esposito
General Manager
Alenia (Aeritalia & Selenia) SpA

Preface

Four major techniques can be adopted to effectively suppress interferences in a radar system; they are (1) antennas designed for low sidelobes, (2) sidelobe blanking, (3) sidelobe cancellation, and (4) nulling or sidelobe reduction and super-resolution with adaptive arrays. These techniques form the subject of the book.

Chapter 1 provides a short description of the problem of *electronic counter-countermeasures* (ECCM) in radar systems. The chapter starts with some *electronic warfare* (EW) terminology (Section 1.1), ECCM in fact being a division of EW. Attention is turned to the description of the *electronic-countermeasures* (ECM) (Section 1.2). The ECM threat to a radar system is important to understand in terms of how to react by using an efficient ECCM strategy. The more important ECM tactics and techniques are briefly recalled as well as their effects on radar performance. The major ECCM radar techniques are described in Section 1.3, and are ranked according to the different radar subsystems in which they are located, namely, the antenna, transmitter, receiver, and signal processor.

After giving an overview of the general principles of ECCM, the remainder of the book is devoted to a detailed discussion of the specific techniques adopted at the antenna level. The description concentrates on the sidelobe ECCM techniques, which have such wide popularity among radar engineers. The great number of technical papers published on these topics allows the presentation of an accurate description of the working principles and processing schemes, and makes possible the performance evaluation to include a wealth of charts, tables, and other numerical data. The purpose is to provide essential information for professionals seeking quick answers for on-the-job problems.

Chapter 2 addresses the low and *ultra-low sidelobe antenna* (ULSA) technology, which is of increasing importance in high performance radar systems, particularly those operating in heavy clutter and jamming. Radars equipped with antennas having receiving patterns characterized by low sidelobe levels accept less clutter, chaff, and jamming power. Low sidelobes are also beneficial in the transmission pattern to reduce clutter illumination, to counteract ESM, and to reduce the effectiveness of antiradiation weapons. Section 2.2 illustrates, with numerical examples, the role played by the low-sidelobe antenna technology in radar systems for defense applications. The key developments that allow significant reduction in sidelobes are considered, and general guidelines for designing ULSAs are described in Section 2.3. The application of these principles to reflector antennas and array antennas are illustrated in the Sections 2.4 and 2.5, respectively. Subsequently, Section 2.6 provides a detailed mathematical analysis of the effects of random errors on the average sidelobe level, random errors being the factors that ultimately limit the achievable minimum value of sidelobes. Finally, the advanced technique of *digital beam-forming* (DBF), which is of paramount importance for modern phased array radars, is extensively described in Section 2.7.

Chapter 3 is completely devoted to the description of the *sidelobe blanker* (SLB) system and evaluation of its performance. The SLB is able to reduce the deleterious effects of an impulse jammer entering the radar through the antenna's sidelobes (Section 3.1). The working principle of SLB and the processing schemes are described in Section 3.2. The performance of the SLB is mainly expressed in terms of probability of blanking the jammer in addition to the usual probability of false alarm and probability of detection (Section 3.3). Design criteria help in the selection of the major system parameters such as the gain of the auxiliary antenna as well as the detection and blanking thresholds (Section 3.4).

Chapter 4 is a thorough review of the sidelobe canceler (SLC) systems. The SLC allows the suppression of high-duty-factor jammers entering in the radar through the antenna sidelobes. The relevance of the SLC system concept transcends the specific application of jammer cancellation. In fact, SLC introduces the field of radar engineering to the key topics of adaptivity, spectrum analysis, and optimum signal processing. These concepts evolve from the academic world to become selling points of the current technology in radar systems.

More specifically, Chapter 4 describes the working principles (Section 4.1), implementation schemes (Section 4.2), and evaluation of performance (Section 4.3). The description of adaptation algorithms includes such recent findings, as the so-called *data domain techniques* (i.e., Givens rotations, Householder reflections). They lead to numerically efficient processing schemes, which can be mapped onto parallel processors, such as systolic arrays, for real-time operation. Performance, which is expressed in terms of *cancellation ratio,* is evaluated for narrowband and wideband jamming. Practical limitations of the system performance are also taken into account. They include the effects of mismatching in the main and auxiliary channels and the effects of the difference between the main-beam sidelobes and the auxiliary patterns. Also considered are space-time processing methods to compensate for the mismatching sources and a trade-off analysis between steady-state cancellation and convergence time of the adaptive systems.

The concept of adaptive arrays, which is a generalization of SLC, is discussed at length in Chapter 5. Adaptive arrays are still in the experimentation phase, but they promise to become the backbone of future radar systems. The concepts described in Chapter 5 involve the basic principles and practical means of jammer cancellation and enhancement of useful target signals (Sections 5.2 and 5.3). Subsequently, the formation of multiple beams is briefly considered as an additional jammer suppression technique (Section 5.4). If the direction from which the jamming is expected to come is known *a priori*, deterministic spatial filtering can be applied (Section 5.5). As a tentative technique to reduce the hardware associated with adaptive beam-forming, the phase-only nulling technique is explored (Section 5.6). This is a means of generating nulls by controlling just the phase values of the array weights while maintaining the amplitude coefficients at predetermined levels. The chapter ends with the concept of super-resolution, which allows separation, tracking, and classification of closely spaced targets and jammers. This technique, widely applied in seismology, astronomy and other fields of science, is in its infancy in radar technology. Nevertheless, experimentation demonstrates that there is an advantage in using super-resolution as compared to conventional, Rayleigh-limited, spectral analysis. Section 5.7 presents several currently available methods, and Section 5.8 illustrates experimental systems and realizations with adaptive arrays.

Finally, a comprehensive bibliography includes the last twenty years of open technical literature. The majority of references are from journals, books, and conference proceedings, which are likely to be found in a large library. Many papers are from IEEE and IEE journals and conference proceedings. The references are listed in alphabetical order and according to the date of issue.

Radar system and antenna engineers, university students, and experts in ECM and ECCM will benefit from reading this book. The reader is assumed to be familiar with basic radar concepts, probability, stochastic process, and linear prediction theories, matrix computations, and linear algebra. The book can be used for university courses or postgraduate courses for radar practitioners. Finally, the book can be used as a reference for practicing engineers.

The author wishes to express his gratitude to several colleagues of the Radar Department of Alenia (Aeritalia & Selenia) SpA for their valuable assistance during the preparation of this book. Dr. L. Timmoneri has been very helpful in the preparation of charts for Chapters 4 and 5. Dr. D. Iovino provided the valuable curves and tables of Chapter 3. The cooperation of Dr. F. Scannapieco, Dr. A. Protopapa, Dr. G. D'Acunzo, Dr. A. DiVito, Dr. R. Pangrazi and Dr. M. DeFazio are also warmly acknowledged. Mr. B. Triola prepared the excellent figures for the book. Special thanks are due to Dr. F.A. Studer, co-author of a number of papers and technical reports on SLC topics.

Finally, the author wishes to acknowledge with thanks Dr. R. Esposito, the general manager of Alenia SpA, and Professor A. Gilardini, who have promoted and encouraged the preparation of this book as the first of a collection of works authored by Alenia engineers. In addition, the author sincerely thanks the director of the Radar and C^2 Division of the Defense Systems Group, Dr. C.A. Penazzi, the Manager of the Radar Factory, Dr. E. Giaccari, and the head of the System Analysis Group, Dr. S. Pardini, for their permission to prepare the material for the book as well as showing their confidence in its quality. The preparation of this book has been encouraged by the author's friend Dr. M.I. Skolnik.

This book has benefited from the great volume of material published in the open technical literature. The author is indebted to the editors of the Institute of Electrical and Electronic Engineers (New York) and of the Institute of Electrical Engineers (London) for permission to reproduce some materials from *Transactions, Proceedings, Records,* and *Digests* of conventions and symposia. Other material has been reproduced from books by permission of Artech House, John Wiley & Sons, Inc. (New York), and other sources duly cited in the text. The author also wishes to express his thanks to the acquisition and editorial staff of Artech House, Inc. for the excellent job producing this book. More specifically, the author wishes to acknowledge the cooperation of Mr. M. Walsh, Ms. P. Ahl, Mr. K. Field, and Ms. S. Brown.

A. Farina
Rome, September 1991

Chapter 1
Introduction to ECM and ECCM Techniques for Radar Systems

Since World War II, both radar and *electronic warfare* (EW) have achieved a very high state of performance. Modern military forces depend heavily on electromagnetic systems for surveillance, weapon control, communication, and navigation. Electronic countermeasures (ECM) are likely to be taken by hostile forces to degrade the effectiveness of electromagnatic systems. As a direct consequence, electromagnetic systems are increasingly equipped with so-called electronic counter-countermeasures (ECCM) to ensure effective use of the electromagnetic spectrum despite the enemy's deployment of EW techniques.

The basic principles used by the radar designer to combat ECM include: (1) actions involved in forcing the noise jammer to spread its power, (2) actions taken to reduce the amount of disturbances entering into the radar, (3) the use of appropriate devices to limit the deleterious effects of jamming in the radar, (4) the options for recognizing and avoiding repeater jamming, and (5) the measures available for operating in the presence of chaff.

This chapter is devoted to a short presentation of ECCM techniques for radar systems subjected to an ECM threat (Farina, 1990). Section 1.1 starts with a review of the terms commonly used in EW and definitions of its main subdivisions. Then, Section 1.2 is devoted to the description of the major ECM techniques and strategies. This is followed (Section 1.3) by a review of several ECCM techniques usually implemented in the various blocks of a radar system, i.e., the antenna, transmitter, receiver and signal processor. Finally, Section 1.4 describes an operational scenario and how it relates to the author's concept of antenna-based ECCM techniques.

1.1 ELECTRONIC WARFARE TAXONOMY AND TERMINOLOGY

Electronic warfare is defined as a military action involving the use of electromagnetic energy to determine, exploit, reduce, or prevent hostile use of the electromagnetic spectrum, and action that retains friendly use of the electromagnetic spectrum (Johnston, 1979). EW is organized into three major categories: *electronic support measures* (ESM), electronic countermeasures, and electronic counter-countermeasures. The definitions of ESM, ECM and ECCM (adapted from (Johnston, 1979)) are listed below.

Electronic support measures is that division of electronic warfare involving actions taken to search, intercept, locate, record, and analyze electromagnetic energy radiated by radars for the purpose of exploiting such radiations in the support of military operations. Thus, electronic support measures provide a source of information with which to conduct electronic countermeasures.

ECM is that division of electronic warfare involving actions taken to prevent or reduce the effective use of the electromagnetic spectrum by radar systems.

ECCM is that division of electronic warfare involving actions taken to ensure the effective use of the electromagnetic spectrum by radars, despite the adversary's use of electronic warfare.

1.2 ELECTRONIC COUNTERMEASURES (ECM)

The objectives of an ECM system are to deny the information (detection, position, track initiation, track update, and classification of one or more targets) that the radar seeks, to surround useful echoes with false target data so that the true information cannot be extracted, or to supply false data so that the information handling system (computer, display, and operator) of the victim radar is impaired.

ECM tactics and techniques may be classified in a number of ways, e.g., by main purpose, whether active or passive, by deployment or employment, according to the platform carrying the ECM device, or by characteristics related to the victim radar (Table 1.1).

The basic ECM purpose is *jamming*. The definition of jamming follows (Van Brunt, 1978, p. 29). The intentional and deliberate transmission of amplitude, frequency, phase, or otherwise modulated, intermittent, continuous wave (CW), or noise signals by either active or passive means for the purpose of interfering with, disturbing, exploiting, deceiving, masking, or otherwise degrading the reception of other signals that are used by radar systems. A jammer is any ECM device that transmits a signal of any duty cycle for the purpose of jamming a radar system (Van Brunt, 1978, p. 85).

Deception is the intentional and deliberate transmission or retransmission of amplitude, frequency, phase, or otherwise modulated, intermittent, or CW signals by either active or passive means for the purpose of misleading in the interpretation or use of information by radar systems (Van Brunt, 1978, p. 54). The categories of deception are *manipulative* and *imitative* (manipulative implies the alteration or simulation of friendly electromagnetic signals to accomplish deception, and imitative

Table 1.1
Taxonomy of ECM Tactics and Techniques

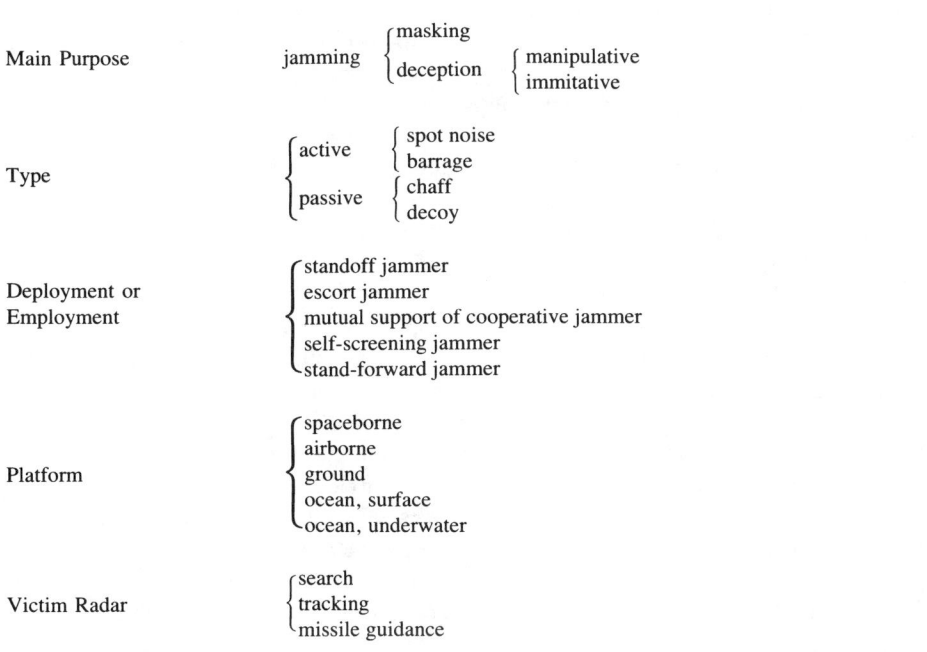

Main Purpose	jamming { masking, deception } { manipulative, immitative }
Type	{ active { spot noise, barrage }, passive { chaff, decoy } }
Deployment or Employment	{ standoff jammer, escort jammer, mutual support of cooperative jammer, self-screening jammer, stand-forward jammer }
Platform	{ spaceborne, airborne, ground, ocean, surface, ocean, underwater }
Victim Radar	{ search, tracking, missile guidance }

consists of introducing radiation into radar channels that imitates a hostile emission). An example of a deception jammer is the *track breaker* type of repeater, which is used against tracking radars. The repeater jammer returns an amplified version of the signal received from the radar. The deception jammer signal, being stronger than the radar's return signal, captures the radar's range gate. The deception signal is then progressively delayed in the jammer using an RF memory, thereby "walking" the range gate away from the actual target. When the range gate is sufficiently removed from the actual target, the deception jammer is turned off, forcing the tracking radar into a target reacquisition mode. "Track breaker" is a general term, which also applies to velocity gate pull-off and various angle deception schemes.

Radio signals generated by special radio transmitters and intended for interfering with or precluding the normal operation of a victim radar system are called *active jamming*. Jamming enters the radar through the antenna's main beam as well as sidelobes. At the input of a victim system, active jamming produces a background which impedes the detection and recognition of useful signals and determination of their parameters. The most common forms of active noise jamming are *spot noise* and *barrage noise*. Spot noise is used when the center frequency and bandwidth of the victim system to be jammed are known and the band is narrow. However, many radars are frequency-agile over a wide band as an ECCM measure against spot jamming. If the rate of frequency agility is slow enough, the jammer can follow the frequency changes and maintain the effect of spot jamming. Barrage or broadband jamming is simultaneously radiated across the entire band of the radar spectrum of interest. This method is used against frequency-agile systems with rates that are too fast to follow, or when the victim's frequency parameters are imprecisely known. Another important reason for using barrage noise jamming is the need for one jammer platform to jam several radars that may be operating in different parts of the jammed band. Barrage noise jammers require considerably more *effective radiated power* (ERP) than does spot noise for equal effectiveness.

Passive ECM is synonymous with chaff, decoys, and other reflectors which require no prime power. Chaff is defined (Van Brunt, 1978, p. 43) as elemental passive reflectors, absorbers, or refractors of radar, communication, and other weapon system radiations that can be floated or otherwise suspended in the atmosphere or exoatmosphere for the purpose of confusing, screening, or otherwise adversely affecting the victim electronic system. Examples are metal foils, metal-coated dielectrics (aluminum, silver, or zinc over fiberglas or nylon being the most common), aerosols, string balls, rope, and semiconductors. The basic properties of chaff are effective scatter area, the character and time of development of a chaff cloud, the spectra of the signals reflected by the cloud, and the width of the band that conceals the target. The objective of decoys as passive countermeasures is to deceive the radar by having it recognize decoy skin returns as the returns from real targets.

Concerning the *deployment* or *employment* of ECM, five classes, as shown in Table 1.1, can be identified (Schleher, 1986, pp. 11–15). Standoff jamming (SOJ) is an ECM tactic, where, in a given area, the jamming platform remains close to, but outside of, the lethal range of enemy weapon systems and jams them to protect the attacking vehicles. Standoff ECM systems employ high-power noise jamming, which must penetrate through the enemy system's antenna receiving sidelobes at long ranges. A typical standoff radar jammer employs an ESM system with direction-finding capability to locate the radar being jammed, a computer to assess the relative threat of the located emitters and to allocate the available jamming resources, and multiple jamming transmitters coupled through directive antennas to provide high ERP throughout the bands of interest.

An *escort jamming platform* accompanies the vehicles and jams the victim radars. The jamming platform must have the same flight characteristics as the strike vehicles to be effective. The effectiveness of an escort jammer is expected to be greater than that of standoff jammers because of its closer proximity to the victim systems. Escort jamming is generally used when the strike aircraft do not have enough available power, space, or payload to protect themselves. When escort jamming is employed, special care must be taken to protect the support aircraft because, if it is destroyed, the whole strike force is thus unprotected.

Mutual-support or *cooperative ECM* involves the coordinated conduct of ECM combat elements against hostile acquisition and weapon-control radars. One advantage of mutual support jamming is the greater ERP available from a collection of platforms in contrast with a single platform. However, the real value of mutual-support jamming is in the coordinated tactics that can be employed. A favorite tactic employed against tracking radars, for example, is to switch between jammers located on separate aircraft within the radar's beamwidth. This "blinking" introduces artificial glint into the radar tracking circuits, which, if introduced at the proper rate (typically 0.1 to 10 Hz), can cause the radar to break angle track. In addition, blinking has the desirable effect of confusing radiation-homing missiles, which might be directed against the jammer radiations.

A *self-screening jammer* (SSJ) is used to protect the carrying vehicle against enemy weapon systems. This situation stresses the capability of an ECM system relating to its power, signal processing, and passive location capabilities.

Stand-forward jamming is an ECM tactic whereby the jamming platform is located between victim radar systems and strike vehicles. The stand-forward jammer is usually within the lethal range of enemy weapon systems for a considerable time. Therefore, only the use of relatively low-cost *remotely piloted vehicles* (RPVs) is practical. The RPVs can assist strike aircraft or missiles in penetrating radar-defended target areas by jamming, ejecting chaff, dropping expandable jammers or decoys, acting as decoys themselves, and performing other related ECM tasks.

Before closing this section, we should mention the problem of qualifying the efficacy of an ECM. The jammer's efficacy depends on many factors, including the features of the victim radar. There does not seem to be a figure of merit that incorporates the effectiveness, either of a specific ECM type or a combination of a number of different ECM techniques. The jammer capability can be characterized by its noise spectral density (p_j), antenna gain (G_j), and the bandwidth (B_j). The size of the jammer is therefore determined by the effective radiated power (ERP = $p_j B_j G_j$), which is useful for analytical work, e.g., calculation of radar range under jamming.

1.3 AN OVERVIEW OF ECCM TECHNIQUES

The primary objective of ECCM techniques and strategies applied to a radar system is to allow the accomplishment of the intended radar functions while counteracting the effects of the enemy's ECM. However, the conflict between ECM and ECCM is endless. When the two systems interact, a dynamic situation arises: ECM disturbs the radar, which, in turn, triggers ECCM, and so on. The limiting factors in the battle are time and cost. Included in these parameters are technological improvements and conceptual advances that control the battle's evolution.

There are two broad categories of ECCM, *electronic* and *operational*. The former includes electronic techniques implemented in the radar subsystems, and the latter refers to the way of selecting the proper combination of operational modes of the radar in response to a specific ECM threat.

The benefits of using ECCM techniques may be summarized as (Johnston, 1979, p. 13): (1) prevention of radar receiver saturation; (2) *constant-false-alarm rate* (CFAR), thus avoiding saturation of the radar data processor and display system; (3) enhancement of *signal-to-jamming ratio* (S/J); (4) *discrimination* of directional interference; (5) *rejection* of false targets, (6) *maintenance of target tracks*; and (7) *counteraction* of ESM. These benefits are obtained by implementing a number of ECCM techniques in the different sections of the radar. The ensuing description is limited to the major ECCM techniques (Table 1.2); for more details, see (Farina, 1990).

Table 1.2
Major ECCM Techniques

Radar Subsystems	ECCM Techniques
Antenna	High Directive Antennas
	Multiple Beams
	Low Sidelobes
	SLB
	SLC
	Adaptive Arrays
	Random Scanning of Main Beam
Transmitter	Radar with Large ERP
	Management of Power in Time and Space
	Frequency Agility and Diversity
	Intrapulse Coding
	PRF Jitter, Stagger, Coding
	AFS
	Use of Millimeter Waves (MMW)
Receiver	Dual-Frequency Conversion
	Large Dynamic Range Receivers
Signal Processing	Digital Coherent and Adaptive Moving Target Indicator (MTI)
	CFAR Detectors
	Pulsewidth and Pulse Repetition Frequency (PRF) Discriminator

1.3.1 Antenna-Related ECCM

The first and probably most important area of the radar to be considered for incorporation of ECCM is the antenna. Because it represents the transducer between the radar and the environment in which it must work, the antenna is the first line of defense against jamming. The directivity of the antenna in the transmission and reception modes allows space discrimination to be used as an ECCM approach. Techniques for generating radar space discrimination include low sidelobes, sidelobe blanking, sidelobe cancelers, adaptive array systems, beamwidth control, antenna coverage, and scan control (Johnston, 1979, pp. 195–224).

Smaller antenna beamwidth, or correspondingly higher antenna gain, can be employed to spotlight a target and to achieve *burn-through jamming*. An antenna with multiple beams (Palumbo, 1989; Giaccari and Penazzi, 1989) can be used to delete the beam containing a jammer while still maintaining detection capabilities for the remaining beams. Increased angular resolution, with respect to that related to the width of the main beam, of close flying targets and jammers can be obtained by adaptive array antennas.

If an air defense radar must operate in a severe ECM environment, degradation in detection range over large portions of the scan volume can occur because of jamming entering the sidelobes, although the main beam is pointing elsewhere. On transmission, the energy radiated into spatial regions outside the main beam is subject to reception by enemy radar *early warning* receivers or *antiradiation missiles* (ARM). The geometry of ARM type engagements is such that the detection of radar emissions through the sidelobes is necessary to provide timely guidance information. This is particularly true for scanning radars, the main beams of which continuously move through the surveillance volume. For these primary reasons, low sidelobes are extremely desirable on both receiving and transmitting. Techniques to achieve low sidelobes are available for reflector type antennas and phased-array antennas as well.

Two additional techniques to prevent jamming from entering by way of the radar's sidelobes are *sidelobe blanker* (SLB) and *sidelobe canceler* (SLC). SLB operates against pulsed interferences, and SLC is valuable in filtering high-duty cycle jamming signals. SLC has a disadvantage as compared to the use of a low-sidelobe antenna. The SLC needs one *degree of freedom* (DOF) for each *stand-off jammer* (SOJ) to be effective. The low-sidelobe antenna design, however, has no such limitation on the number of jammers that it can handle. Nonetheless, SLC and SLB can be adopted by any type of radar already developed, whereas an ultra-low sidelobe antenna needs the redesign of a new, costly antenna. An interesting generalization of SLC is the adaptive array system concept, which provides more degrees of freedom in nulling sidelobes as well as main-lobe jammers.

Certain deception jammers depend on anticipation of the beam scan or on knowledge or measurement of the antenna scan rate. Random electronic scanning effectively prevents this deception type of jammer from synchronizing to the antenna scan rate, thus defeating it. Although adding complexity, cost, and possibly weight to the antenna, the control of beamwidth, coverage, and scan is a valuable and worthwhile ECCM feature of all radars.

1.3.2 Transmitter-Related ECCM

The different types of ECCM are related to the proper use and control of the power, frequency, and waveform of the radiated signal (Johnston, 1979, pp. 151–188). One "brute force" approach to defeating a noise jammer is to increase the radar's transmitter power. This technique, when coupled with "spotlighting" the radar antenna on the target, results in an increased ERP value, thus giving a significant increase in the radar's detection range. However, such an approach is not effective against chaff, decoys, repeaters, spoofers, and so on. Also, high ERP has the disadvantage of making the radar more vulnerable to detection and location by ESM and ARM. Reduction or management of transmitted power when less power is required for target detection is one technique commonly used in *low probability of intercept* (LPI) radar against ARM (Schleher, 1986, pp. 171–178; Farina and Galati, 1985).

More effective is the use of complex, variable, and dissimilar transmitted signals that place the maximum burden on the ECM. One way is to change the frequency of the radar transmitter in the frequency-agility or frequency-diversity modes. Frequency agility usually refers to the radar's ability to change the transmitter frequency on a pulse-to-pulse or batch-to-batch basis. The batch-to-batch approach allows the system to perform doppler processing, which is not compatible with frequency agility on a pulse-to-pulse basis. Frequency diversity refers to the use of several complementary radar transmissions at different frequencies, either from a single radar (e.g., a radar having stacked beams in elevation by employing different RF signals on each elevation beam) or several radars. The objective of frequency agility and diversity is to force the jammer to spread its energy over the entire agile bandwidth of the radar, thus causing a reduction of the jammer density and resulting ECM in its degraded effectiveness (Van Brunt, 1978, pp. 217–229).

Frequency agility and diversity techniques represent a form of *spread-spectrum* ECCM in which the information-carrying signal is spread over as wide a frequency (space, time) region as possible to reduce detectability and make jamming more difficult. Spread-spectrum features associated with LPI radars pertain to the realm of *radar waveform design*. Waveform coding includes PRF jitter, PRF stagger, and coding, and, perhaps, shaping of the transmitted radar pulse. All these techniques make deception jamming or spoofing of the radar difficult because the enemy should not know or anticipate the fine structure of the transmitted waveform. Intrapulse coding to achieve *pulse compression* may be particularly effective in improving target detection capability by radiation of a larger amount of average radar power without exceeding peak power limitations within the radar and by improving range resolution (larger bandwidth), which, in turn, reduces chaff returns and resolves targets to a higher degree. Because the peak power is lower, a pulse compression radar is also less detectable by ESM equipment than is a conventional pulsed radar.

Some advantage can be gained by including the capability of examining the jammer signals, finding holes in its transmitted spectrum, and selecting the radar frequency with the lowest level of jamming. This approach is particularly useful against pulsed ECM, spot noise, and nonuniform barrage noise. Its effectiveness primarily depends on the extent of the radar agile bandwidth and the acquisition speed and frequency tracking of an "intelligent" jammer. A technique suited to this purpose is referred to as *automatic frequency selection* (AFS) and is described in detail in (Strappaveccia, 1987).

Another method, which is employed to reduce the effect of main-beam noise jamming, raises the transmitter frequency to narrow the antenna's beamwidth. This restricts the sector that is blanked by main-lobe jamming and also provides a strobe in the direction of the jammer. Strobes from two or three spatially separated radars allow the location of the jammer.

The rationale of raising the transmitter frequency leads to the millimetric-wave radar system concept (Skolnik, 1990; Bodnar, 1987). MMW radars typically operate around 35 and 94 GHz. They offer increased angular resolution as compared to microwave radars and potentially wide transmission bandwith. However, their operation is generally restricted to short-range applications due to the considerable attenuation of MMW, which occurs in the lower atmosphere. The MMW radar could make use of the propagation attenuation to achieve covert operation.

1.3.3 Receiver and Signal Processing-Related ECCM

Because the bandwidth at the receiver input would be broad to accommodate the system's frequency agility, the receiver should use a dual-frequency conversion procedure. In addition, an image rejection filter and high dynamic IF mixers are worth having.

To avoid receiver saturation due to jamming, which would result in the virtual elimination of target information, wide dynamic range receivers are adopted. Linear receivers are recommended and old-fashioned logarithmic receivers are ruled out for radars that require high levels of clutter attenuation.

Digital coherent signal processing greatly alleviates the effects of clutter and chaff. This is motivated by the use of coherent doppler processing techniques, such as fixed or adaptive MTI (Farina, 1987a).

A disadvantage of MTI processors is the relatively large amount of pulses (up to ten or more) that must be transmitted at a stable frequency and PRF. A responsive jammer could measure the frequency of the first transmitted pulse and then center the jammer to spot jam the following pulses. Also, the requirement for a stable PRF precludes the use of pulse-to-pulse jitter, which is one of the most effective techniques against deception and camouflage jammers, which rely on anticipating the radar transmitter's pulse. In addition, high-performance MTI operation requires a wide-dynamic-range linear receiver, which precludes the use of the hard or soft limiting associated with a number of ECCM fixes (e.g., Dicke-Fix (Schleher, 1986)) against certain forms of jamming.

Noncoherent devices are also required because of the limited degree of clutter, chaff, and jammer suppression achieved in practice by coherent devices so that the residual interference is still a significant source of false alarm. Among the noncoherent devices worth mentioning are the CFAR and the pulsewidth and PRF discriminator, which are effective against pulsed jammers.

Dicke-Fix is a type of CFAR used to counter high rates of swept-frequency CW jamming and swept spot-noise jammers. The Dicke-Fix receiver concept requires the use of a coherent hard limiter, which preserves the phase of the signal while keeping the amplitude at a constant value. The coherent limiter is inserted upstream of the pulse compression filter in a radar that uses phase-coded signals. In reception, the jammer and target signals are chopped in amplitude. The preservation of the target signal phase coding allows the integration of target energy by means of the pulse compression filter matched to the phase code. The Dicke-Fix processing scheme suffers from two limitations. The first is related to the detection loss experienced when the target does not compete with the jammer. The other disadvantage is the masking effect of a weak target signal that is sufficiently close to a strong target in range (compared with the spatial extension of the code).

The pulsewidth discrimination circuit measures the width of each received pulse. If the received pulse is not of approximately the same width as the transmitted pulse, the signal is rejected and not passed to the signal processor or display. Similar concepts apply to discrimination based on PRF. A pulsewidth discrimination technique can help in rejecting chaff. In fact, echo returns from chaff corridors are much wider than the transmitted pulse.

1.4 FOCUS OF THE BOOK

The purpose of this section is to focus the attention of the reader on the main problem considered in this book and the preferred, proposed technical solutions. Figure 1.1 is a pictorial illustration of an operational scenario, showing a radar, a target T_1 in the main antenna beam, and a far-off platform T_2. The SOJ, on-board the platform T_2, tends to mask the presence of the incoming aircraft T_1 by injecting in the radar receiver, through the radar antenna pattern, high duty-cycle interfering signals. In addition, the target T_1 may also carry an SSJ for self-protection purposes. A radar *plan position indicator* (PPI) display would appear completely blurred to an operator. The operator would not be able to detect any target and to measure the target polar coordinates. The situation depicted in the figure is very basic. More challenging scenarios should include numerous aircraft with adequate ECM, including the presence of chaff clouds.

We have already indicated that the antenna is the first line of defense of the radar and adequate techniques should be adopted at this level to achieve a consistent reduction of the jamming signals. This book describes a number of antenna-based

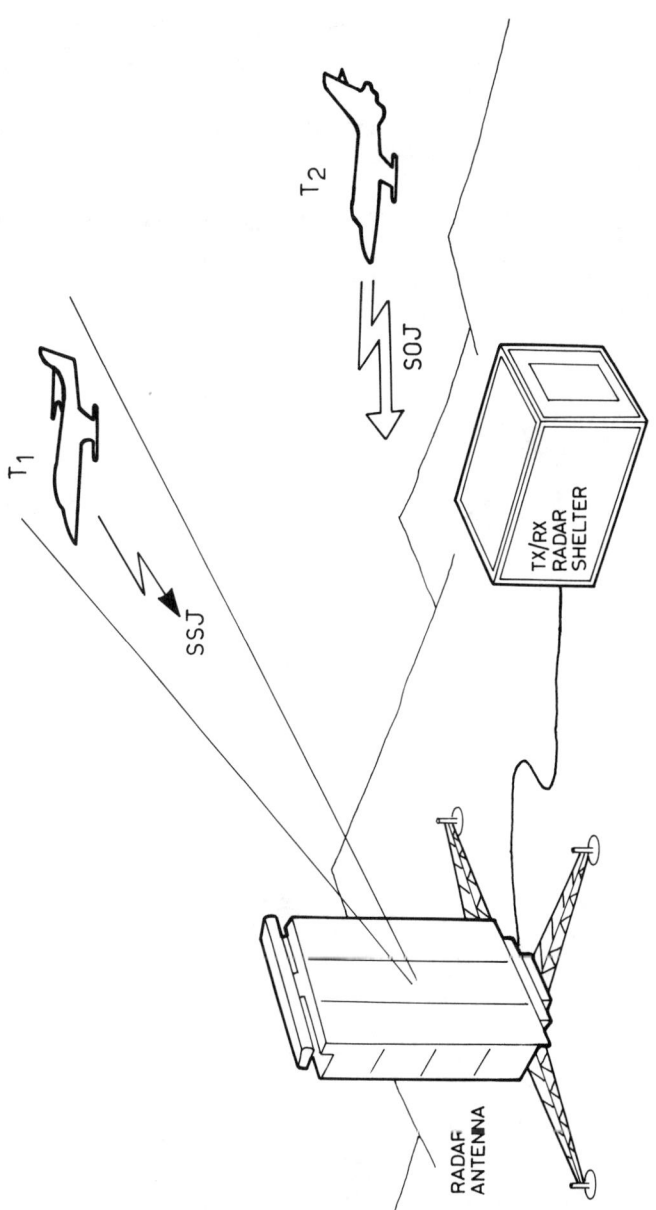

Figure 1.1 An operational scenario.

electronic counter-countermeasures that are deemed necessary to thwart the ECM action, thus making possible the detection, tracking, and classification of targets in the area of radar surveillance.

The following techniques are considered in detail in this book:

- Ultra-low sidelobe antenna, described in Chapter 2, to counteract sidelobe jammers;
- Sidelobe blanker (SLB), analyzed in Chapter 3, to null out pulsed jammers;
- Sidelobe canceler (SLC), the theme of Chapter 4, to filter high-duty-cycle jammers;
- Adaptive arrays, extensively described in Chapter 5, representing the technique at the forefront of research and development efforts with promises of superior performance against sidelobe and main beam jammers.

Chapter 2
Low Sidelobe Antennas

2.1 INTRODUCTION

This chapter addresses low and ultra-low sidelobe antenna technology, which is of increasing importance in high performance radar systems, particularly those operating in heavy clutter and jamming (Patton, 1980).

Official definitions of low and ultra-low sidelobe antennas are not available at present. Peak sidelobes are usually expressed in decibels relative to the peak of the main-lobe antenna. Sidelobes of -30 dB mean that the highest sidelobe is 30 dB below the main lobe. The average sidelobe level, expressed in dBi (i.e., relative to the isotropic gain, which is the gain of a lossless antenna that radiates uniformly in all directions.), is also important because it describes the sidelobe pattern over a wide angular interval. As an example, if 10% of the radiated power is in the sidelobes, the average sidelobe level is -10 dBi; if 1% of power is in the sidelobes, the average sidelobe is -20 dBi, and so on. Following the proposed numerical values given in (Barton, 1988, p. 188 and Schrank, 1988, p. 341), we assume that low-sidelobe antennas are those having the peak sidelobe as 30 to 40 dB below the main-lobe peak, and ultra-low sidelobe antennas have peak sidelobes lower than -40 dB. In terms of average levels, low sidelobes are defined as between -5 and -20 dBi, while ultra-low sidelobes as being less than -20 dBi. Ordinary peak sidelobes are 13 to 30 dB below the main-lobe peak and from 0 to -5 dBi relative to isotropic for the average sidelobes.

Radars equipped with antennas having receiving patterns characterized by low sidelobe levels accept less clutter, chaff, and jamming power. Low sidelobes are also beneficial in the transmission pattern to reduce clutter illumination, and to counteract ESM and antiradiation weapons. Section 2.2 illustrates, with numerical examples, the role played by the low-sidelobe antenna technology in radar systems for defense applications.

In the mid to late 1960s, the first ULSA was realized at Westinghouse (Baltimore) for the Airborne Warning and Control Systems (AWACS); the key developments that made possible the reduction of sidelobes from -35 to approximately -50 dB will be reconsidered, and general guidelines for the design of a ULSA will be described in Section 2.3. The application of these principles to reflector and array antennas will be illustrated in the Sections 2.4 and 2.5, respectively. Section 2.6 provides a detailed mathematical analysis of the effects produced by random errors on the average sidelobe level, random errors being the factors that ultimately limit the achievable minimum value of sidelobes. Finally, the advanced technique of digital beam-forming, which is of paramount importance for modern radars, is extensively described in Section 2.7.

2.2 RELEVANCE OF LOW AND ULTRA-LOW SIDELOBE ANTENNAS IN RADAR SYSTEMS

To show the benefit of having a low-sidelobe receiving antenna pattern, consider the effect of noise jamming on radar system performance. An effective and often used type of ECM is the noise jammer (see Section 1.2), which radiates energy over a bandwidth equal to or greater than the instantaneous bandwidth of the radar. The effect of this type of ECM is to increase the system noise temperature, T_s, to some new value T'_s, so that

$$T'_s = T_s + T_j \tag{2.1}$$

where T_j is the noise temperature produced by the jammer. The end result is that the useful signal is submerged by the interference.

Realistically, the radar may be required to operate in an isotropic noise background caused by the presence of a number of jammers operating within the radar's receiver passband. Such a noise background will affect the radar in both the mainlobe and sidelobe regions. Background radiation of this type is usually given in terms of *power spectral density* (PSD) in W/Hz and the noise temperature of the jammer is simply

$$T_j = \text{PSD}/k \tag{2.2}$$

where k is Boltzmann's constant. The effect of barrage jammers dedicated to a specific radar can be determined in the following manner. The noise power, P_n, entering the radar through the antenna sidelobes is equal to

$$P_n = \frac{P_j G_j}{4\pi R_j^2} A_{SL} \qquad (2.3)$$

where P_j is the jammer power, G_j is the jammer gain, R_j is the radar-to-jammer range, and A_{SL} is the equivalent aperture area of the radar sidelobes. The area is determined by the sidelobe gain, G_{SL}, so that

$$A_{SL} = \frac{G_{SL}\lambda^2}{4\pi} \qquad (2.4)$$

which, if substituted in (2.3), gives

$$P_n = \frac{P_j G_j G_{SL}\lambda^2}{(4\pi R_j)^2} \qquad (2.5)$$

Thus, the noise temperature of the jammer is

$$T_j = \frac{P_n}{B_j k} = \frac{P_j G_j G_{SL}\lambda^2}{B_j k (4\pi R_j)^2} \qquad (2.6)$$

where B_j is the jammer bandwidth. We immediately can see that a reduction in the sidelobe level, G_{SL}, of the antenna pattern corresponds to a proportional reduction of P_n. The effective radar range in the presence of a jammer is calculated by substitution of the system noise temperature, T'_s of (2.1), with T_j given by (2.6), in the well known radar equation (Barton, 1988).

To be more effective, consider the following three numerical examples. The first example shows the ratio between the radar range R_1 in the presence of a jammer divided by the clear radar range R_0. The radar operates with a wavelength of 10 cm, has a main-beam peak gain of 30 dB, is characterized by a receiver noise figure of 3 dB, and a receiving loss of 3 dB. The jammer, operating at a distance R_j from the radar, has an effective radiated power spectral density of 100 W/1 MHz. By definition, the ERP is the product of the transmitter power by the antenna gain of the jammer (see Section 1.2). Figure 2.1 shows the ratio R_1/R_0 in percentage as a function of the sidelobe level of the radar antenna along the jammer direction, and having R_j as parameter.

In the second example shown in Figure 2.2, the quantity R_1/R_0 is drawn as a function of the radar-jammer distance R_j, having the sidelobe level SLL of the radar antenna as parameter. The other radar parameters are the same as the previous example.

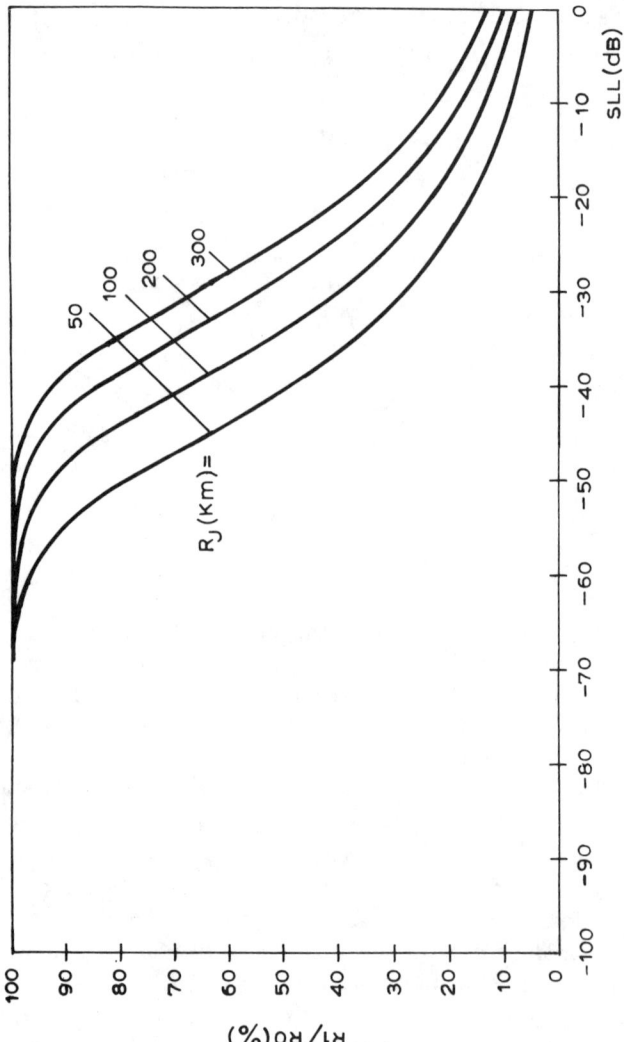

Figure 2.1 Reduction of radar range under jamming *versus* the sidelobe level of the radar antenna. The parameter of the curves is the jammer distance from the radar.

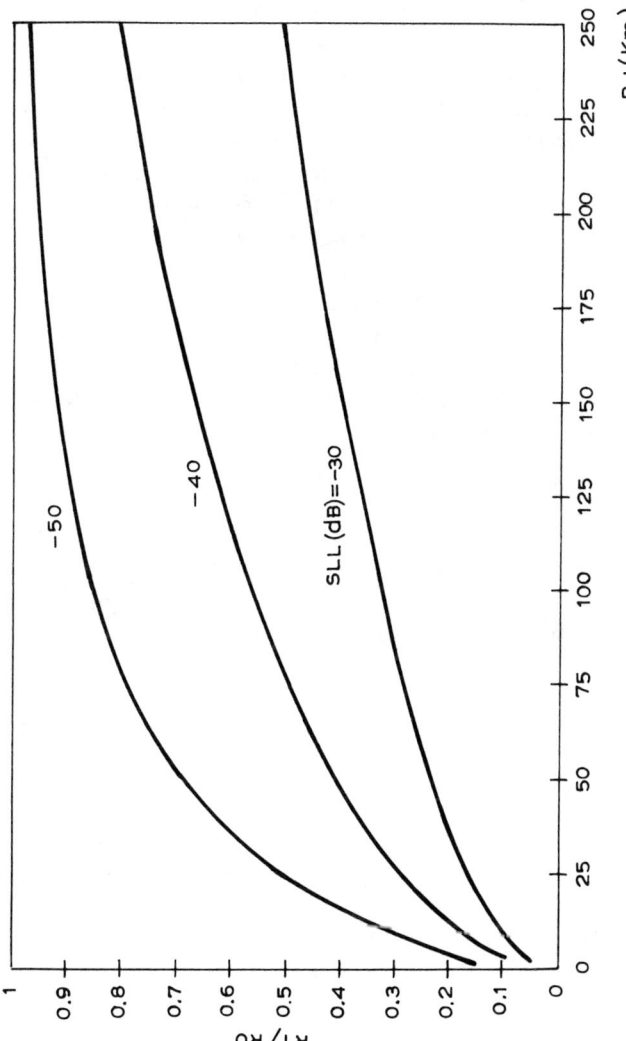

Figure 2.2 Reduction of the radar range under jamming *versus* the jammer distance from the radar. The parameter of the curves is the sidelobe level of the radar.

A more complex jamming scenario is considered for the third example. There are five jammers operating at an altitude of 10 km; their distance from the radar is 100 km. The jammer azimuth positions are $-20°$, $-10°$, $0°$, $20°$, and $30°$, respectively. The radar is a phased array operating at 10 cm wavelength, with a pencil beam of 3° azimuth and 6° elevation beamwidth, and the corresponding gain of 30 dB. The antenna, which is sited 10 m above ground level, has a tilt of 10° in elevation. The sidelobe levels of the antenna in the azimuth-elevation plane are shown in Figure 2.3. Note that, in the interval (0°, 5°) in azimuth and (0°, 10°) in elevation, the antenna pattern is Gaussian shaped. In the remaining sectors of the plane, the sidelobes are stepwise-constant. The radar noise figure and receiver loss amount to 3 dB each. Let us now consider what happens when a target is present at certain elevation and azimuth angles, θ_{el} and θ_{az}, respectively. We assume that the phased-array beam is electronically steered along the target direction. Figure 2.4 illustrates the parameter R_1/R_0 as a function of the target azimuth angle, θ_{az}, having the target elevation angle θ_{el} as parameter. We can see that, when the target elevation is 5°, in practice, the jammers go through the main lobe of the radar. As the target elevation angle increases, the jammers are received by the radar through the antenna sidelobes, and their deleterious effects are mitigated.

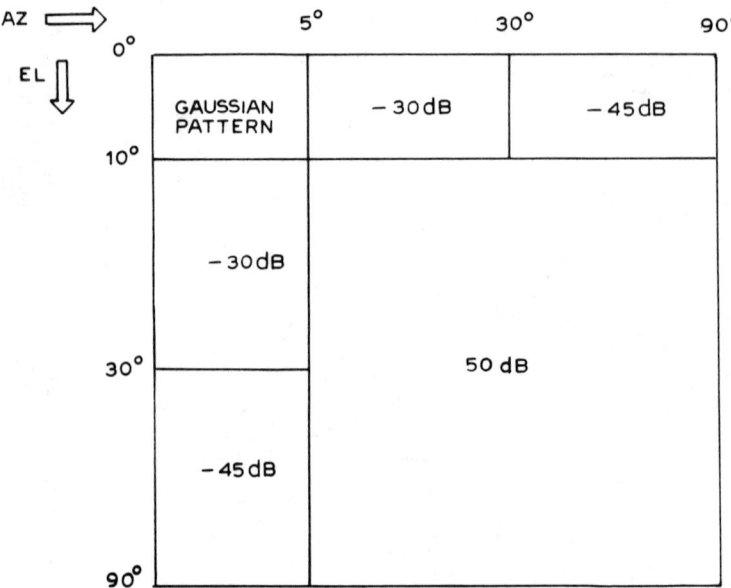

Figure 2.3 Sidelobe level of the antenna in the azimuth-elevation plane.

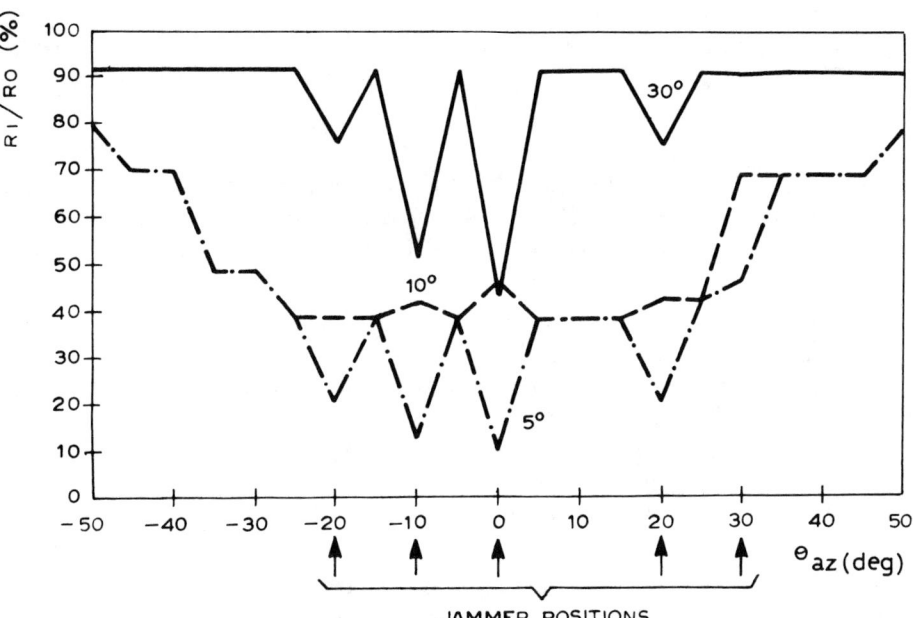

Figure 2.4 Radar range reduction against five jammers.

2.3 BASIC PRINCIPLES TO ACHIEVE LOW AND ULTRA-LOW SIDELOBE ANTENNAS

2.3.1 General Guidelines

The tapering functions of antenna aperture to achieve sidelobes below −40 dB have been well known for many years. They are Dolph-Chebyschev, Taylor, Hamming, Blackman, Bickmore-Spellmire, and Kaiser-Bessel. C.L. Dolph published his paper, titled "A Current Distribution for Broadside Arrays which Optimizes the Relationship between Beam Width and Side-Lobe Level," in June 1946 (Dolph 1946). Nevertheless, the application of this and other related theories was not successful until the mid 1960s. This was due to many reasons, such as (Schrank, 1988):

- The inadequate accuracy of amplitude and phase tapering across the antenna aperture necessary to achieve the theoretical low sidelobe patterns predicted by the tapering function;
- The incomplete understanding of the effect of mutual coupling between the radiating elements of an array;
- Techniques for affordable precision machining had not yet been developed;
- The development of pattern test ranges suitable for measuring low sidelobes was in its infancy.

In the AWACS program an S-band array of waveguides with 4000 radiating slots achieved sidelobes of -50 dB. The antenna was electronically scanned in elevation and mechanically rotated in azimuth. Key techniques and technologies adopted in the AWACS program were (Schrank 1988):

- Ferrite phase shifters, for scanning the beam in elevation, which achieved rms phase accuracy of less than 1°;
- A computer-aided design program for slotted waveguide arrays in which mutual coupling was taken into account;
- Computer-aided machining for milling the slots with exactitude at their precise locations and with their proper dimensions;
- Network analyzers and range equipment for high accuracy testing.

The basic lesson to be learned from this experience is that achieving low and ultra-low sidelobes requires providing the proper aperture distributions in amplitude and phase, and carefully controlling and compensating for the many error sources in the design, fabrication, assembly, and siting of the antenna (Skolnik, 1990, Section 7.5). This apparently simple rule is not as easy to apply as it appears due to the forthcoming considerations. Nevertheless, ULSA technology is increasingly required and implemented in modern radars (Adam, 1988, p. 26).

A low-sidelobe design increases the antenna cost and makes manufacturing difficult. For instance, assuming that digital phase control is used, additional bits of control may be required to reduce phase quantization errors. Increasing the amplitude weighting, conversely results in a less efficient use of the antenna aperture. Therefore, more elements must be added to the antenna array to achieve the required gain if severe amplitude weighting is used.

A very low sidelobe antenna, when tactically deployed, may not maintain the same sidelobe pattern measured on an antenna test range. This is so because nearby structures or obstacles can cause an increase in the sidelobe levels in certain directions. Typical rotating antennas for land-based applications have their sidelobes also determined by reflections from local terrain, buildings, and other structures. Similar difficulties are encountered in shipboard antennas, or aircraft and by radomes. Consequently, the antenna design must consider not only the sidelobes to be achieved by the isolated antenna, but also those achieved in a typical installation. Efforts should not be made to achieve very low sidelobes if they are degraded by the environment surrounding the installation (Billetter, 1989).

In summary, we conclude with the following points:

1. The first requirement in the design of a ULSA antenna is to choose the deterministic amplitude weighting for the aperture to achieve the desired antenna sidelobe levels. The aperture weighting causes a loss of antenna gain, on the order of 1 to 2 dB, and a broadening of the main beam;
2. The second requirement concerns the manifold design and manufacturing errors, which must be consistent with the desired design sidelobe level;
3. A system-level trade-off is usually made to balance the performance versus the cost to produce the design.

2.3.2 Aperture Tapering Functions for Low Sidelobes

As is well known, the antenna radiation pattern is the Fourier transform of the complex distribution of the antenna aperture illumination. If the amplitude distribution is uniform across the antenna and the phase distribution is linear, although it may be with a tilt angle relative to the antenna aperture, the radiation pattern is a "sinc" or $(\sin x)/x$ function with the main lobe steered at that tilt angle. The pair of first sidelobes, occurring at 1.6 θ_a (where θ_a is the half-power beamwidth), will be -13.6 dB from the main-lobe peak. The sidelobe peak amplitudes will not fall off below -30 dB until the tenth sidelobe, occurring at $\theta = 11.9 \, \theta_a$ (Barton, 1988, p. 150).

To achieve sidelobes lower than -13 dB, a tapered amplitude distribution is necessary, which results in a wider beamwidth and a correspondingly lower directive gain. The reduced gain can be expressed in terms of *taper loss* or its equivalent *aperture efficiency*. For example, an ultra-low sidelobe level of -50 dB decreases the efficiency with respect to a uniformly illuminated antenna by about 1.5 dB and widens the main lobe by a factor on the order of 2. Various amplitude distributions, such as Dolph-Chebyschev and Taylor, minimize the widening of the main beam and the reduction of antenna gain to obtain low sidelobes. If the optimum tapering is exactly implemented across the antenna aperture, the corresponding low sidelobes will inevitably be achieved due to the Fourier transformation relationship previously mentioned. However, if amplitude or phase errors exist, the resulting pattern will be degraded and the sidelobes will be higher (Schrank, 1988).

The optimum tapering function is defined as one that produces the narrowest beamwidth, measured between the first null on each side of the main beam, with no sidelobes higher than the stipulated level. Dolph solved this problem for a linear array of discrete elements by using Chebyschev polynomials. If the number of elements becomes infinite while element spacing approaches zero, the Dolph pattern becomes the optimum continuous line-source aperture, called the *Chebyschev pattern,* given by (Bodnar, 1987, p. 529):

$$E(u) = \cos \sqrt{u^2 - A^2} \qquad (2.7)$$

where

$$u = \frac{\pi L}{\lambda} \cos \theta$$

$$A = \operatorname{arccosh}(R)$$

and L is the antenna length, λ is the wavelength, θ is the azimuth angle, and R is the main-lobe-to-sidelobe voltage ratio. The corresponding antenna beamwidth is $\beta \lambda / L$, where β, called the beamwidth constant, is a complex function of R. The Chebyschev pattern provides a useful basis for comparison with other patterns, although it is physically unrealizable. In fact, all the sidelobes have the same amplitude value and the corresponding aperture distribution has an impulse (edge brightening) at both ends of the aperture (Bodnar, 1987, p. 529).

Taylor has developed a method for avoiding the above problems by arbitrarily closely approximating the Chebyschev pattern with a physically realizable pattern. The approximation was very close in the main-beam region, but let the wide-angle sidelobes decay in amplitude. Taylor used a closeness parameter $\bar{n}$ in the analysis. As $\bar{n} \to \infty$, the Taylor pattern approaches the Chebyschev pattern. By using the largest $\bar{n}$ that still produces a monotonic aperture distribution, we obtain the beamwidth constant β and the aperture efficiency η of Figure 2.5. Note that the η value of the Chebyschev pattern is not shown because it is very low. Additionally, notice that the beamwidth from the Taylor distribution is almost as narrow as that of the Chebyschev pattern while still producing excellent aperture efficiency (Bodnar, 1987, p. 530). Several other aperture tapering functions are known such as the $\cos^n(\cdot)$, Hamming, and many others (Harris, 1978). However, the Taylor distribution is becoming the most widely used. A second type of aperture optimization of interest is the one which minimizes the integrated (average) sidelobe level. This is accomplished by the Kaiser-Bessel aperture weight (Harris, 1978, p. 73) and may be more appropriate where the primary objective is to minimize total interference which is uniformly distributed in the sidelobe region. An example of various radiating patterns is given in Section 2.7.2.

2.4 REFLECTOR ANTENNAS WITH LOW AND ULTRA-LOW SIDELOBES

The conventional reflector antenna, comprising a doubly curved paraboloid and a horn feed in front of the reflector, is designed to give a reasonable compromise

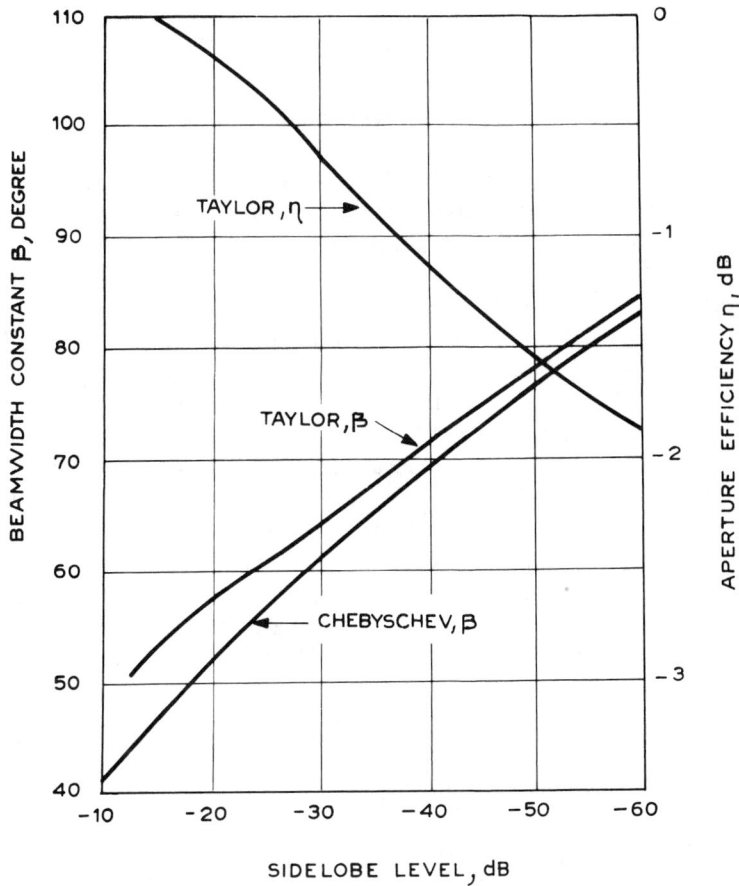

Figure 2.5 Aperture efficiency η and beamwidth constant β versus sidelobe level for line-source aperture distributions. (Adapted from Bodnar, 1987, pp. 531–532, *Principles and Applications of Millimeter-Wave Radar*, N. C. Currie and C. E. Brown eds., Artech House, Norwood, MA, 1987.)

between efficiency (i.e., η = 50–60%), first sidelobe level (−25 dB from mainlobe peak), and spillover lobes (−35 dB).

The basic limitations of this antenna type in terms of the achievement of low sidelobes are the following:

1. *Aperture blockage*—The placement of the feed and its support structure results in blockage of part of the energy radiated by the aperture. The pattern of the blocked aperture is estimated as the unblocked reflector pattern minus the pattern produced by the feeding structure. The blockage results in a gain reduction

and an increase of the sidelobe level of the unblocked reflector. To give an idea of the sidelobe level limitation produced by the blockage, an example shows that the ratio A_b/A (where A_b is the optical shadow area of the feed structure and A is the reflector area) should be on the order of 0.01 to reach a sidelobe level of -40 dB (Bodnar, 1987, p. 534).

2. *Feed with simple structure*—The feed, which is typically a simple waveguide horn, produces a cosine-shaped amplitude distribution, whereas the amplitude taper required for low sidelobes has inflected shape at the edges of the aperture. In other words, the tapering function should have a number of derivatives equal to zero at the edges of the aperture, which is not the case of a simple horn.

3. *Spillover*—Part of the power radiated by the feed misses the reflector and produces a pattern in the rear part of the reflector. Other causes are the reflector edge effects: diffraction and currents circulating around the edges onto the rear surface (Barton, 1988, p. 158).

4. *Reflector surface tolerance*—Loss caused by surface tolerance is normally small, except for antennas with very large L/λ (beamwidth less than 0.5°). The reflector surface tolerance affects both the phase and amplitude distributions in the aperture plane, thus degrading the sidelobe level. The nature of sidelobe perturbation depends on the amplitude of the deviation of the physical surface with respect to the desired theoretical surface, the spatial correlation of the deviations, and the level of the theoretical sidelobes. As has been shown for conventional antennas, the peak-to-peak error from the actual surface and the theoretical surface should be no more than 0.1λ to maintain sidelobe levels in the -25 to -30 dB range (Bodnar, 1987, p. 525). To reach -40 dB, the surface tolerance should be on the order of 0.06λ. Typical digitally controlled milling machines, capable of accuracies of ± 0.001 inch peak, are able to provide good surface accuracy compatible with low-sidelobe requirements. Additional errors, resulting when joining together the panels of a reflector, provide sidelobe level perturbations comparable to or greater than those previously mentioned.

The design techniques to obtain reflector antennas with low sidelobes are those which eliminate aperture blockage, use special feeds for the proper antenna aperture illumination, use absorbers and shielding to reduce the spillover, and use adequate milling machines to obtain good accuracy of the reflector surface. Some of these techniques are now briefly described.

To circumvent the blockage of a front-fed paraboloid, an advisable approach is to adopt an offset-fed paraboloid (Bodnar, 1987, p. 544), or, alternatively, an axisymmetric dual-reflector Cassegrain antenna using the polarization twist technique at the main reflector to avoid blockage by the grating subreflector (Barton 1988, p. 161).

Complex feeds, such as corrugated horns and focal-plane arrays, are designed to approximate Taylor or similar illumination functions. These oversized feeds must be offset to avoid blockage. An example is represented by a parabolic cylinder and a constrained line feed, which produce the low-sidelobe illumination function in the azimuth plane. The paraboloid is wide enough to minimize the spillover. The paraboloid is fed from a position offset in elevation to avoid blockage (Barton, 1988, p. 193).

A number of reflector edge treatments are used to reduce spillover and diffraction backlobes. Lining the rim of the dish with microwave absorbers provides some minor improvement. Castellating the rim significantly improves the front-to-back radiation ratio. Backlobe suppression with spillover reduction is achieved with the absorber-lined tunnels. These are cylindrical tubes of metal or dielectric lined on the inside with microwave absorbers to catch the spillover radiation and reduce the edge diffraction effects. Better than 60 dB front-to-back radiation ratios are reported (Schrank, 1988, p. 351).

In conclusion, we can state that to offset reflector antennas may have equal sidelobes performance to the slotted arrays in only a few special cases where counteracting measures, i.e., absorbing materials, *et cetera*, can be implemented. In general, most radars involve a scanning antenna, where the environment and the installation limitations would make such measures impractical. A reflector antenna can be made to approach the array performance in a few special cases, such as fixed services and point-to-point communication links.

2.5 ARRAY ANTENNAS WITH LOW AND ULTRA-LOW SIDELOBES

The degradation of sidelobe levels of an array antenna is due to mutual coupling of radiating elements, systematic (or correlated) errors, and random errors (Schrank, 1988). Mutual coupling can be compensated for by means of proper computer design programs used to determine the design parameters for the array radiating elements (slot, dipoles). Systematic errors, such as array non-flatness, must be controlled by careful fabrication processes. Systematic errors generally affect near-in sidelobes or give rise to periodic lobes. Random errors generally affect the far-out sidelobe level, and hence the average sidelobe levels of the antenna. Random errors are the factors that ultimately limit the sidelobe performance of an antenna (Schrank, 1988).

A list of error sources that have to be considered in low sidelobe array antenna design includes:

1. The phase and amplitude errors of the illumination of individual radiating elements, resulting from the feed network and the phase shifters;
2. The mechanical deviation in the element location;
3. The mutual coupling effects;
4. The failed elements.

A detailed analysis of the degradation of the average sidelobe level due to random amplitude and phase errors and the failure of radiating elements is described in Section 2.6. To give an idea of the quantitative effect of random errors, consider the following simple equation (Schrank, 1988, p. 343; Barton, pp. 157, 183):

$$\overline{SL} \text{ (dBi)} = 10 \log_{10} \pi(\sigma_\phi^2 + \sigma_a^2) \tag{2.8}$$

which states the relationship between the average sidelobe level, $\overline{SL}$, relative to isotropic and the standard deviations, σ_ϕ and σ_a, of the phase and amplitude errors, respectively. It is seen that to have -10 dBi average sidelobe level, the phase-only tolerance required amounts to $10°$ rms, while the amplitude-only tolerance required is 1.5 dB rms. The proof of an equation similar to (2.8) will be given in Section 2.6.

Ultra-low sidelobes can be achieved in slotted waveguide arrays because accurately milling the slot in the correct position and with the required dimensions is possible. The transmission line between slots is a very short, simple, and precise piece of waveguide. In addition, because the slots are fed serially, the design errors occur gradually over the length of the array (Schrank, 1988, p. 345). Arrays using phase shifters at each radiating element for electronic scanning in azimuth and elevation, are more difficult to design for low and ultra-low sidelobe performance. The requirement of -20 dBi average sidelobe or -45 dB for peak sidelobe is difficult to meet, even using phase shifters with several bits (Barton, 1988, p. 189).

Arrays having mechanical rotation in azimuth and electronic scanning in elevation may offer ULSA performance in the azimuth plane. The tapering for the antenna pattern in the azimuth plane may be obtained by *constrained feeding* (the signal is confined to waveguide, stripline, or coaxial feed networks with discrete branches connecting to the elements) or by *space feeding* (the signal is radiated from a horn and propagates as a wave through air or a continuous dielectric medium to reach the elements (Barton, 1988, p. 175). Constrained feed arrays may produce the precise illumination taper for a ULSA. The suspended stripline feed allows the achievement of the amplitude and phase accuracy required for ULSA performance. This precision cannot be achieved by using printed circuit board techniques because of dielectric constant variations inherent in large boards. An entirely new situation exists in this case, as compared to the slotted waveguide arrays, because the path to adjacent elements can pass through entirely different couplers and lines, although path lengths are designed to be equal. Errors are no longer gradual and exceptional precision is required to the many individual components in the path to a radiator to achieve ULSA (Schrank, p. 348). An example of a successful realization of horizontal networks of a planar array antenna using air-dielectric stripline is described in (Giaccari and Penazzi, 1989, p. 59), where -37 dB near-in sidelobes and -50 dB far-out sidelobes are shown to be obtained with this and additional advanced manufacturing techniques.

Another interesting realization using this system is the rotating planar (5m × 10m) array operating at UHF as described in (Carlson, et al., 1990). The antenna is made up of 14 air-dielectric stripline distribution networks that divide the energy from the feed point for each row to the row's 24 radiating elements. The stripline consists of parallel ground planes of aluminium with a center conductor running between them. Dielectric spacers support the center conductor, which is separated above and below from the ground planes by air gaps. Air-dielectric stripline was used for the distribution network to minimize losses and to avoid the illumination errors that result from nonhomogeneity of solid dielectrics. The azimuth pattern of the array results form an accurate tapering which provides 29 dB of gain above the isotropic level. The required sidelobe performance is a value more than 50 dB below the main beam peak or 21 dB below the isotropic level (-21 dBi).

Space-feed arrays may also provide the precise illumination taper in the azimuthal plane for ULSA performance due to the advances in feed horns and focal-plane array feeds. Space feeds have a major advantage in cost, as their power division occurs in the air. A disadvantage is that the feed horn structure, which illuminates the array, may produce spillover.

Modern stacked-beam antennas use planar arrays with an elevation beamforming matrix, while the antenna is still mechanically rotating in azimuth (Giaccari and Penazzi, 1989). In some systems, amplifiers are used upstream the RF matrix to ease the matrix design since moderate losses can be tolerated without affecting the system noise temperature. However, the gain and phase stability of these amplifiers is critical in establishing the elevation sidelobe levels. In addition, the failure of one row amplifier introduces large elevation sidelobes, even though it may have a small effect on the gain of each elevation beam channel (Barton, 1988, p. 197).

Digital beam-forming is the latest technology in flexibility of receiving arrays (Barton, 1988, p. 206; Farina and Galati, 1985, p. 266). Briefly, the signals from the array elements are down converted to a proper IF band and then converted from analog to digital form. Afterwards, a number of beams are digitally formed by linear combinations with proper weights of the received digital samples. The signals relevant to each beam are subject to further processing, such as doppler filtering and CFAR thresholding. Considering ULSA requirements, note that the entire error budget is allocated to the receivers and analog-to-digital converters; in fact, there are no phase shifters. Many bits are required to keep the sidelobe level at a reasonable level. The cost in receiver implementation and digital throughput is a large price to pay for the resulting flexibility offered by the DBF technology (Barton pp. 206–207). The topic of DBF is extensively described in Section 2.7.

In conclusion, the future challenge of low-sidelobe technology can be recognized in the design and fabrication of two-dimensional scanning, active aperture arrays with low sidelobes. Also, combining low sidelobes with adaptive, conformal, and DBF capabilities is quite interesting and attractive from an operational viewpoint (Schrank, 1988, p. 366).

2.6 ANALYSIS OF RANDOM ERROR EFFECTS ON AVERAGE SIDELOBE LEVEL OF ARRAYS

This section examines the effects of random errors on the sidelobe level of the radiation pattern of a phased-array antenna. The array errors that result from feed network, phase shifters, mechanical location, the orientation of radiating elements, and quantization phenomena can be taken into account by means of a phase error and an amplitude error for each element in the array. The analysis presented here considers the errors to be random variables, as they create random sidelobes added to the ideal antenna pattern. In a more general treatment, deterministic or correlated errors should also be considered; in this case higher sidelobes should be expected but only at a limited number of locations.

Two technical approaches are available in the literature; the first evaluates the average sidelobe behavior (Wirth, 1989; Groger *et al.*, 1989), while the second calculates the probability density function (pdf) of the sidelobe level. Both approaches will be described in this section, although the second technique is recommended when accurate analysis and design are required. In fact, operation requirements are correctly specified in terms of a histogram of sidelobe level rather than average sidelobe level. This is particularly true for a low-sidelobe antenna. For ECCM applications, to specify a very low value for the average sidelobe level is not enough. The antenna may have a few high peak sidelobes, making the radar vulnerable to jammers.

Consider a linear array having the following ideal radiation pattern:

$$f_0(u) = \sum_{i=1}^{N} a_i \exp\left[-j\frac{2\pi}{\lambda} x_i u\right] \qquad (2.9)$$

where

$u = \cos\theta$, with θ being the azimuth angle from the array broadside,
$N =$ the number of array elements,
$x_i =$ the coordinate of the ith radiating element,
$a_i =$ the weight amplitude applied at the ith radiating element.

Assume to have amplitude and phase errors, δa_i and $\delta \phi_i$, on the ith radiating element. The errors on the whole number N of radiating elements are independent and have a constant value of the pdf in the following intervals:

$$|\delta a_i| \leq \Delta a \quad i = 1, 2, \ldots, N$$
$$|\delta \phi_i| \leq \Delta \phi \quad i = 1, 2, \ldots, N \qquad (2.10)$$

The array pattern $f(u)$, subject to the amplitude and phase errors, is

$$f(u) = \sum_{i=1}^{N} (a_i + \delta a_i) \exp\left[-j\left(\frac{2\pi}{\lambda} x_i u + \delta\phi_i\right)\right] \quad (2.11)$$

which is a random variable with its own mean, variance, and pdf. The mean value of the array pattern is

$$E\{f(u)\} = \sum_{i=1}^{N} a_i e^{-j(2\pi/\lambda)x_i u} E\{e^{-j\delta\phi_i}\} + \sum_{i=1}^{N} E\{\delta a_i\} e^{-j(2\pi/\lambda)x_i u} E\{e^{-j\delta\phi_i}\} \quad (2.12)$$

The previous equation becomes

$$E\{f(u)\} = f_0(u) \frac{\sin \Delta\phi}{\Delta\phi} \approx f_0(u) \quad (2.13)$$

because $E\{\delta a_i\} = 0$, where the second equality holds for not very large $\Delta\phi$ values. The squared value of the array radiation pattern is

$$|f(u)|^2 = \sum_{i=1}^{N} (a_i + \delta a_i)^2 +$$

$$\sum_{i=1}^{N}\sum_{k\neq i}^{N} (a_i + \delta a_i)(a_k + \delta a_k) e^{-j(2\pi/\lambda)u(x_i-x_k)} e^{-j(\delta\phi_i-\delta\phi_k)} \quad (2.14)$$

The corresponding mean value is

$$E\{|f(u)|^2\} = \sum_{i=1}^{N} [a_i^2 + E\{\delta a^2\}] + \left(\frac{\sin \Delta\phi}{\Delta\phi}\right)^2 \cdot \sum_{i=1}^{N}\sum_{k\neq i}^{N} a_i a_k e^{-j(2\pi/\lambda)u(x_i-x_k)} \quad (2.15)$$

The variance of the array radiation pattern results to be

$$\text{var}\{f(u)\} = E\{|f(u)|^2\} - |E\{f(u)\}|^2 \quad (2.16)$$
$$= \frac{N}{3}\Delta^2 a + \left[1 - \left(\frac{\sin \Delta\phi}{\Delta\phi}\right)^2\right]\sum_{i=1}^{N} a_i^2$$

which is dependent on the maximum values of the amplitude and phase errors. The *relative sidelobe level* (RSL) is defined as

$$\text{RSL} = \frac{\text{var}\{f(u)\}}{|f_0(u=0)|^2} = \frac{\text{var}\{f(u)\}}{\left(\sum_{i=1}^{N} a_i\right)^2} \qquad (2.17)$$

$$= \frac{N\Delta^2 a}{3\left(\sum_{i=1}^{N} a_i\right)^2} + \left[1 - \left(\frac{\sin \Delta\phi}{\Delta\phi}\right)^2\right] \frac{\sum_{i=1}^{N} a_i^2}{\left(\sum_{i=1}^{N} a_i\right)^2}$$

By introducing the antenna efficiency η (note that $\eta = 1$ when any $a_i = 1$) (Carlier, 1979):

$$\eta = \frac{\left(\sum_{i=1}^{N} a_i\right)^2}{N \sum_{i=1}^{N} a_i^2} \qquad (2.18a)$$

and with the following approximation:

$$\left(\frac{\sin \Delta\phi}{\Delta\phi}\right)^2 \approx 1 - \frac{\Delta^2 \phi}{3} \qquad (2.18b)$$

the previous equation becomes

$$\text{RSL} = \frac{1}{3\eta} \frac{\Delta^2 a}{\sum_{i=1}^{N} a_i^2} + \frac{1}{3\eta N} \Delta^2 \phi \qquad (2.19a)$$

which, introducing the relative amplitude error $\Delta a' = \Delta a / a_i$ independent of the array element index i, gives

$$\text{RSL} = \frac{1}{3\eta N} [(\Delta a')^2 + \Delta^2 \phi] \triangleq \sigma^2 \qquad (2.19b)$$

This equation shows that the relative sidelobe level is limited by the maximum amplitude and phase errors to a quantity σ^2, which decreases as the number of radiating element is increasing.

Table 2.1 shows the average sidelobe level due to random errors induced by the quantization of amplitude and phase element weights. In fact, with a phase shifter having P bits, for instance, the phase can be set to the desired value with a residual error:

$$\Delta\phi = \pm\frac{\pi}{2^P} \quad \text{(peak phase error)} \tag{2.20}$$

$$\sigma_\phi = \frac{\pi}{\sqrt{3}\,2^P} \quad \text{(root mean square phase error)} \tag{2.21}$$

Similar considerations apply to the amplitude quantization. The exercise refers to a planar array with a number N of radiating elements taken as a parameter of the table, the antenna efficiency is $\eta = 0.5$, corresponding to a certain tapering of the antenna illumination; the mean amplitude excitation is $\bar{a} = 0.5$ and $a_{\max} = 4\bar{a}$. The table indicates the effect of the amplitude and phase errors due to quantization with a limited number of bits. As anticipated, an increase of the radiating elements is of great benefit to achieve low sidelobes, although a coarse quantization is applied. In fact, 4 bit quantization of the amplitude weight corresponds to a quantization step of 6.25% and a value of $\Delta a' = 0.0625\sqrt{2} = 0.0887$, while 4 bit quantization of

Table 2.1
Average Sidelobe Level due to Errors by Quantization for a Planar Array
$\eta = 0.5$, $a_{\max}/\bar{a} = 4$, $\bar{a} = 0.5$ (from Wirth, 1989; Groger et al., 1989)

Number of Elements	1000	5000	10,000
Amplitude Errors			
2-bit	−40.7 dB	−47.7 dB	−50.7 dB
3-bit	−46.8 dB	−53.8 dB	−56.7 dB
4-bit	−52.8 dB	−59.8 dB	−62.8 dB
Phase Errors			
2-bit	−33.8 dB	−40.8 dB	−43.8 dB
3-bit	−39.8 dB	−46.8 dB	−49.8 dB
4-bit	−45.8 dB	−52.8 dB	−55.8 dB
Element Failure K/N			
1%	−46.9 dB	−53.9 dB	−56.9 dB
10%	−36.9 dB	−43.9 dB	−46.9 dB

the phase weight gives $\pm 11.25°$ (0.19634 rad) of quantization step. Application of (2.19b) gives the numerical values in Table 2.1.

Examine the relative sidelobe level when there are failures of radiating elements. At first, consider the case of the failure of just one radiating element, for example, the jth. The RSL is as follows:

$$\text{RSL} = \frac{(a_j)^2}{\left(\sum_{i=1}^{N} a_i\right)^2} \tag{2.22}$$

which is essentially the ratio between the squared weight amplitude of the jth failed element divided by $|f_0(u=0)|^2$. The mean value of RSL due to the failure of one generic radiating element is

$$\text{RSL} = \frac{\sum_{i=1}^{N}(a_i)^2}{N\left(\sum_{i=1}^{N} a_i\right)^2} = \frac{1}{N^2 \eta} \tag{2.23}$$

where the second equation is obtained by introducing the definition of antenna efficiency η (2.18a). Note that the RSL value now depends on $1/N^2$ in contrast to the dependence of $1/N$ due to random errors on the weights. Finally, if K out of N radiating elements fail, the corresponding RSL value is

$$\text{RSL} = \frac{K}{N} \frac{1}{\eta N} \tag{2.24}$$

where (K/N) is the fraction of failed elements. Table 2.1 also illustrates the effects on the sidelobe level of a certain percentage of failed elements. For instance, a 1% element failure limits the maximum achievable sidelobe level to -54 dB for an antenna having 5000 elements. Increasing the overall number of radiating elements causes an improved value of sidelobe level.

Let us now present the pdf of the antenna sidelobes. This topic has been studied at length by many authors. See, for example, (Ruze, 1966; Carlier, 1979; Hsiao, 1984a, b). These papers show that the statistical distribution of the antenna radiated power is as follows:

$$p(r) = \frac{2r}{\sigma^2} \exp\left[-(a^2 + r^2)/\sigma^2\right] I_0(2ar/\sigma^2) \tag{2.25}$$

where r^2 is the actual radiated power $|f(u)|^2$ by the antenna, a^2 is the error-free radiated power $|f_0(u)|^2$, σ^2 is the array error power as given by (2.19b), and $I_0(\cdot)$ is the modified Bessel function of the first kind. Remember that the same pdf describes the amplitude of a sinusoidal function plus a random noise having a Gaussian pdf for the in-phase and quadrature components. In fact, the actual array pattern is the error-free array pattern (2.9), which is the sum of cosine-sine terms, degraded by amplitude and phase errors for each cosine-sine term. Equation (2.25) may be integrated to produce the amount of sidelobe level above any given level. For instance, Figure 2.6 shows the sidelobe levels having 10% probability of being exceeded in a real array, as a function of theoretical sidelobe levels, for several values of error power σ^2. As a numerical application, consider the case of -40 dB error-free sidelobe level. The actual sidelobe level (which will be exceeded with probability 0.1 in a real array) will be -39 dB for $\sigma^2 = -60$ dB, -37 dB for $\sigma^2 = -50$ dB, -34 dB for $\sigma^2 = -40$ dB, and so on. Also, note that, as the design sidelobe level is reduced, the actual sidelobes tend to the value of σ^2 given by (2.19b). The reduction in design sidelobe level (that of the unperturbed radiation pattern $f_0(u)$) is usually

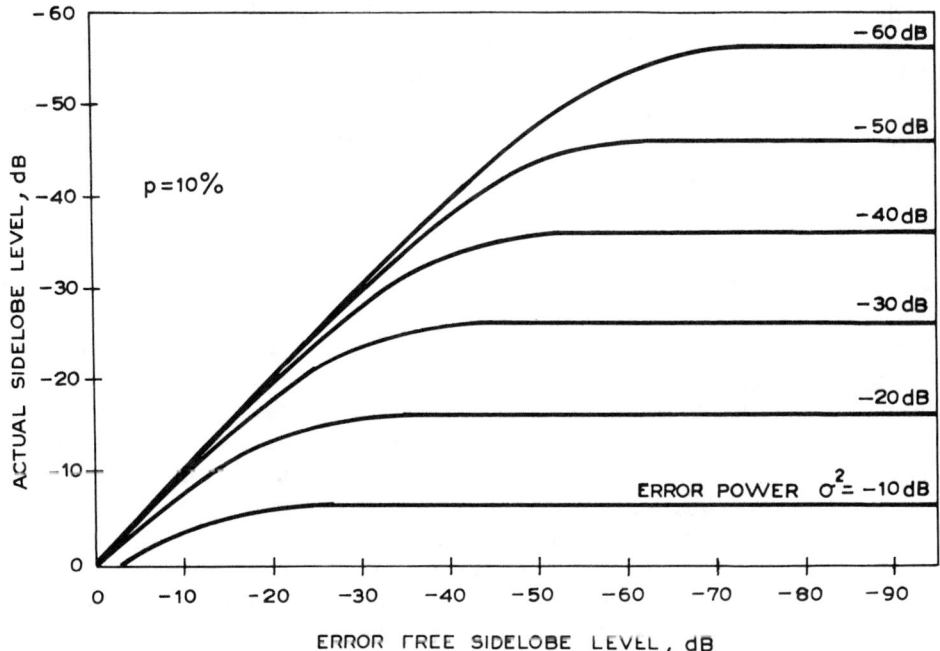

Figure 2.6 Sidelobe level exceeded in 10% of cases for various excitation errors, as a function of the error-free sidelobe level. (From Carlier, *The Marconi Review,* Second Quarter, 1979, pp. 119–134.)

accompanied by an increase in beamwidth, and hence a reduction in illumination efficiency, η, which, in turn, increase the achievable value σ^2. Thus, an optimum value of design sidelobe level exists for most arrays and is on the order of σ^2 (Carlier, 1979).

2.7 DIGITAL BEAM-FORMING (DBF) TECHNIQUE

The purpose of this section is to provide design criteria for new ULSA antennas as opposed to how it has been done in the past. The technology is available or soon to be for a number of narrow-bandwidth (i.e., a few megahertz) array applications, having the digital ports that are one-dimensional, e.g., in azimuth, where this type of flexibility is most needed. Digital technology is moving so fast that much wider bandwidths may be viable in the not too distant future. Many of the very advanced array-related ECCM techniques and numerous solution algorithms have been very predominant in the technical literature since the late 1970s. While the cancellation functions which use this technology will be well described in Chapter 5, the DBF concept and DBF specific hardware issues will be discussed in this section. Specifically, the basic DBF terminology and concepts will be introduced in Section 2.7.1. Subsequently the various beam-forming techniques will be the topic of Section 2.7.2. The analog-to-digital (A/D) conversion schemes are the subject of Section 2.7.3. In Section 2.7.4, a detailed analysis of the performance limitations of DBF is presented. The procedures to calibrate the DBF are illustrated in Section 2.7.5. The two closing sections deal with technological issues and existing DBF systems, respectively.

2.7.1 Basic Concepts

The term "digital beam forming" is taken to refer to a processing structure that accepts the digital signals from an array antenna and performs spatial processing on them. More precisely, an array antenna samples the electromagnetic waves at many locations in the antenna's aperture. Subsequently, the signals from each receiving element are individually converted to complex digital numbers at megahertz rates and transferred to a high-speed digital processor. The end-product of this processing is a set of beams differently oriented in space, each beam giving access to a number of range and doppler cells. Here, DBF is considered for the receiver mode.

A generic DBF comprises an array of receiving antennas, receiver modules, analog-to-digital converters (ADCs), a digital beamformer, a controller, and a calibration unit (Figure 2.7). The receiver module is a complete heterodyne receiver performing the functions of frequency down-conversion, filtering, and amplification to a power level commensurate with A/D conversion. The ADC plays a key role because it determines the dynamic range of the DBF. The dynamic range represents

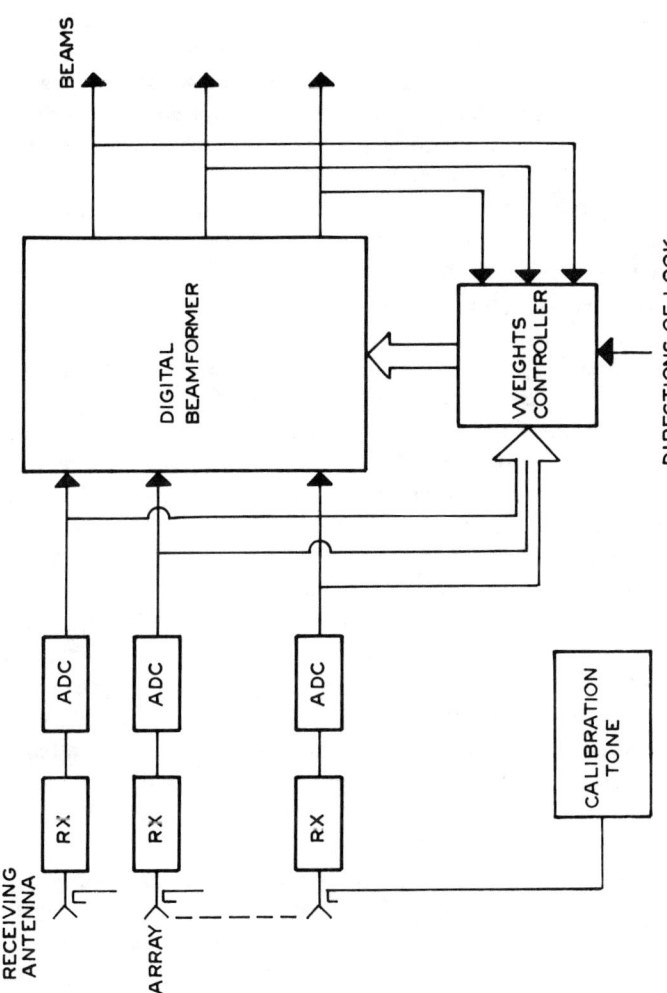

Figure 2.7 Basic scheme of a DBF.

the range of signal strength over which the system will perform as expected. For a DBF, the dynamic range increases by 6 dB per bit of the ADC and by the number, N, of parallel receiving channels, expressed in dB (Steyskal, 1987).

The digital beamformer, a very fast parallel processor, should have capability of billions of operations per second. It forms multiple beams by finding the inner product of the set of received samples from the array elements and the sets of weights that differently shape the beams originating from the antenna. The weights are generated by a separate controller. The controller calculates the weights on the basis of the desired look directions and depending on the amount of interference received at the input channels (*open-loop adaptivity*). *Closed-loop adaptivity* methods foresee the use of the residual disturbance at the output beams to determine the weights. The computational load on the weights controller is also of interest, depending on the particular control algorithm that is implemented (see Chapters 4 and 5). Roughly speaking, we can say that the load is proportional to N^3, which is the number of multiplications required for inversion of N-dimensional interference covariance matrix. More efficient methods may require N^2 multiplications, or even less. These operations, however, need to be performed only at a rate of change of the external environment, e.g., one to one hundred times per second. Thus, the total load (typically, several million operations per second, MOPS), for practical applications, is not trivial, but certainly possible. Consequently, the hardware implementation of the controller is envisaged as a programmable processor having a collection of software-coded algorithms for linear and nonlinear processing of array data. Finally, the calibration unit is needed to guarantee the matching of the receiving channels, and thus obtaining low sidelobes.

The DBF processing scheme contrasts with traditional phased-array beamforming, shown in Figure 2.8, where the signals gathered by the array are properly shifted in phase, summed in an analog device, down-converted to the proper frequency bands, and successively transformed into digital words. By this well proven technique, a limited number (less than 10, in general) of contemporary beams can be formed at the RF or IF stage with passive phased-array antennas and analog technology. The disadvantages of this technique are related to the difficult control of the sidelobe level, high loss, absence of individual beam shape control, complex construction, and corresponding heavy equipment.

The DBF technique of Figure 2.7 allows, in principle, to form far more than 10 contemporaneous beams. Closely spaced beams without penalty in signal-to-noise ratio (SNR) can be obtained because the SNR values have already been established in the low-noise amplifiers preceding the digital beamformer. Additional advantages of this technology are the separate control of each beam and the possibility to shape the antenna pattern in range and azimuth in a deterministic or adaptive fashion. The antenna pattern can be shaped as a function of range to place nulls in range cells containing clutter and chaff by using different weight sets (for the combination of

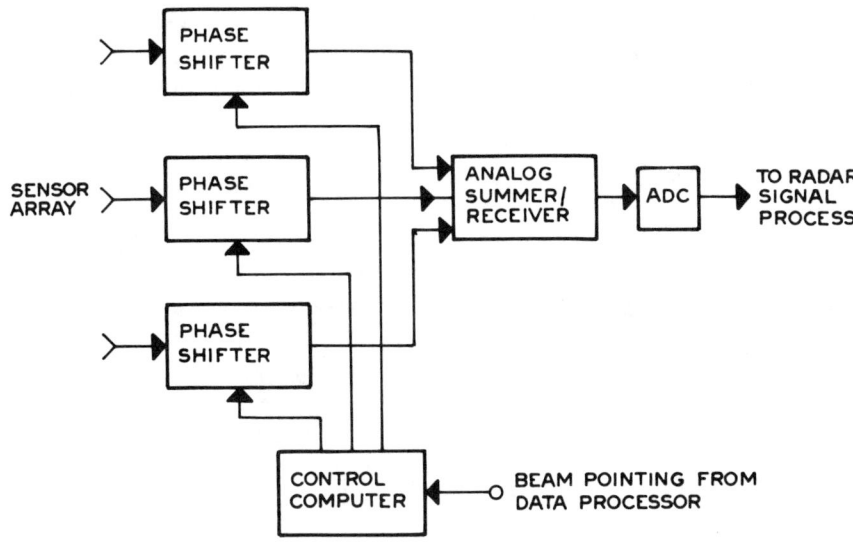

Figure 2.8 Traditional phased-array beam-forming.

phased-array elements) in different range cells. The antenna pattern can also be adaptively controlled to reduce the effect of directional jammers with unknown characteristics. The weights control law involved in the jammer suppression is quite complex, and can only be efficiently implemented in digital form. Moreover, the null depths are not affected by amplitude and phase errors in antenna elements and receiver channels because they can be compensated by means of calibration techniques implemented in the DBF. This approach avoids the requirement of very tight absolute tolerances in receiver channels. In addition to the use of the internal calibration tone indicated in Figure 2.7, another calibration technique applies a test signal to the input of the antennas by an auxiliary antenna placed, for instance, in the near-field of the array, but outside the normal scanning range of the beams. The signals produced at the receiver outputs are used to modify the weights of the array elements (see Section 2.7.5 for details). Moreover, the on-line calibration allows for very low sidelobes in the antenna pattern due to the compensation, which can be carried out on the different sources of sidelobe increments, such as errors in the aperture illumination, component fabrication tolerance, mismatching of receiving channels, scattering, and multipath of the environment in which the antenna is deployed.

The main disadvantage of the DBF technique relates to the system complexity, which increases with the signal bandwidth, number of array elements, and number of simultaneous formed beams. Other limitations are the availability and cost of

ADCs, frequency and dynamic-range characteristics of ADCs, and the throughput (data rate) required for the digital beamformer. Partial implementation is possible by digitizing at the subarray level, as shown in Figure 2.9, thus reducing the system complexity and cost.

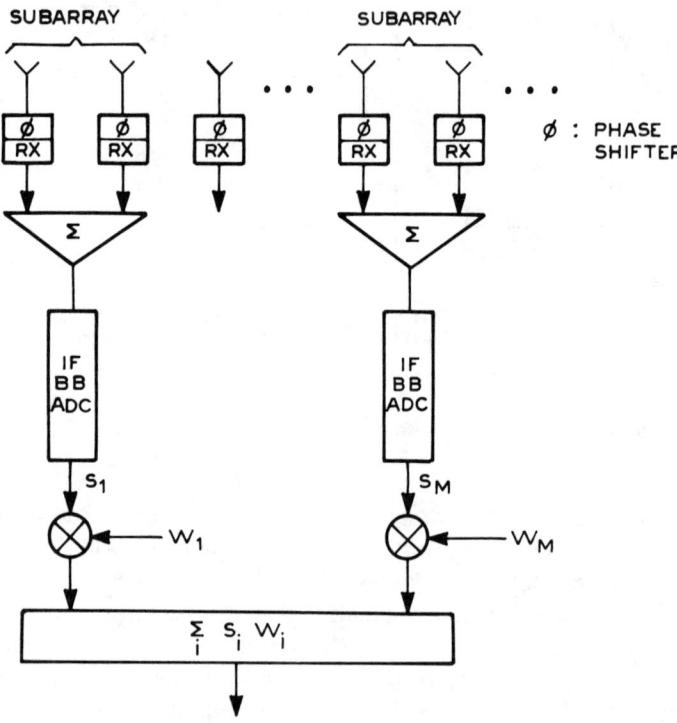

Figure 2.9 Beam-forming at subarray level.

2.7.2 Beam-Forming Techniques

The DBF is a powerful tool to generate a cluster of beams. A modern beam-forming technique is based on the frequency-domain approach. Due to the linearity of the beam-forming process, the Fourier transform $B(\omega, \theta_1)$ of the beamformer output is obtained by the Fourier transform $S_n(\omega)$ ($n = 1, 2, \ldots, N$) of the sensor outputs as follows:

$$B(\omega, \theta_1) = \sum_{n=1}^{N} a_n S_n(\omega) e^{-j\omega t_n(\theta_1)} \tag{2.26}$$

where $\omega = 2\pi f$ denotes the angular frequency, a_n indicates the weighting coefficient and $t_n(\theta_1)$ is the delay required to steer the beam in the direction θ_1. Equation (2.26) applies to a linear array. For narrowband signals, the beam pattern $B(\omega, \theta_1)$ may be synthetically indicated $B_1(\theta)$, where the angular frequency ω has been suppressed, the angular direction of the beam has been put as a subscript, and θ indicates the angular coordinate of the beam. For an array with N receiving elements and interelement distance d, the previous expression (2.26) becomes

$$B_1(\theta) = \sum_{n=1}^{N} a_n s_n e^{-j2\pi f(d/c)\sin(\theta-\theta_1)} \qquad (2.27)$$

where c is the light speed. In the digital beam-forming technique, the received signals $s_n\{n = 1, 2, \ldots, N\}$, the weighting a_n, and the phase shifts are represented by digital words. The summation is also performed digitally.

When a large number of similar beams are to be generated, the fast Fourier transform (FFT) algorithms (noteworthy is the well known Cooley-Tukey radix 2) is recommended to reduce the computational load of operations, as the one indicated by (2.27). The number of real multiplications for an N-point FFT is on the order of $[2N(\log_2 N - 3) + 8]$ (Barton, 1980; Steyskal, 1987) if N is a power of 2, against $4N$ real multiplications to form just one beam with the discrete Fourier transform (DFT) approach. The FFT is so efficient that, as long as the number of elements is less or equal to 2000, fewer operations are needed to compute the entire set of N FFT beams than to compute four customized beams. The computation efficiency of the FFT is gained at the expense of a reduced flexibility in controlling the individual beam response (i.e., all the patterns are identical and the beam spacing is determined by orthogonality constraints).

A linear array is capable of forming beams in one dimension, whereas a planar (rectangular) array is capable of forming beams in two dimensions by applying the proper phase at each element. For a rectangular array of MN elements, the two-dimensional (2D) beam-forming operation essentially consists of a two-dimensional FFT, which can be obtained by taking an N-point FFT on each of the M rows followed by an M-point FFT on each of the N columns. The fan-shaped beam formed in the first FFT is narrowed to a pencil beam on the second FFT. The basic processing scheme is shown in Figure 2.10.

To have an idea of the processing requirements for multiple digital beamformers, consider a 2D array of 16×16 antenna elements used to form 256 beams in azimuth and elevation directions. If the instantaneous signal bandwidth is 1 MHz, the 256 beams must be formed in each time resolution cell of 1 μs. This means that sixteen 16-point FFTs have to be completed in 1 μs because of the pipelining of the data from row to column FFT. Therefore, the maximum time allotted for each 16-point FFT operation is 62.5 ns. Consequently, the digital beamformer processor of Figure 2.7 should be able to perform one real multiplication every 1.56 ns, i.e., 0.64

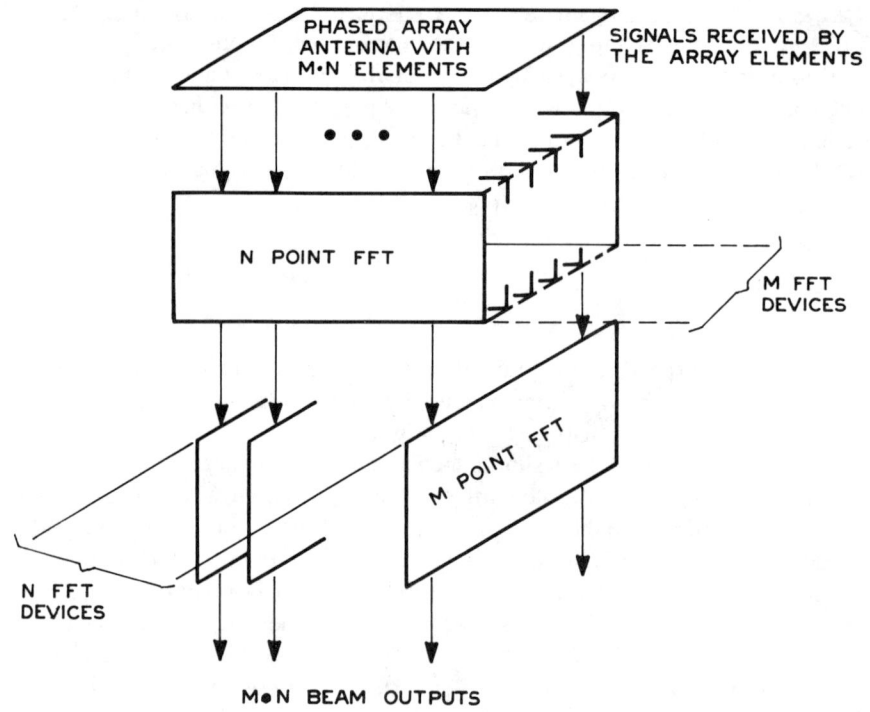

Figure 2.10 DBF with FFT processors.

$\times \ 10^9$ operations/s. Steyskal and Rose (1989) report that a very fast beamforming processor is in construction. The expected performance of the processor is on the order of 25×10^9 operations/s.

An alternative technique related to the formation of custom beams, each with an *ad hoc* shape and pointing direction, uses the DFT, where the weights of the array signals are individually selected to produce a specific radiating polar diagram. In this way, any number of closely spaced, low-sidelobe beams can be formed without degradation in SNR. This contrasts with analog multiple-beamformers (Butler matrix, Rotman lens), where the output beams must be orthogonal to avoid loss of signal power. A noticeable advantage of the method is the possibility of selecting and applying *ad hoc* weighting functions (e.g., Kayser-Bessel, Dolph-Chebyschev, Gaussian, *et cetera*) to obtain low and even ultra-low sidelobes. Examples of the theoretically achievable patterns for a linear antenna of 16 elements 0.5λ far apart are shown in Figure 2.11 (DiVito and Iovino, 1987). In particular, Figure 2.11(a) sketches, as a reference, the uniform weighting, whereas Figures 2.11(b) to 2.11(d) refer to the Dolph-Chebyschev, Gaussian, and Kaiser-Bessel weighting functions. Sidelobes 40

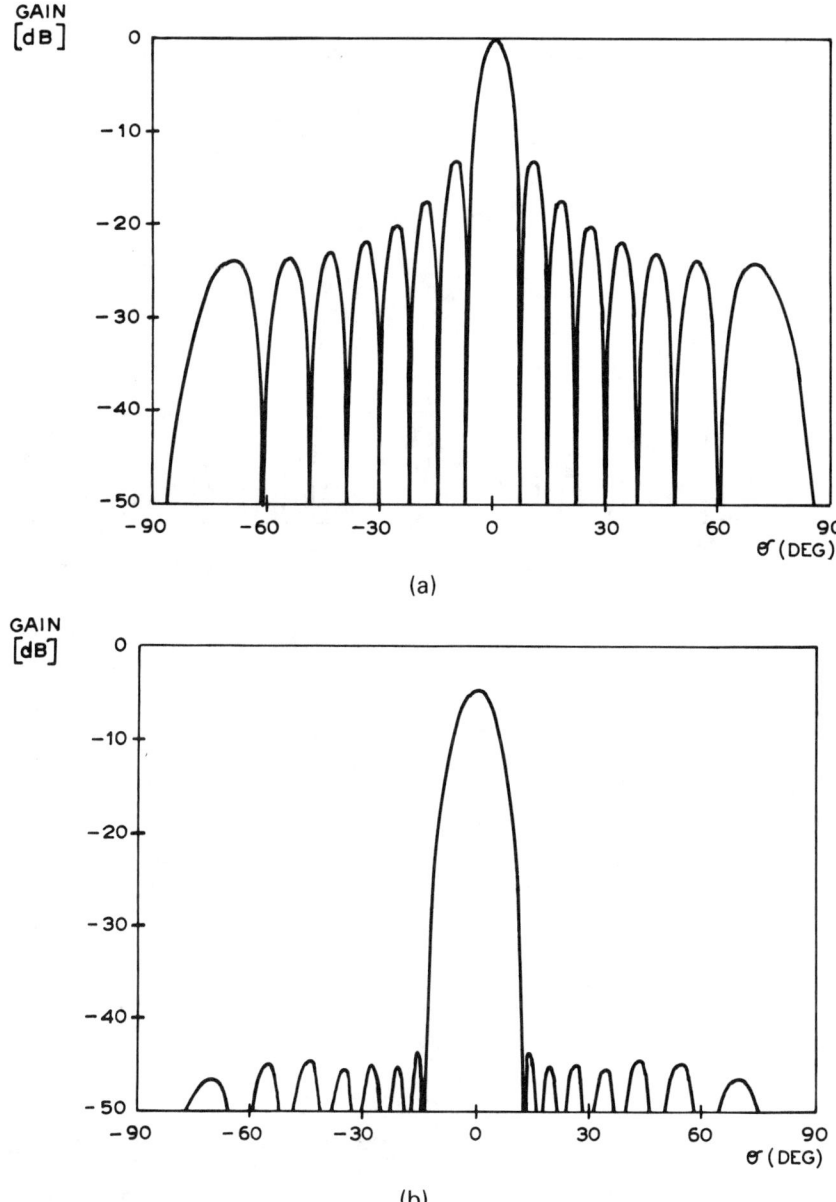

Figure 2.11 (a) Beam pattern, uniform weighting; (b) beam pattern, Dolph-Chebyschev (40 dB) weighting; (c) beam pattern, Gaussian weighting with parameter $\alpha = 2.5$; (d) beam pattern, Kaiser-Bessel weighting with parameter $\alpha = 1.7$.

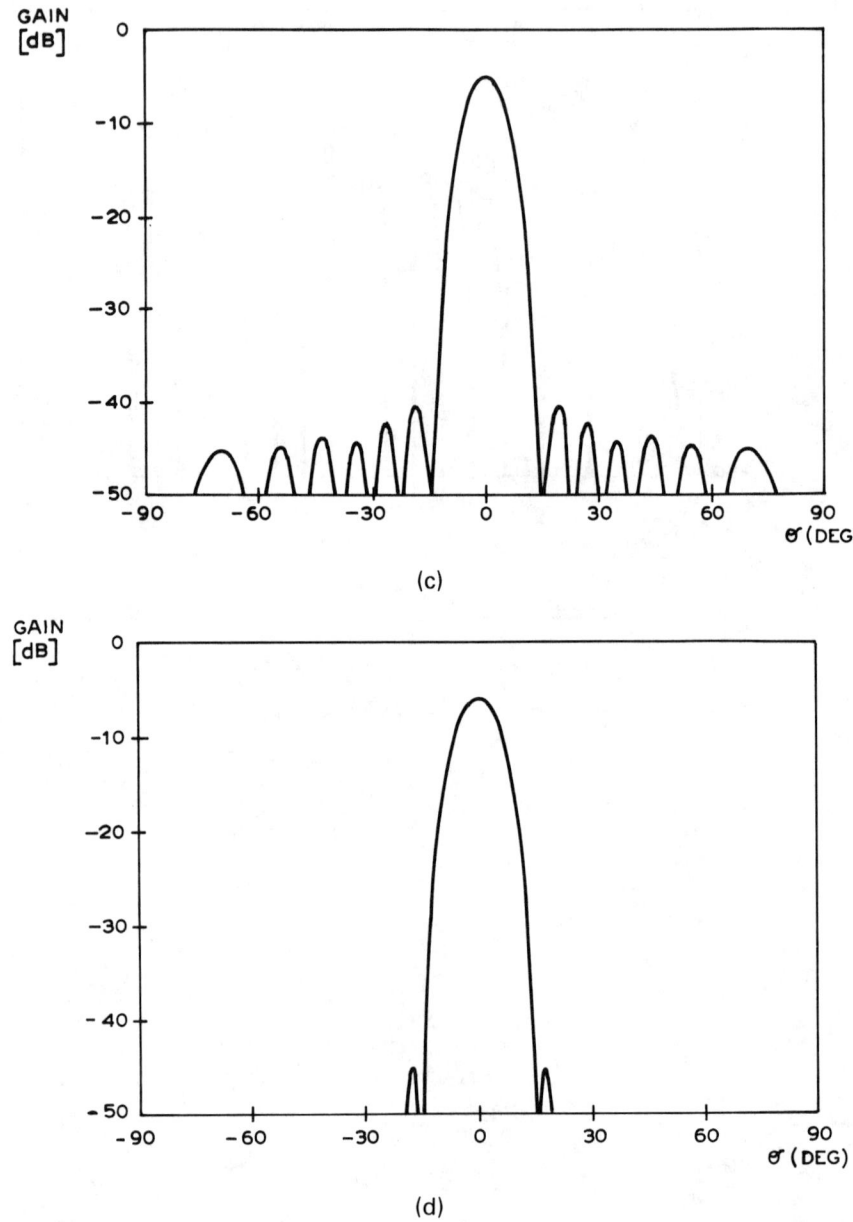

Figure 2.11 continued.

dB down from peak are achieved by these weightings at expense of a broadening of the main beam. To maintain angular resolution while reducing sidelobes results in increased aperture sizes with concomitant cost increases. As an example, reduction of sidelobes from −25 dB to −50 dB requires a 30% increase in elements to maintain constant angular resolution (Loomis, 1990).

The technique of customized beams is computationally intensive for the digital beamformer. The processing scheme for this technique is sketched in Figure 2.12. We note that the number of beams that can be obtained is not strictly related to the number of antenna elements as is the case for the previous FFT approach.

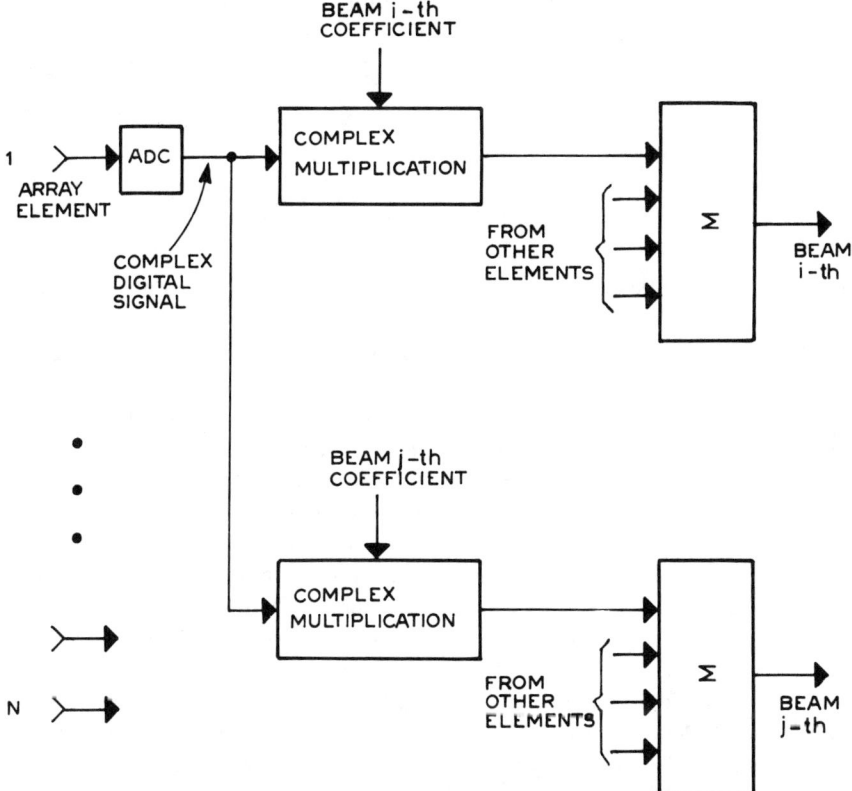

Figure 2.12 Digital formation of customized beams.

2.7.3 Analog-to-Digital Conversion Options and Related Processing Schemes

The bandpass signal $s(t)$ received by an element of the array antenna is

$$s(t) = I(t)\cos 2\pi ft + Q(t)\sin 2\pi ft \tag{2.28}$$

where f is the carrier frequency, $I(t)$ and $Q(t)$ are the in-phase and quadrature components of the baseband modulating signal. The useful information $I(t)$ and $Q(t)$ must be extracted from the RF signal and represented by digital words. This section illustrates two technical approaches: (1) conversion to baseband and digitization, and (2) direct sampling and digitization of the IF signal.

Coventional radars use a pair of analog mixers to extract the $I(t)$ and $Q(t)$ signals prior to digitizing the two baseband channels. The mixers, amplifiers, and low-pass filters of Figure 2.13 need an accurate alignment to ensure matching of their phase and gain responses across the signal bandwidth. The two baseband channels are digitized by two ADCs which should sample the modulation envelopes at the same instant of time. The Figure 2.13 also shows the presence of a digital phase shifter, which can be implemented by means of the following numerical algorithm, requiring four real multiplications:

$$I'(n) = I(n) \cos\theta - Q(n) \sin\theta$$
$$Q'(n) = I(n) \sin\theta + Q(n) \cos\theta \qquad (2.29)$$

where θ is the phase shift value to be applied to the array element. An amplitude weighting also may be applied by modifying the equation.

The required sampling rate is derived by looking at the Figure 2.14. In particular, Figure 2.14(a) sketches a typical radar pulse with the 3 dB spectrum width of B. In the baseband conversion (Figure 2.14b), the highest significant frequency is approximately $0.7B$ for this example, and thus the Nyquist criterion requires a minimum sampling rate of $1.4B$ in each of the two baseband channels (Barton, 1980).

An alternative to this method is to use a single, fast ADC to sample directly the IF waveform. Mixing, low-pass filtering, and other processing functions are then performed digitally. This approach offers better matching of the phase and amplitude responses of the in-phase and quadrature components than with the approach using dual ADCs. For instance, there are no I and Q video amplifiers to align and dc offsets are inherently low. The corresponding processing scheme is that shown in Figure 2.15. The signal samples $s(n)$ are multiplied by the samples of the local oscillator. This corresponds to sampling the signal $s(n)$ at a time instant in which only one component (either $I(n)$ or $Q(n)$) is different from zero. In the successive time instant, the other component will become different from zero. Because the in-phase and quadrature components are sampled at different instants of time, they have to be aligned in time by interpolation filters, usually implemented as a finite impulse response (FIR).

To quantify the required sampling rate, consider Figure 2.14(c), which shows the signal spectrum down-converted to an IF different from zero. It is preferable to minimize the IF so that the sampling rate and hence the conversion rate required of the ADC are also minimized. However, the "folding" of the spectrum around zero

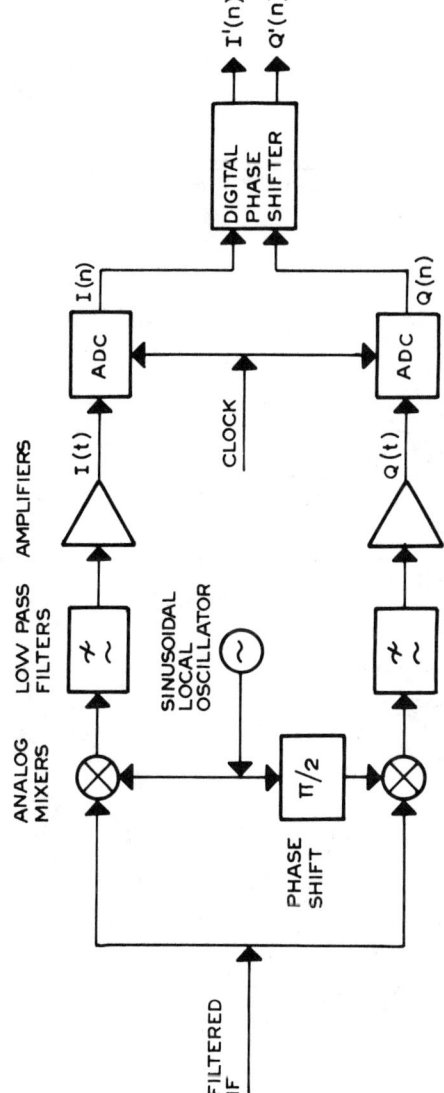

Figure 2.13 Scheme of baseband digital conversion in a receiving channel of a digital beamformer.

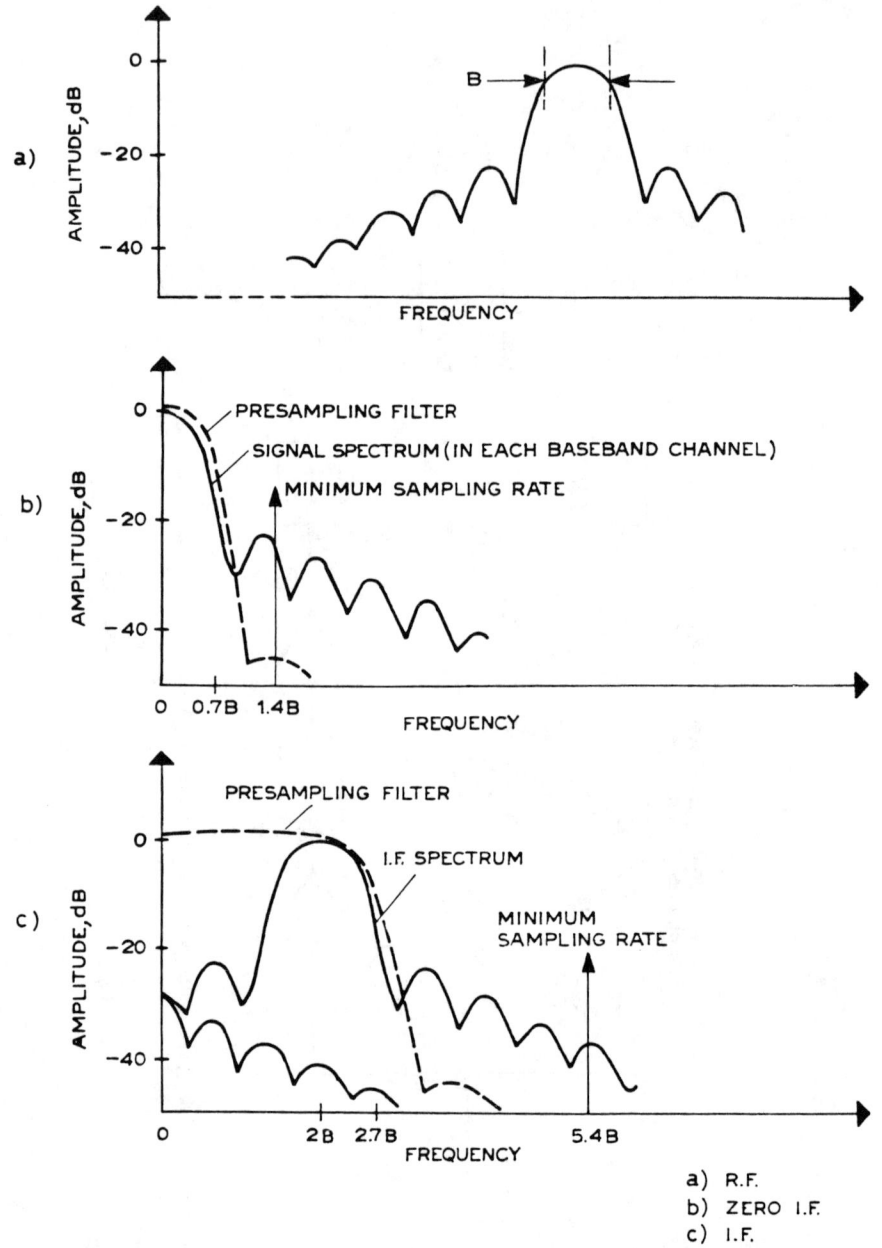

Figure 2.14 Sampling a typical radar pulse spectrum. (From Barton, *IEE Proc.*, Vol. 127, Pt.F, No. 4, August 1980, pp. 266–277.)

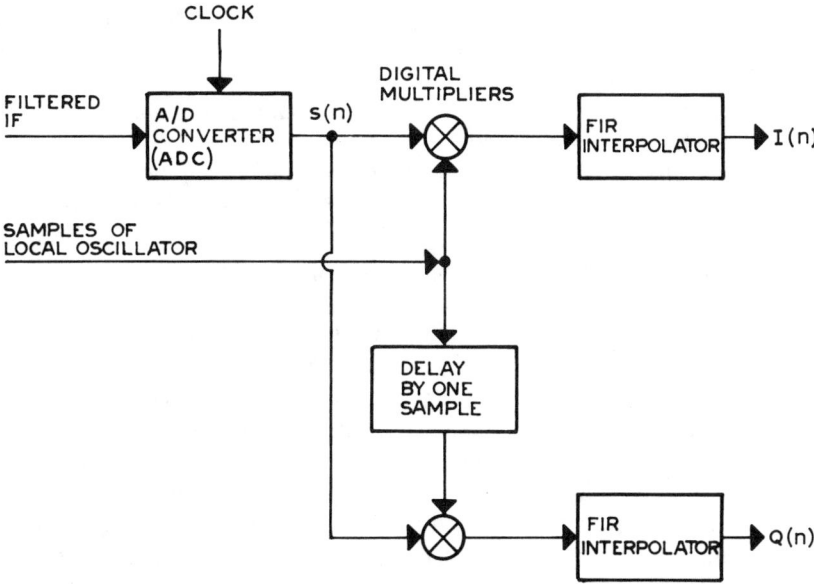

Figure 2.15 Scheme of IF Digital conversion in the receiving chain of a digital beamformer.

frequency causes spurious contributions to occur in the bandpass of the IF filter, which will add to the main spectrum. To keep the distortion small, the relative level of the folded spectrum within the IF filter should be less than about −40 dB. Having chosen an IF equal to $2B$ for the example and assuming that the IF filter has a bandwidth of significant response of $1.4B$, by application of Nyquist's criterion for adequately sampling the signal spectrum, the minimum required sampling frequency, being twice the upper frequency limit of the useful spectrum, is $2[(2 + 1.4/2)B] = 5.7B$ (Barton, 1980). The practical applicability of the digitization at IF is related to the availability of sufficiently fast ADC devices with the required number of bits for dynamic-range purposes. An experimental prototype, described by Jackson and Matthewson (1986), uses a seven-bit monolithic bipolar flash converter with sampling rate of 12 MHz, four times the IF center frequency of 3 MHz. The authors report a measured gain match across the frequency band within 0.1 dB and the phase orthogonality being everywhere less than 0.2° from the ideal 90° split.

Another experimental realization is described in (Steyskal and Rose, 1989). It concerns a receiver module for DBF that is divided into three subsections: RF (C-band input to 45 MHz output), IF (45 MHz input to 500 kHz output), and digital (500 kHz input to 12-bit in-phase and quadrature-phase outputs at 500 kHz data rate). The IF sampling incorporates a single ADC capable of capturing the IF signal and performing the analog-to-digital conversion at four times the signal bandwidth. The

I and Q components are then generated digitally through an FIR filter. This technique is intended to eliminate dc offset errors and to maintain orthogonality based on the digital word size used in calculating the I and Q signals. The ADC selected for the receiver operates at a frequency of 2 MHz, which results in a 500 kHz I and Q data rate with a 400 kHz signal bandwidth.

A further realization of digital baseband quadrature sampling is the one described in (Teitelbaum, 1991). It refers to a rotating UHF planar array consisting of a vertical stack of 14 corporate-fed rows. MIT Lincoln Laboratory is developing this experimental array to adaptively suppress, with digital techniques, jammers in the elevation sidelobes. Each of the antenna rows feeds a channel receiver and an ADC. Since the accuracy of conventional analog-quadrature video detectors would limit the nulling performance, signals are digitized at 1 MHz intermediate frequency and digitally processed to baseband quadrature channels. The IF signals in each channel are oversampled at a 4 MHz rate to recover the in-phase and quadrature-phase components. Because of the 4:1 oversampling, the process of digitally mixing to baseband in quadrature channels becomes trivial, requiring only multiplication by $+1$, -1, and 0. A FIR low-pass filter is used to eliminate the negative frequencies sideband of the sampled waveform as well as any DC offset introduced by the ADC. The sampled waveform is then decimated by a factor of 4 to obtain a 1 MHz complex sampling rate.

2.7.4 Performance Limitations in a Digital Beam-Forming Network

In this section, we examine the effects of the errors present in the various sections of the receiving channels in a digital beam-forming network. In particular, we try to answer the following relevant question. How closely does each receiving channel need to match each other in phase and amplitude to achieve low and even ultra-low antenna sidelobes? The analysis is undertaken for the linear array of N receiving elements shown in Figure 2.16. The figure illustrates the composition of each receiving channel, which is the cascade of (1) the RF low-noise amplifier; (2) the down-conversion from RF to IF, the bandpass filter, and the IF amplifier; (3) the baseband conversion and phase detection to produce the in-phase (I) and quadrature (Q) channels; (4) the A/D conversion; and (5) the digital network for the formation of M beams. We should note that the number M of beams is generally different from the number of receiving antennas. In this general case, the digital beam-forming is operated by a suitable (M, N) matrix **W** of weights, as follows:

$$\mathbf{B} = \mathbf{WS} \qquad (2.30)$$

where **B** is the M-dimensional vector of the formed beams, and **S** is the N-dimensional vector of samples by the N receiving channels.

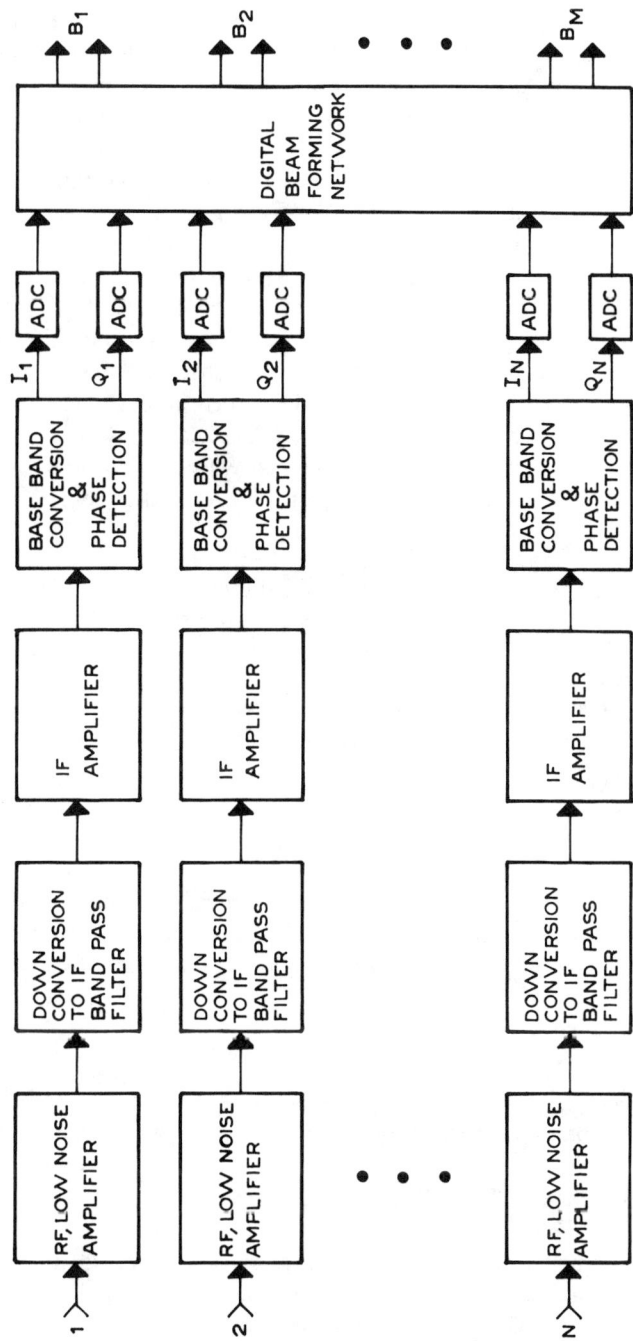

Figure 2.16 Digital beam-forming processing scheme at baseband.

In the event of $M = N$ the beam-forming network could be efficiently implemented by the FFT algorithm; in the general case of $M \neq N$, the weighting matrix **W** is derived by resorting to the proper algorithms as discussed in Section 2.7.2.

In the present analysis, the following errors are considered: (1) the amplitude and phase mismatching errors introduced by the receiving element and the blocks up to the baseband conversion; (2) the gain and phase imbalance errors introduced by the phase detector; (3) the quantization errors of digitized signals, (4) the quantization errors of beam-forming weights; and (5) the thermal noise (DiVito and Iovino, 1986). For sake of simplicity, the mutual coupling between the receiving elements is not taken into account; however, the correction of this effect is discussed in Section 2.7.5. In the following, the analysis will be made by modeling the errors as random variables. Other analyses are based on the use of deterministic models for the errors; for instance, in (Picardi, 1988) the errors are considered to be sinusoidal functions so that the paired echoes theory is applied.

A plane wave, impinging on the array antenna with a direction of arrival of θ degrees from the array boresight, generates at the Kth receiving element the signal:

$$s'_K = \exp j(2\pi ft - K2\pi \frac{d}{c} f \sin\theta)$$
$$= \exp j(2\pi ft + K\gamma), \quad K = 1, 2, \ldots, N \qquad (2.31)$$

where f is the RF carrier, d is the interelement distance, and $\gamma \triangleq - 2\pi(d/c)f \sin\theta$. Ideally, the signal at the input of the digital beam-forming network would be an exact reproduction of

$$s_K = A \exp(K\gamma), \quad K = 1, 2, \ldots, N \qquad (2.32)$$

where A performs the overall amplification of the processing chain. The set of N samples of $s_K(K = 1, 2, \ldots, N)$, when combined with the proper set of weights $W_{K,l}(K = 1, 2, \ldots, N; l = 1, 2, \ldots, M)$, produces the lth beam:

$$B_l = \sum_{K=1}^{N} W_{K,l} s_K, \quad l = 1, 2, \ldots, M \qquad (2.33)$$

By using the vector notation, this equation can be rewritten as follows:

$$B_l = \mathbf{W}_l^T \mathbf{S}, \quad l = 1, 2, \ldots, M \qquad (2.34)$$

where $\mathbf{W}_l$ is the N-dimensional vector of weights related to the lth formed beam. The whole set of M formed beams can be synthetically written as in (2.30), where the M rows of the matrix $\mathbf{W}(M,N)$ are the M vectors $\mathbf{W}_l(l = 1, 2, \ldots, M)$ of the previous expression (2.34). In reality, the signal at the input of the phase detector appears as follows:

$$s''_K(t) = A(1 + \alpha_K) \exp[j(2\pi f_{IF}t + \psi_K + K\gamma)] + n_K \exp(j2\pi f_{IF}t) \quad (2.35)$$

where α_K and ψ_K are the gain and phase errors introduced by the whole receiving chain up to the input of the phase detector, and n_k is the thermal noise, which is assumed to be white Gaussian distributed with zero mean and variance $2\sigma^2$. The in-phase and quadrature signals at the output of the phase detector suffer for the amplitude (β_K) and phase (ε_K) error imbalance:

$$s'''_K = I_K + jQ_K \quad (2.36a)$$

$$I_K = A(1 + \alpha_K)(1 + 0.5\beta_K) \cos(K\gamma + \psi_K + 0.5\varepsilon_K) + n_{I_K} \quad (2.36b)$$

$$Q_K = A(1 + \alpha_K)(1 - 0.5\beta_K) \sin(K\gamma + \psi_K - 0.5\varepsilon_K) + n_{Q_K} \quad (2.36c)$$

where, for the sake of simplicity, the imbalance errors have been divided in equal parts among the two channels. Owing to the limited amount of phase errors (ideally, $\psi_K \pm 0.5\varepsilon_K \approx 1°$), the following approximations can be retained:

$$\begin{aligned}\cos(K\gamma + \psi_K + 0.5\varepsilon_K) &\approx \cos K\gamma - (\psi_K + 0.5\varepsilon_K) \sin K\gamma \\ \sin(K\gamma + \psi_K - 0.5\varepsilon_K) &\approx \sin K\gamma + (\psi_K - 0.5\varepsilon_K) \cos K\gamma\end{aligned} \quad (2.37)$$

Consequently, the mathematical expressions of I_K and Q_K can be rewritten as follows:

$$\begin{aligned}I_K &\approx A \cos K\gamma + AC_{1K} \sin K\gamma + AC_{2K} \cos K\gamma + n_{I_K} \\ Q_K &\approx A \sin K\gamma + AC_{3K} \cos K\gamma + AC_{4K} \sin K\gamma + n_{Q_K}\end{aligned} \quad (2.38)$$

where:

$$\begin{aligned}C_{1K} &= -(\psi_K + 0.5\varepsilon_K)(1 + \alpha_K)(1 + 0.5\beta_K) \\ C_{2K} &= (1 + \alpha_K)(1 + 0.5\beta_K) - 1 \\ C_{3K} &= (\psi_K - 0.5\varepsilon_K)(1 + \alpha_K)(1 - 0.5\beta_K) \\ C_{4K} &= (1 + \alpha_K)(1 - 0.5\beta_K) - 1\end{aligned} \quad (2.39)$$

A compact expression of s'''_K is

$$s'''_K = I_K + jQ_K = s_K + \Delta s_K + n_K \quad (2.40)$$

with Δs_K being:

$$\Delta s_K = 0.5[(C_{2K} + C_{4K}) + j(C_{3K} - C_{1K})]s_K$$
$$+ 0.5[(C_{2K} - C_{4K}) + j(C_{1K} + C_{3K})]s_K^* \quad (2.41)$$

where the asterisk stands for a complex conjugate. The next operation on the received signal is the quantization of I_K and Q_K, which is done by $(b + 1)$ bits and a quantization step of $q = 2^{-b}$. A simple analysis assumes that the quantizations error is modeled as an additive white noise, uncorrelated with the signal, and evenly distributed in $[-0.5q, 0.5q]$. The mean value of the quantization error is zero, while the variance is $q^2/12$. Consequently, the overall noise power on the in-phase or quadrature channel is

$$P_n = \sigma^2 \left[1 + \frac{(q/\sigma)^2}{12}\right] \quad (2.42)$$

A sensitivity loss is defined as follows:

$$L = 10 \log_{10}\left[1 + \frac{(q/\sigma)^2}{12}\right] \quad (2.43)$$

By choosing a number of bits such that the quantization step is comparable to the thermal noise level, i.e., $q = 2\sigma$, the loss L amounts to 1.25 dB.

The last error to consider is due to the quantization of the weights $W_{K,l}$ ($K = 1, 2, \ldots, N; l = 1, 2, \ldots, M$). The mathematical expression (2.34) of the lth beam can be rewritten as follows:

$$B_l = \sum_{K=1}^{N} (W_{K,l} + \Delta W_{K,l})(s_K + \Delta s_K + n_K)$$
$$= (\mathbf{W}_l^T + \Delta \mathbf{W}_l^T)(\mathbf{S} + \Delta \mathbf{S} + \mathbf{n})$$
$$= \mathbf{W}_l^T \mathbf{S} + \mathbf{W}_l^T \Delta \mathbf{S} + \mathbf{W}_l^T \mathbf{n} + \Delta \mathbf{W}_l^T \mathbf{S} + \Delta \mathbf{W}_l^T \Delta \mathbf{S} + \Delta \mathbf{W}_l^T \mathbf{n} \quad (2.44)$$

where vector symbols have been introduced in the second and third rows of the equation.

As before, the error $\Delta W_{K,l}$ is also random with zero mean value and variance $q^2/12$. For simplicity, the same number of bits have been taken for quantizing the weights and the I and Q signal components. Owing to the uncorrelated nature and zero mean values of the errors, the mean and mean squared values of the beam B_l are calculated as follows:

$$E\{B_l\} = \mathbf{W}_l^T \mathbf{S} \tag{2.45}$$

$$E\{|B_l|^2\} = |\mathbf{W}^T\mathbf{S}|^2 + E\{|\mathbf{W}_l^T \Delta \mathbf{S}|^2\} + E\{|\mathbf{W}_l^T \mathbf{n}|^2\} + E\{|\mathbf{S}^T \Delta \mathbf{W}_l|^2\} \tag{2.46}$$

The mean value of the lth beam is equal to the ideal pattern and the mean square value of the beam is calculated by disregarding the second-order terms, i.e., those containing the products between errors. In the following, we provide an explicit expression of the mean square value of the lth beam. These expressions fulfill our purpose:

$$|\mathbf{W}_l^T \mathbf{S}|^2 = |B_l|^2 \tag{2.47a}$$

$$E\{|\mathbf{W}_l^T \Delta \mathbf{S}|^2\} = \sum_{K=1}^{N} \sum_{i=1}^{N} W_{K,l} W_{i,l}^* E\{\Delta s_K \Delta s_i^*\} = A^2 \sigma_T^2 \sum_{K=1}^{N} |W_{K,l}|^2 \tag{2.47b}$$

with

$$\sigma_T^2 \approx \sigma_\alpha^2 + \sigma_\psi^2 + \frac{1}{4}(\sigma_\beta^2 + \sigma_\varepsilon^2) \tag{2.47c}$$

$$E\{|\mathbf{W}_l^T \mathbf{n}|^2\} = \sum_{K=1}^{N} \sum_{i=1}^{N} W_{K,1} W_{i,l}^* E\{n_K n_i^*\} = \left(2\sigma^2 + \frac{q^2}{6}\right) \sum_{K=1}^{N} |W_{K,l}|^2 \tag{2.47d}$$

$$E\{|\mathbf{S}^T \Delta \mathbf{W}_l|^2\} = \sum_{K=1}^{N} \sum_{i=1}^{N} s_K s_i^* E\{\Delta W_{K,l} \Delta W_{i,l}^*\} = \sum_{K=1}^{N} |s_{K,l}|^2 \frac{q^2}{6} = NA^2 \frac{q^2}{6} \tag{2.47e}$$

Note that (2.47b) requires some tedious algebraic calculation, and (2.47e) provides for the signal quantization error. In summary, the expression of the mean square value of the lth beam is as follows (DiVito and Iovino, 1986):

$$E\{|B_l|^2\} = |B_l|^2 + \left[A^2\left(\sigma_\alpha^2 + \sigma_\psi^2 + \frac{\sigma_\beta^2}{4} + \frac{\sigma_\varepsilon^2}{4}\right)\right.$$
$$\left. + \left(2\sigma^2 + \frac{q^2}{6}\right)\right] \sum_{K=1}^{N} |W_{K,l}|^2 + NA^2 \frac{q^2}{6} \tag{2.48}$$

where σ_α and σ_ψ are the standard deviations of the gain and phase errors of the receiving chain to the phase detector input, and σ_β and σ_ε are the standard deviations of the amplitude and phase error imbalance of the phase detector. The terms that sum to the ideal value $|B_l|^2$ have the effect of filling up the nulls of the ideal pattern of the lth beam and limiting the achievable peak-to-sidelobe ratio.

Now let us apply the previous equation to the following practical example (DiVito and Iovino, 1986). Consider a linear array with $N = 16$ receiving channels and the following error budget:

$$\sigma_\alpha = 2.3 \times 10^{-3} \text{ (0.2 dB)}, \quad \sigma_\psi = 0.1°, \quad \sigma_\beta = 11 \times 10^{-3} \text{ (0.1 dB)}, \quad \sigma_\varepsilon = 0.5°$$

$$b = 7 \quad \text{and} \quad \sigma = 0.5q$$

In the case of uniform weighting, we obtain these numerical values:

$$|B_l|^2_{\max} = A^2 N^2$$
$$E\{|\mathbf{W}_l^T \Delta \mathbf{S}|^2\} = \sigma_T^2 A^2 N = 5.8\ 10^{-4} A^2 N$$
$$E\{|\mathbf{W}_l^T \mathbf{n}|^2\} = \frac{2}{3} N q^2 = \frac{2}{3} N 2^{-2b}$$
$$E\{|\mathbf{S}^T \Delta \mathbf{W}_l|^2\} = \frac{NA^2}{6} 2^{-2b} \tag{2.49}$$

where σ_T^2 is defined in equation (2.47c).

To see the relevance of the three error terms, we compare them to the maximum value of the ideal term:

$$E\{|\mathbf{W}_l^T \Delta \mathbf{S}|^2\}/|B_l|^2_{\max} = -44.4 \text{ dB}$$
$$E\{|\mathbf{W}_l^T \mathbf{n}|^2\}/|B_l|^2_{\max} = -55.9 \text{ dB}$$
$$E\{|\mathbf{S}^T \Delta \mathbf{W}_l|^2\}/|B_l|^2_{\max} = -62.1 \text{ dB} \tag{2.50}$$

These terms do not influence the sidelobe level for the case of uniform weighting (with the first sidelobe 13 dB down from the peak), but they would be important when amplitude tapering is applied to achieve very low sidelobes. Among the three terms, the first one, concerning the amplitude and phase mismatch to the phase detector input, is the more relevant. So, a calibration procedure must be devised to reduce the importance of these errors. We note that halving the error standard deviations would result in a 6 dB reduction of the term $E\{|\mathbf{W}_l^T \Delta \mathbf{S}|^2\}$. The last two terms are dependent on the number of bits; therefore, they can be lowered by an appropriate increase of the bit number. However, the convenience of using a large number of bits should be considered in relation to the capability of amplitude and phase matching of the channels.

2.7.5 Calibration Procedures

To obtain accurate beam shapes and low sidelobes with a multiple receiver array, we need to correct for the gain and phase errors mentioned in the preceding subsection. If the system is not so aligned, the weighting coefficients applied to the incoming signals are in error due to the misalignment of the channels and the beam pattern thus is degraded.

Calibration can be incorporated in a DBF because differences in the analog front ends can be compensated digitally. Calibration also is required for DBF systems where each channel contains a complete microwave receiver, all of which must be amplitude and phase matched over the entire signal bandwidth. Any differences will preclude precise pattern sidelobe control and open-loop nulling. Calibration may be performed in two ways (Steyskal, 1987; Steyskal and Rose, 1989; Wardrop, 1985).

Direct calibration uses a near-field or a far-field source, and measures the phase and gain errors in each channel with respect to a reference channel. By taking the reciprocals of these (complex) error terms, we can produce a set of correction coefficients, which, when applied in the beam-forming process, align the signals into an effective plane wavefront. Methods that use an external signal are sensitive to multipath propagation and the configuration of the antenna site.

The array also may be indirectly calibrated by using test signals injected into the channels via couplers situated between each antenna element and its receiver (see Figure 2.7). However, the distribution network of the internally injected signal will not produce identical signals in each channel, and so the network itself must be calibrated by using an external source. This, however, needs to be performed only once. Methods based on internal test signals calibrate the receiver chain, except for the passive antenna elements.

Steyskal and Rose (1989) describe an array calibration experiment performed with a 32-element linear array operating in C-band. The array has a self-calibration system that monitors the 32 receiving channels for subsequent digital correction of channel imbalance. The basic component is a high-precision, bidirectional loop, which distributes the pilot calibration tone to each elemental receiver. Before digital correction, the channel varied by ± 2 dB in amplitude, and the phase had a random distribution. After compensation, authors report that the root mean square (rms) errors were about 0.26 dB in the amplitude and $2.5°$ in phase, which is commensurate with a -40 dB sidelobe level.

Due to mutual coupling effects, the elements in a small array have different radiation patterns, and this may preclude precise pattern control and low sidelobe levels. However, DBF can correct for these effects and lead to improved pattern quality. The problem and the principle for its correction can be explained as follows (Steyskal, 1987). The signal at the output of the individual antenna element has one dominant component due to the direct incident plane wave, and several lesser components due to scattering of the incident wave at the neighboring elements. Each scattered component can be expressed in terms of the plane wave incident at the scattering element and a coupling coefficient to the receiving element. Thus, the set $\{s_n, n = 1, 2, \ldots, N\}$ of N element output signals is linearly related to the N values $\{s_{dn}, n = 1, 2, \ldots, N\}$ of the incident plane wave at the antenna element apertures. The former represent the known, coupling perturbed signals; the latter constitute the unknown, desired signals. Defining the corresponding N-dimensional column vectors $\mathbf{s}$ and $\mathbf{s}_d$, we have

$$\mathbf{s} = \mathbf{C}\mathbf{s}_d \qquad (2.51)$$

and, consequently,

$$\mathbf{s}_d = \mathbf{C}^{-1}\mathbf{s} \qquad (2.52)$$

Thus, the coupling correction is accomplished by multiplying the received signal $\mathbf{s}$ by the inverse coupling matrix $\mathbf{C}^{-1}$. In practice, the coupling coefficients C_{mn} are obtained from measurements, and the inverse matrix $\mathbf{C}^{-1}$ is computed off-line and then applied to the digitized array signals. Steyskal (1987) reported that the technique applied to an eight-element array reduces the initial pattern sidelobe level of -20 dB to -30 dB.

2.7.6 Implementation of DBF

As shown by the Figure 2.7, a generic DBF comprises the antenna elements, receiver modules, ADCs, digital beamformer, weight controller, and calibrator. Each subsystem requires a specific technology. At present, three technologies have been identified as critical to the development of DBF for future radar systems, (Valentino, 1984; Steyskal and Rose, 1989):

1. Gallium arsenide (GaAs) microwave monolithic integrated circuits (MMICs);
2. Very large scale integration (VLSI) and very high speed integrated circuits (VHSIC);
3. Fiber optics.

These technologies will allow DBF to meet the data-handling and processing requirements, given the size and power constraints of future radar systems, in a cost effective manner.

GaAs MMICs can be extensively used in the receiver module for front-end amplification and mixing, while silicon (Si) integrated circuits (ICs) can be used to implement the IF functions. Either Si or GaAs technologies can be used for the ADC, which is the most challenging device to implement. In fact, the future required sampling rate is on the order of tens to hundreds of MHz, and the dynamic range is commensurate to 8-10 bits. The MMICs will be printed on GaAs crystalline wafers to produce compact, active microwave circuits at reasonable cost. Receiver modules and ADCs should easily fit into the space provided by the array antenna.

The design of an effective digital processor with the capacity and speed required by a DBF radar is limited by today's semiconductor chips' capabilities. High-speed and submicron sized printed circuit technology will enable fabrication of devices with greater circuit densities, lower power consumption, and faster data rates. New fabrication techniques in the area of VLSI logic circuits should provide the speed and logic circuit density necessary for practical design and construction of the beamformer and weight controller processor. The development of VHSIC (US program), VHPIC (very high performance integrated circuits, UK program), and the like are also critical because a DBF radar's effectiveness depends on rapid formation of simultaneous multiple beams and signal processing. This capability imposes a very high processing load on the beamformer.

Fiber optics offer a means of simplifying the data transfer and distribution lines between the DBF and the receiver modules. This technology will be applied at the interface between the ADC outputs, DBF, and distribution of local signals to each of the elemental receivers. Because many wires are required to make these connections, containing them can be difficult. The ultra-high-bandwidth capabilities of fiber optic cable will enable multiple signals to be carried on one line, resulting in a simple and practical packaging solution.

2.7.7 Existing DBF Systems

This subsection describes the DBF systems developed worldwide, as they appear in the open technical literature (e.g., Steyskal, 1987). The earliest DBF system is the ELRA phased-array radar in Germany (Wirth, 1989). It has separate, circular transmitting and receiving arrays (300 elements, 1:8.2 fill factor; 768 elements, 1:6.25 fill factor, respectively) and operates at S-band. Element calibration is performed via a fixed probe in the near-field. The outputs of the double-conversion receivers are combined into 48 subarrays with digital outputs, from which a number of beams are formed and the entire beam cluster is scanned by analog phase shifters.

Several experimental DBF systems have been reported from the UK. An eight-element array, intended to demonstrate rejection of barrage jamming, is discussed in (Barton, 1980) and experimental results on main-beam nulling are presented. Another example is that developed by GEC Research Laboratories, known by the acronym GASP (GEC Array Signal Processor test bed) (Wardrop, 1985). This is a narrow-band, high-dynamic-range system, which operates with a 32-element S-Band linear array, to be used for research into multistatic systems as well as for investigating array signal processing algorithms. The system is capable of generating up to six simultaneous beams. The ADCs perform a conversion every 3 μs to a 12-bit accuracy, giving a dynamic range of some 60 dB with an IF bandwidth of 330 kHz.

A comprehensive suite of software tools has been developed for alignment and diagnostic purposes (Wardrop, 1985).

In the United States, one large system is the experimental over-the-horizon (OTH) radar, the CONUS B-OTH by General Electric Company. This is reported to be a bistatic system operating between 6 and 20 MHz with a bandwidth of about 100 kHz. On receiving, up to 82 elements are used to form four simultaneous beams that cover the transmitted beamwidth. Adaptive nulling and on-line receiver error correction are provided. Another DBF radar receiver array with eight elements at X-band, with double-conversion receivers, analog I and Q channels, and eight-bit ADC is reported in (Loomis, 1990; Steyskal, 1987). Finally, Steyskal (1987) reported that Rome Air Development Center tested a linear, 32-element array at C-band with triple-conversion receiving modules and 0.5 MHz bandwidth. The array incorporates self-calibration via a high precision distribution network to inject pilot tones at the array element front-ends. This array provides a means of obtaining high-quality low-sidelobe patterns. We should mention the experimental work done recently in Japan (Inatsune, et al., 1989). The system is a planar array of 8 × 8 radiating elements at X-band. The IF is 30 MHz and the IF bandwidth is 5 MHz. The ADC operates with 8 bits in the IF stage at the 10 MHz rate. The digital beamformer is implemented by a minicomputer. Several algorithms have been tested, such as those related to the DFT and FFT. Weighting functions like the Dolph-Chebyschev and Taylor with −40 dB have been successfully tested.

2.8 CONCLUDING REMARKS

This Chapter has discussed the low and ultra-low sidelobe antenna techniques. We have indicated, with numerical examples, the practical relevance of having such techniques installed in defense radar when operating in the presence of jamming sources (Section 2.2). Subsequently, the basic principles to achieve low and ultra-low sidelobes have been presented (Section 2.3) and applied to the reflector antennas (Section 2.4) and array antennas (Section 2.5). A detailed analysis of the average sidelobes generated by random amplitude and phase errors and failed elements in an array antenna has been presented in Section 2.6. The advanced technique of DBF also has been extensively described (Section 2.7).

There is no doubt that ultra-low sidelobe antennas are highly advantageous for many types of radar such as ground-based, shipborne, airborne, and even spaceborne. The basic principles are also well established. In practice, the implementation of such antenna types requires modern procedures for design, fabrication, and testing of the system. Future development of the low-sidelobe antenna techniques can be expected in the areas of the phased array with electronic scanning in both angles, in the DBF, and in the conformal arrays. Concerning the DBF, several experimental systems have demonstrated the feasibility of this technology. A problem with DBF is the large amount of parallel data signals to process at any instant of time. The further advance of technology should make this problem affordable.

Chapter 3
Sidelobe Blanking System

3.1 INTRODUCTION

When a signal enters the radar receiver and exceeds the detection threshold, the signal is assumed to have been intercepted by the radar antenna's main beam, i.e., the signal is received only when the antenna is pointed in the direction of the signal source. Nevertheless, a very strong signal may enter the antenna sidelobes, thus being interpreted as a main-beam signal and resulting in a major angle error. Because of this vulnerability in the angle measurement system, radar designers put considerable effort into decreasing the radar's sensitivity to signals outside the antenna main beam by design and fabrication of antennas with very low sidelobes (Chapter 2). ULSA obviously is not the panacea for all radar systems. We should remember that the cost could be considerably high and low sidelobes are traded-off with a defocusing of the main-beam pattern. Suppression techniques that are not so costly are envisaged for a specific class of disturbances.

Sidelobe interference of interest in this chapter for suppression purposes is of impulse type, i.e., characterized by a low duty cycle. For instance, a radar signal intercepted by an ESM device (see Section 1.2) can be properly delayed (e.g., into the next radar interpulse period) and transmitted into the radar sidelobes, thus appearing as a target at different range and azimuth values than the true target. Of course, multiple false targets in range can be injected into the radar sidelobes.

To counter such low-duty-cycle sidelobe jammers, we may employ the *sidelobe blanker* (SLB), which is used to turn off the radar receiver output when an unwanted signal (i.e., interference pulse, strong target echo, clutter echo, *et cetera*) appears in the radar antenna sidelobes. One way of dealing with these unwanted signals is to provide the radar with an auxiliary (also referred to as *guard*) channel. In essence, this consists of a separate receiver, the input of which is supplied by a small antenna (e.g., dipole, horn, *et cetera*) mounted close to the radar antenna (see Figure 3.1).

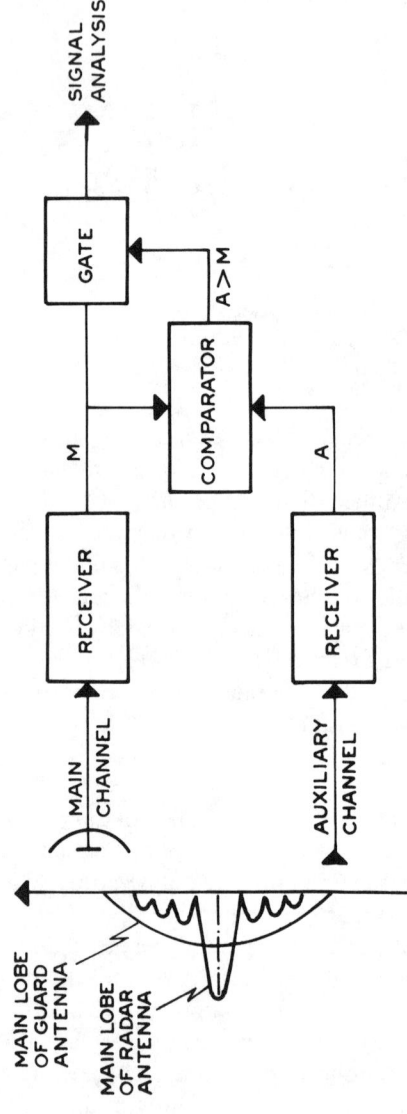

Figure 3.1 Simplified block diagram of the sidelobe blanker.

The width of the guard's main lobe is sufficient to encompass the entire region illuminated by the radar antenna's sidelobes, and the gain of the guard's main lobe is greater than that of any of the sidelobes. Any detectable target in the radar antenna's sidelobes will therefore produce a stronger output from the guard receiver than that from the main receiver. However, because the gain of the radar antenna's main lobe is much greater than that of the guard antenna, any target in the radar antenna's main lobe will produce a much stronger output from the main receiver than that from the guard receiver. Consequently, by comparing the outputs of the two receivers and inhibiting the output of the main receiver when the output of the guard receiver is stronger, we can prevent any targets that are in the sidelobes from entering the signal analysis circuits and appearing on the radar display. An example of the practical effectiveness of the SLB device is presented in the literature, where the *plan position indicator* (PPI) display is shown for a radar that is subject to pulsed interferences, but with and without the SLB (Johnson and Stoner, 1978).

An approximation to the guard test could be mechanized by comparing the sum-and-difference signal magnitudes in a monopulse system, as illustrated in Figure 3.2. This procedure is less effective than the guard processing since the difference pattern does not form a perfect omnidirectional pattern over the principal sum channel sidelobe region. On the other side, the technique of Figure 3.2 is easily mechanized in a monopulse radar without the need of additional hardware.

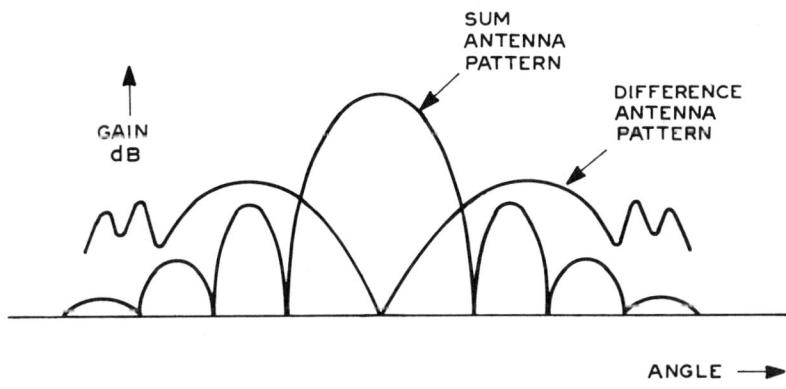

Figure 3.2 SLB implemented by using sum-and-difference antenna patterns in a monopulse system.

Unfortunately, the ECM designer sometimes can counter the SLB technique. The radar system engineer also should be aware of this aspect before deciding to use the SLB technique in a specific radar system. The ECM transmitting antenna can operate with orthogonal polarization to that of the radar receiving antenna. Polarization is defined as the orientation of the electric vector of a radiated signal; orientations include vertical, horizontal, circular, and elliptical. As shown in Figure 3.3 for a typical search radar, a signal injected into the antenna at a polarization other than the antenna's design polarization can disrupt the antenna pattern in a way that destroys the criteria required for effective comparison of signals in the main and auxiliary antennas. This is because the two antennas (main and auxiliary) are physically different. In most cases, the radar antenna and the SLB antennas are sufficiently unlike one another, and so they produce different polarization characteristics that make them vulnerable to cross-polarization techniques. In Figure 3.3, the solid line patterns represent the properly polarized case for a typical radar antenna; the dotted lines represent the cross-polarized case (i.e., the ECM transmitting antenna's polarization is orthogonal to the radar and auxiliary receiving antenna's polarization). For the co-polarized case, the relationship of the two patterns is correct for proper blanking operation; for the cross-polarized case, the patterns show a reversal of roles for the two antennas. For a cross-polarized signal in the sidelobes, the main antenna provides more gain than that in the auxiliary antenna. Therefore, signals arriving via the antenna sidelobes with a polarization orthogonal to the design polarization of the antenna will pass through to the radar detection circuits. We should point out that the cross-polarization technique is effective if the ECM transmitted signal is orthogonal or nearly orthogonal to the radar receiving antenna. However, ECM receivers detect and measure the polarization of the radar transmitting antenna, which may or may not have the same polarization as the radar receiving antenna (Chrzanowski, 1990).

The SLB should be able to produce the maximum (in percent) sidelobe blanking with the least loss of radar sensitivity for the main-beam targets. *Percentage*

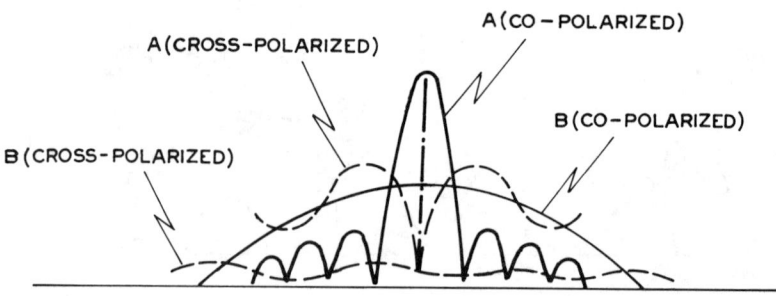

Figure 3.3 Polarization patterns of the main (a) and auxiliary (b) antennas. (From Chrzanowski, *Active Radar Electronic Countermeasures*, Artech House, Norwood, MA, 1990.)

blanking is defined as the portion of sidelobe disturbances uniformly distributed in angle that are prevented from being processed as targets. *Radar sensitivity reduction* is defined as the increase in SNR necessary for a given main-beam target to achieve the same detection probability as without the blanker. The presence of the SLB may, in fact, reduce the probability of detection of a target in the radar's main beam. This is because there is a finite probability at any given time that thermal noise in the blanker channel output will have a larger amplitude than signal-plus-noise at the radar receiving antenna's output. This could result in blanking of a signal that should successfully exceed the false-alarm-rate threshold in a radar without the SLB. This phenomenon is illustrated by Figure 3.4, showing a small target (i.e., with low SNR) that is blanked, although it is in the radar main beam, because the blanker channel noise exceeds the signal-plus-noise in the receiver radar. This phenomenon reduces the radar detection probability (Arancibia, 1978). Another phenomenon producing a detection loss occurs when a main-beam target is blanked because it occupies the range cell of an unwanted interference in the radar sidelobes. Therefore, proper design of the SLB involves a trade-off between the desired confidence with which sidelobe targets can be reliably rejected and the loss in detectability of main-lobe target (i.e., relative to a single-channel system), which the radar designer is willing to accept. The trade-off is influenced by the gain margin of the auxiliary antenna with respect to the radar sidelobes.

The SLB system has been studied in detail (Maisel, 1968; Harvey and Wood, 1980; O'Sullivan, 1987; O'Sullivan et al., 1989; Arancibia, 1988; Skolnik, 1990, pp. 17.11–17.16). All the references are devoted to the analytical or experimental evaluation of SLB performance. Maisel (1968) provides a calculation of the detection probabilities of the SLB for a useful target in the main lobe and for jammers in the sidelobes of the radar antenna by assuming nonfluctuating pulsed signals in the radar. However, we should note that these two quantities are not sufficient to describe completely the performance of the SLB, new operating characteristics will be introduced in the Section 3.3 for proper design of the SLB system.

The SLB scheme presented in Harvey and Wood (1980) differs from that of Maisel (1968) for the presence of a more complex blanking logic. The analysis is more complete as well because, in addition to the previously mentioned probabilities of detection, the detection loss incurred when the SLB is switched on is also evaluated in the absence of interference. Applications of the major analytical results are made for two types of interference: *noncoherent* pulse interference (defined as a train of interference pulses that are not synchronized to the radar PRF) and *coherent* pulse interference (consisting of a train of pulses that are synchronized with the transmitted pulses in such a way that the interference pulses enter the radar at constant apparent range on successive PRIs).

The analysis in (O'Sullivan, 1987; O'Sullivan *et al.*, 1989) extends the pioneering work of (Maisel, 1968) along two directions: (1) a *processing configuration* incorporating CFAR devices in the two channels, offering superior performance when

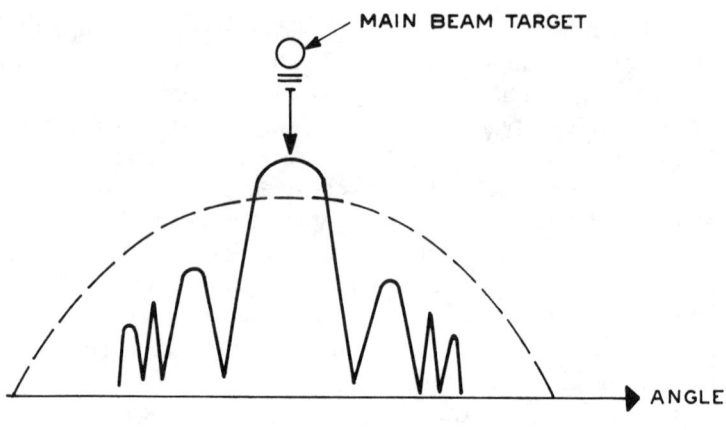

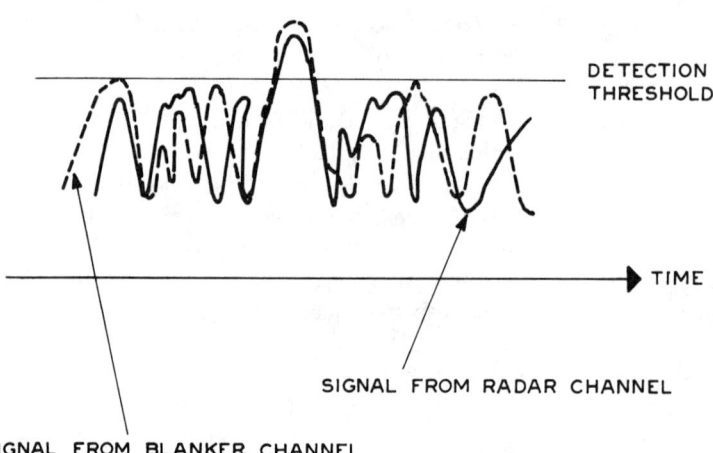

Figure 3.4 Radar sensitivity reduction due to the SLB presence. (Adapted from Arancibia, *Microwave Journal,* Vol. 21, No. 3, March 1978, pp. 69–73.)

the noise background against which the system operates is nonstationary, and (2) a *performance evaluation* considering fluctuating target models (i.e., Swerling I and III) in addition to Swerling 0.

Here, we prefer to consider the simple SLB scheme proposed by Maisel (1968); however, the performance evaluations are completely reworked to account for a number of probabilities and the detection loss that were not adequately considered. The working principle and the basic scheme of the system is described in Section 3.2, while the performance evaluation and design criteria are illustrated in Sections 3.3 and 3.4.

3.2 WORKING PRINCIPLE AND PROCESSING SCHEMES

In this section, we reconsider the SLB system's working principle and describe its configuration and processing scheme in more detail. We know that the basis of operation is extremely simple. Figure 3.5 depicts the radiation patterns of the main antenna (with maximum gain, G_t) and a low-gain auxiliary antenna. We assume that the gain, G_A, of the auxiliary antenna is higher than the maximum gain G_{sl} of the sidelobes of the radar antenna. Figure 3.6 shows the system block diagram with the two receiving channels (one for the radar antenna and the other for the auxiliary antenna), blanking logic, and usual radar detection threshold on the single pulses delivering signals to subsequent processing circuits (e.g., noncoherent integration of target echoes and thresholding) for further signal analysis. The SLB compares, at each range bin, each pulse received and processed (logarithmically detected in the

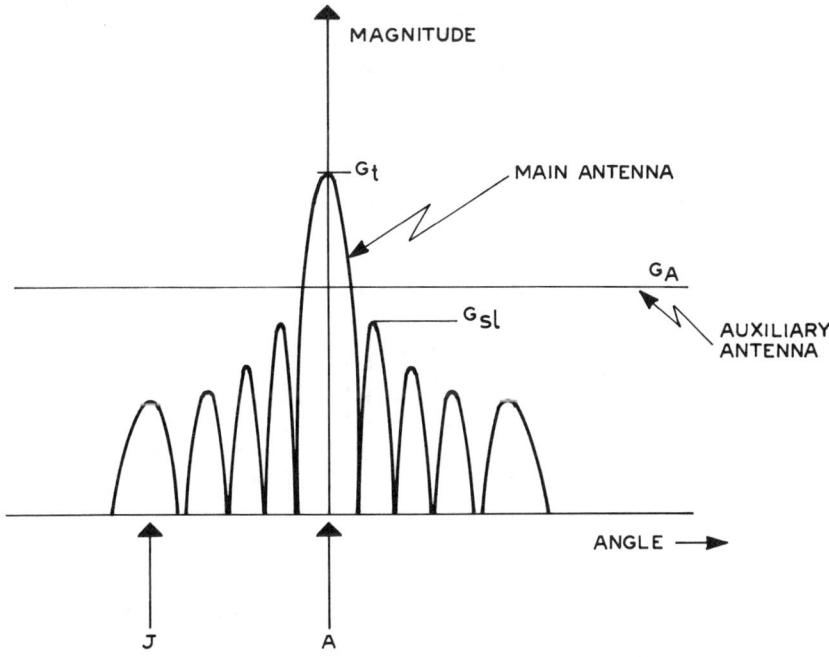

Figure 3.5 Main and auxiliary antenna patterns for SLB.

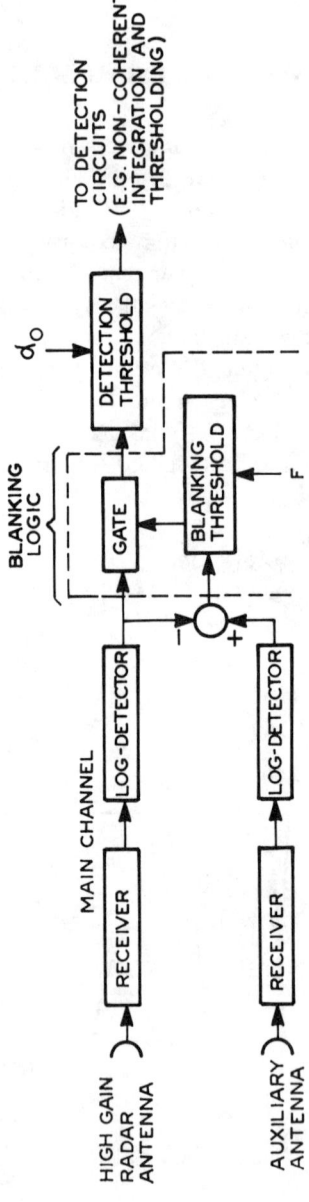

Figure 3.6 Basic sidelobe blanking system. (From Maisel, *IEEE Trans. on Aerospace and Electronic Systems*, Vol. AES–4, No. 2, March 1968, pp. 174–180, © 1968 IEEE.)

Maisel (1968) scheme) by the two parallel channels. Thus, the SLB decides whether to blank the main radar channel on a single sweep basis and for each range bin. A target "A" situated in the main lobe will result in a large signal in the main receiver channel and a smaller signal in the auxiliary receiver channel. A proper blanking logic allows this signal to pass by comparing the difference signals of the two channels with a suitable threshold, F. Because the main channel's signal is sufficiently higher than the auxiliary channel's signal, that of the main channel goes through the gate and then undergoes the usual thresholding to ascertain whether a target has been detected by the radar. Targets or jammers "J" situated in the sidelobes give small main channel signals but larger auxiliary channel signals so that these targets are suppressed by the blanking logic, i.e., the gate of Figure 3.6 interrupts the flow of the radar signal toward the radar detection thresholding.

The gain of the auxiliary antenna is set everywhere to a value higher than the radar sidelobes. However, the relative gain between the main and auxiliary antennas is influenced by several factors (e.g., polarization, multipath, electromagnetic isolation between the antennas) so that the relative gain may vary considerably. If the main antenna sidelobes are high, to provide sufficient broad-beam auxiliary gain in the region of the principal radar sidelobes may be impossible with a simple antenna. If a main antenna possesses good sidelobes, SLB is not rendered unnecessary because there is always some power level at which interference pulses may enter the radar. The possession of good sidelobes, however, makes the SLB task much easier because the auxiliary antenna design problems are lessened (Harvey and Wood, 1980).

Low-sidelobe antennas and ULSAs for radars have sidelobes below the isotropic gain. In this case, the auxiliary antenna may be omnidirectional. This is the hypothesis assumed in Figure 3.5 and in the mathematical development of Section 3.3. However, many practical antenna designs result in sidelobes well above the isotropic gain, which implies that the auxiliary antenna must have greater than unity gain and thus cannot be omnidirectional. (This general situation is illustrated in the drawing of the auxiliary patterns of Figures 3.1 to 3.4). The auxiliary antenna can be a dipole behind the reflector of a radar antenna, as illustrated in Figure 3.7, or a horn to cover the first higher sidelobes of the radar antenna. If the radar is equipped with a phased-array antenna, the auxiliary can be a subgroup of elemental apertures distributed over the array as mentioned in (Chin et al., 1989).

A phenomenon that causes a departure of the SLB system from the desired ideal behavior is due to the mismatching of the main and auxiliary channels. Major sources of mismatching are in the RF sections, where the connection between the radar antenna and the remaining system is by means of a waveguide, whereas the connection for the auxiliary antenna is normally by means of a coaxial cable. In addition, the auxiliary channel being for receiving only, does not need RF protection circuits. Matching conditions have to be guaranteed over the entire radar bandwidth,

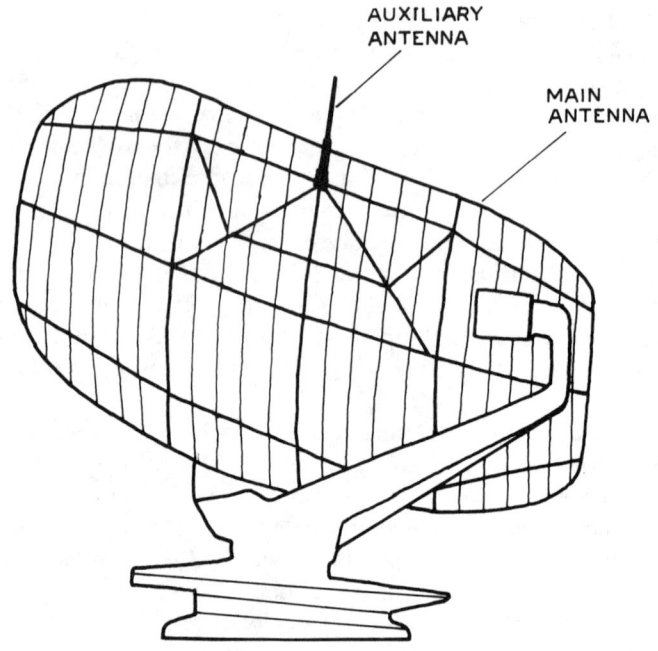

Figure 3.7 Location of the auxiliary antenna with respect to a reflector type of radar antenna.

even in frequency-agility conditions. Less severe are the matching conditions in the IF sections. The matching condition of the two channels is a critical point; it ensures that the two compared signals differ only for the two different antenna patterns and not for other spurious phenomena.

The presence of receiver noise in each channel is an additional cause for departure from the desired ideal behavior and requires, by proper choice of SLB antenna gain and threshold levels (α_0, F, see Figure 3.6), obtaining a satisfactory compromise between three conflicting requirements. Namely, for a given false-alarm probability, we seek to maximize probability of detection on the main lobe, to minimize probability of detection in the sidelobes, and to minimize the size of the SLB antenna (Maisel, 1968).

Another topic is related to the compatibility between the SLB and other signal processing techniques. Usually, the radars are equipped with MTI (moving target indication) or equivalent devices for clutter filtering in the doppler frequency domain. MTI processing can be inserted before or after the SLB. Figure 3.8 shows the processing scheme depicting the MTI upstream from the SLB. Note that the radar and auxiliary receiving channels perform the same functions: (1) RF conversion; (2) IF conversion; (3) baseband (BB) conversion; (4) analog-to-digital conversion; (5) MTI

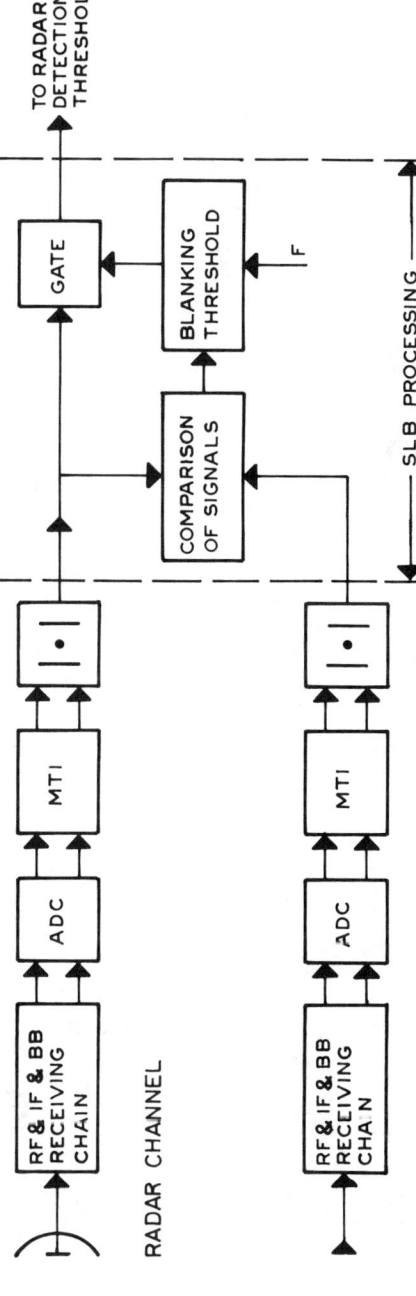

Figure 3.8 MTI processing upstream of the SLB logic.

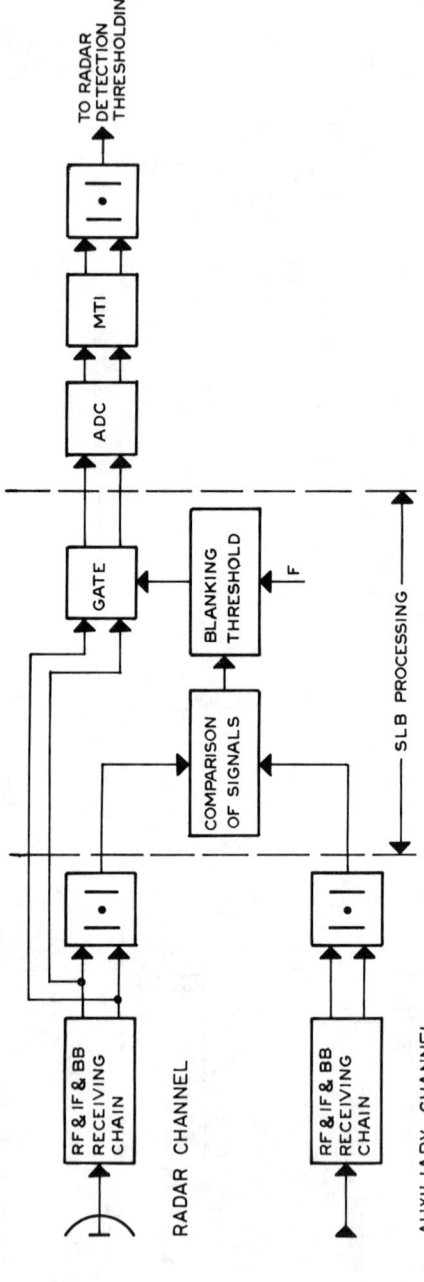

Figure 3.9 MTI processing downstream of the SLB logic.

processing, and (6) envelope detection. The signals produced in this way are compared to ascertain if the radar signal envelope is larger than that of the auxiliary signal. If so, the radar signals proceed toward the usual radar detection logic; otherwise, the signal flow is interrupted by the gate. The processing scheme for insertion of the MTI downstream from the SLB is shown in Figure 3.9. Only if the radar signal envelope is larger than the auxiliary signal envelope, the (I, Q) components of the radar signals are subject to the MTI processing and detection thresholding. The two processing options of Figures 3.8 and 3.9 have their distinct conveniences and drawbacks. However, the scheme of Figure 3.8 is the favored, although it requires two MTIs instead of one, as in Figure 3.9. The main reason for this choice is due to the poor performance of the scheme of Figure 3.9. In fact, strong and range-extended clutter in the sidelobes may give rise to too many blanking commands, which may produce the cancellation of a target signal in the main beam at the same range cell of the clutter echoes (see Figure 3.10).

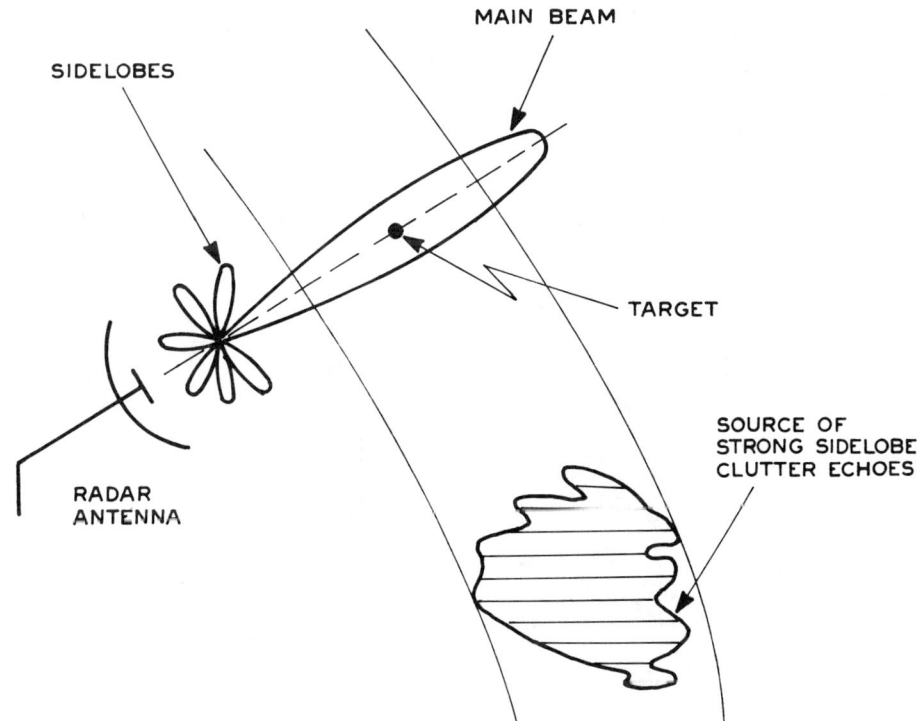

Figure 3.10 Main-beam target competing with strong range-extended sidelobe clutter. The target range is in the range interval of the clutter.

The SLB is also well suited to work in conjunction with the SLC system (see Chapter 4). As a matter of fact, one auxiliary antenna of the SLC array may be used when required for SLB purposes. We anticipate the SLC to be able to cancel high-duty-cycle sidelobe interference, whereas it is unable to filter impulse disturbances. The scheme of Figure 3.11 illustrates a possible combination of the SLB and SLC processors. The radar main antenna is surrounded by a number of auxiliary antennas, the majority of which are dedicated to the SLC function, and one auxiliary is devoted to the SLB device. The antennas are equipped with individual, carefully matched receivers. A number of differences exist between the SLC and SLB processors. Specifically, the SLC operates coherently on several range cells, whereas the SLB works on the signal amplitudes in one range cell. The envelope-detected signals emerging from the sidelobe canceler and the SLB auxiliary channel are compared to ascertain

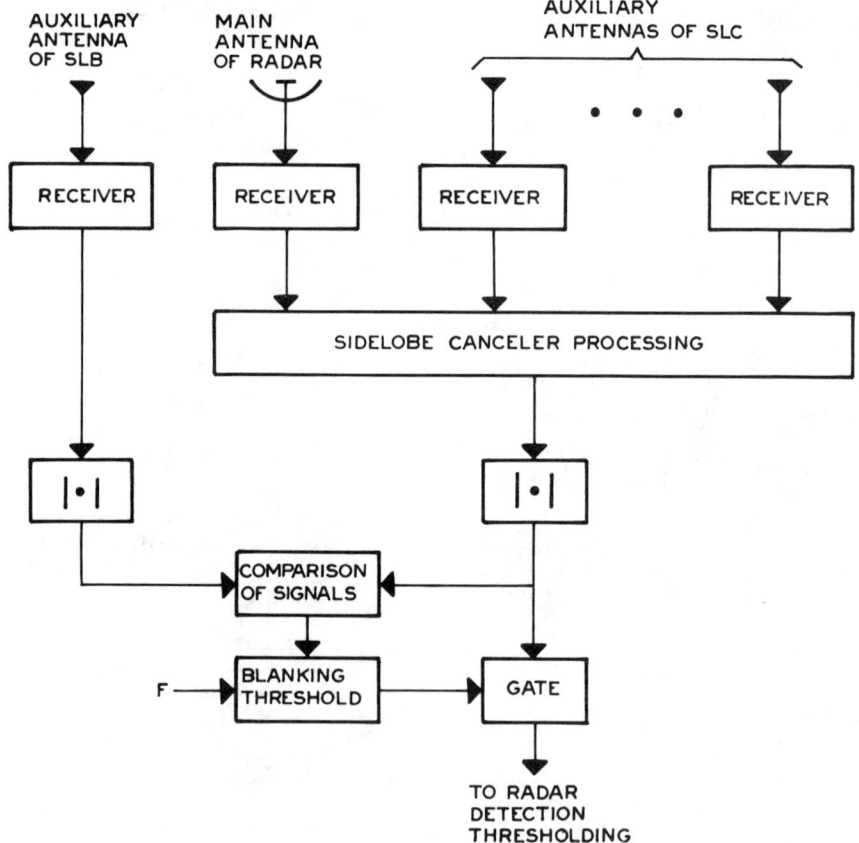

Figure 3.11 Joint use of SLB and SLC systems.

whether impulse disturbances are present in the radar sidelobes. If so, the gate inhibits the flow of the radar signals toward the detection thresholding. Finally, we wish to quote Chin *et al.* (1989), who consider the electromagnetic interference problem between the SLC and SLB antennas. Specifically, the reference considers the case in which the auxiliary antennas are subarrays of the main radar antenna. An interferometric effect results from combining SLB and SLC functions in a single radar, which can severely limit the usefulness of such a combination. Simply, the SLB antenna pattern is highly lobed by the presence of the SLC antennas. Nevertheless, Chin *et al.* (1989) present two techniques that make viable the union of SLB with SLC. The two techniques are named *aperture switching* and *jammer switching* or *beam switching,* respectively. Briefly, aperture switching, the preferred technique, moves the subarray phase centers on a pulse-to-pulse basis over the pulse-burst duration, thus reducing the unwanted interferometric nulls without degrading the jammer nulls.

3.3 PERFORMANCE EVALUATION

In this section, the performance of the SLB is evaluated as a function of (1) the sidelobe level of the radar antenna, (2) gain of the auxiliary antenna, and (3) signal-to-noise and jammer-to-noise power ratios per single pulse. The corresponding curves may be used to select the proper values of the auxiliary antenna gain and the detection and blanking thresholds. The numerical calculations of this section are derived from (Iovino, 1987).

A mathematical scheme equivalent to Figure 3.6 for the calculation of the SLB performance is that shown in Figure 3.12. A blanking signal is generated when the ratio (v/u) between the detected signals in the two channels is greater than the blanking threshold, F. The detection threshold, α_0, is for a single-hit detection with respect to the thermal-noise power.

The performance of the system may be analyzed by looking at the different outcomes obtained as a consequence of the pair (u, v) of processed signals (see Figure 3.13). Three hypotheses are tested: (1) the null hypothesis H_0 corresponding to the presence of noise in the two channels; (2) the H_1 hypothesis pertaining to the target in the main lobe; and (3) the H_2 hypothesis corresponding to target in the main lobe and/or interference signal in the sidelobe region in the same range cell as the target. These three hypotheses correspond to the three regions of the (u, v) plane depicted in Figure 3.13:

N: Detection of a null hypothesis;

D: Detection of a useful target;

B: Blanking command.

The three regions are mathematically defined as follows:

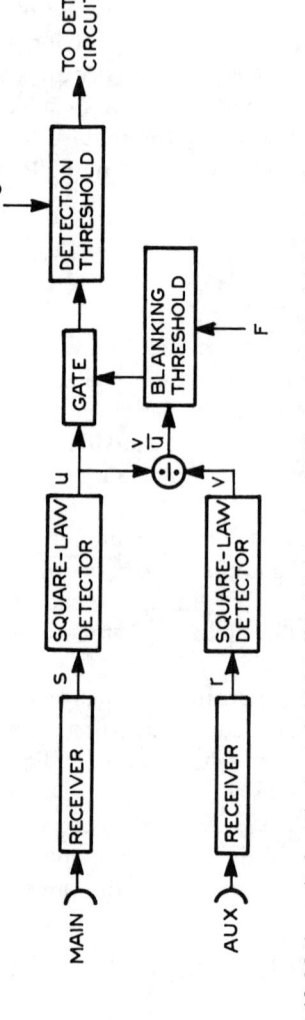

Figure 3.12 Mathematical equivalent of sidelobe blanking system of Figure 3.6. (From Maisel, *IEEE Trans. on Aerospace and Electronic Systems*, Vol. AES-4, No. 2, March 1968, pp. 174–180, © 1968 IEEE.)

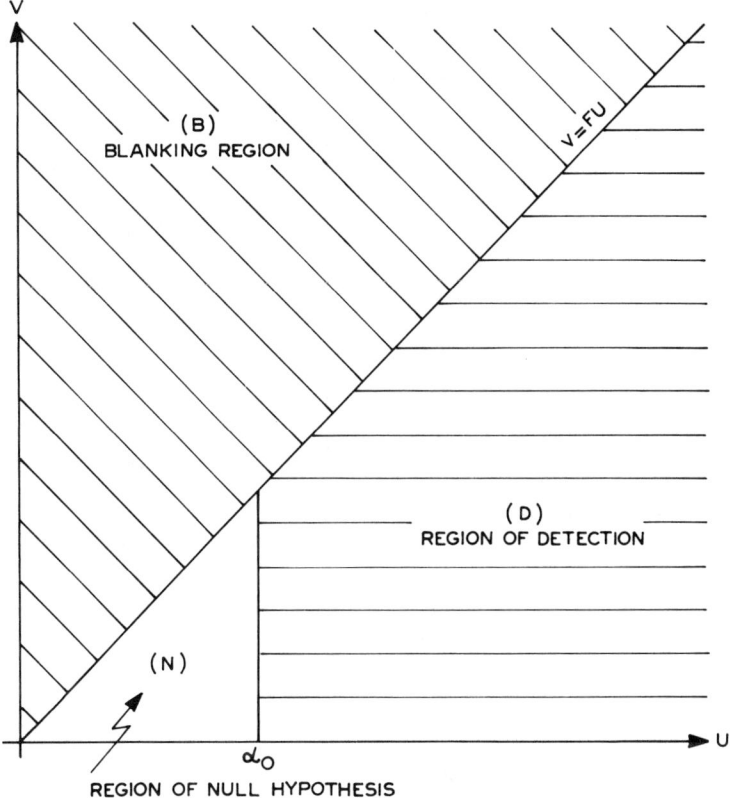

Figure 3.13 The multihypotheses testing problem in the (u, v) plane.

$$N: 0 \leq u \leq \alpha_0 \quad \text{and} \quad 0 \leq v < Fu$$
$$D: u > \alpha_0 \quad \text{and} \quad 0 \leq v < Fu \quad (3.1)$$
$$B: u \geq 0 \quad \text{and} \quad v \geq Fu$$

The presence of the blanking threshold, F, reduces the target detection domain, thus producing a detection loss with respect to the case in which the SLB is not present (i.e., $F = \infty$). The detection loss increases as F decreases. In the following, we will show how to limit the detection loss according to a prescribed jammer blanking probability.

To proceed with the performance evaluation, let us assume the idealized antenna patterns shown in Figure 3.14. Both patterns are normalized to the radar antenna gain, G_t. The sidelobe level of the radar antenna is set to the constant value

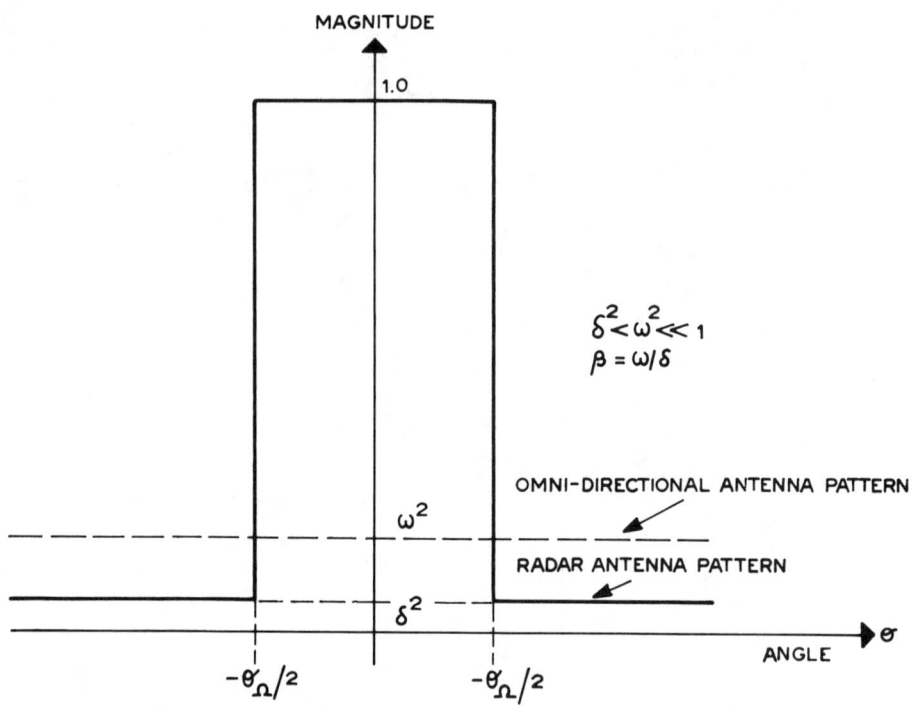

Figure 3.14 Idealized main and auxiliary antenna patterns.

of $\delta^2 = G_{sl}/G_t$; and the width of the main beam is θ_Ω; the gain of the auxiliary antenna is $\omega^2 = G_A/G_t$. The ratio $\beta = \omega/\delta = (G_A/G_{sl})^{1/2}$ is referred to as the *voltage gain margin* of the auxiliary antenna with respect to the sidelobe of the radar antenna. For correct operation of the SLB, the gain margin β should be such that

$$\beta \geq 1 \qquad (3.2a)$$

$$\beta^2 \geq F \qquad (3.2b)$$

where β^2 is the *power gain margin*.

The first inequality represents the concept, already mentioned, that the auxiliary gain should cover the radar sidelobes as much as possible. The second inequality can be understood by taking into account the thermal noise. In fact, if the thermal noise were negligible, the blanking condition would be

$$\frac{v}{u} \geq \beta^2 \qquad (3.3)$$

where the equal sign applies only for the case in which the jammer is received at the peak of the highest sidelobe. Consequently, to select the blanking threshold $F = \beta^2$ would be sufficient. In the presence of thermal noise, a prudent choice is that of (3.2b).

The problem of relating the key system parameters, F, β, ω, and δ also has been explicitly addressed in (O'Sullivan et al., 1989). The inequalities $\omega^2 \leq F \leq \beta^2$ should apply if the mainlobe targets are to fall within the region of detection, D (see Figure 3.13), for any target signal-to-noise power ratio, and if sidelobe targets are to avoid the same region D with high probability. We should note that in the inequality, ω^2 plays the role of the power gain ratio between the auxiliary antennas and the peak of the radar main beam. The precise value of F within the above interval involves a trade-off between detectability loss on main-lobe targets and the reliability of rejecting sidelobe targets. Numerical values of F pertain to the range $\omega^2 \approx 0.001$ and $1 < \beta^2 < 2$.

To proceed with the mathematical details, let us consider the nth azimuthal sample $s(n)$ of the radar signal upstream from the envelope detector (see Figure 3.12). The following expressions apply in relation to the underlying hypotheses:

$$s(n) = \begin{cases} w_s(n), & H_0 \\ w_s(n) + A \exp(j\phi_A), & H_1 \\ w_s(n) + A \exp(j\phi_A) + C \exp(j\phi_C), & H_2 \end{cases} \quad (3.4)$$

where A is the amplitude of a steady-state target received in the main beam, ϕ_A and ϕ_C are the target and jammer phases, respectively, evenly distributed in $[0, 2\pi]$, and C is the amplitude of a jammer received in the sidelobe region. Note that a model different from Swerling 0 for the target and the jammer could be selected, which would involve a retracing of the forthcoming calculations without impairing the validity of the whole approach. The nth sample of the thermal noise, $w_s(n)$, is assumed to be white Gaussian distributed with zero mean and variance $\sigma_s^2 = 2\sigma^2$. The following signal-to-noise and jammer-to-noise ratios can be defined upstream of the square-law detector of the radar channel:

$$\text{SNR} = A^2/2\sigma^2 \quad (3.5a)$$

$$\text{JNR} = C^2/2\sigma^2 \quad (3.5b)$$

As a consequence, they include the antenna gain values and the loss in the receivers. Having assumed the maximum gain of the radar antenna to be unity, apart from the processing loss, the SNR is the value at the radar antenna input. The jammer-to-noise power ratio upstream of the radar antenna is equal to JNR divided by the level of the radar sidelobes δ^2. The radar sidelobes are given by $(\omega/\beta)^2$. As a consequence, a JNR value of -5 dB associated with $\beta^2 = 5$ dB and $\omega^2 = -15$ dB provides a jammer-to-noise power ratio at the antenna input of 15 dB.

A similar mathematical model can be adopted by the nth azimuthal sample $r(n)$ of the signal upstream from the envelope detector of the auxiliary channel. The following equations apply:

$$r(n) = \begin{cases} w_r(n), & H_0 \\ w_r(n) + \omega A \exp(j\phi_A), & H_1 \\ w_r(n) + \omega A \exp(j\phi_A) + \beta C \exp(j\phi_C), & H_2 \end{cases} \quad (3.6)$$

where $w_r(n)$ is the nth sample of thermal noise in the SLB channel, uncorrelated with $w_s(n)$ and having the same variance (i.e., $\sigma_r^2 = \sigma_s^2 = 2\sigma^2$).

The nth sample of the radar signal downstream of the square-law detection is

$$u(n) = |s(n)|^2 \quad (3.7)$$

The corresponding probability density functions (pdf) in the three alternative hypotheses are

$$P_u(u/H_0) = \frac{1}{2\sigma^2} \exp(-u/2\sigma^2) \quad (3.8a)$$

$$P_u(u/H_1) = \frac{1}{2\sigma^2} \exp\left(-\frac{u + A^2}{2\sigma^2}\right) I_0\left(\frac{A\sqrt{u}}{\sigma^2}\right) \quad (3.8b)$$

$$P_u(u/H_2) = \frac{1}{2} \int_0^\infty t J_0(At) J_0(Ct) J_0(t\sqrt{u}) \exp\left(-\frac{\sigma^2 t^2}{2}\right) dt \quad (3.8c)$$

$I_0(\cdot)$ and $J_0(\cdot)$ are the modified Bessel functions of the first kind and zero order. The derivation of the third integral may be found in (Esposito and Wilson, 1973). This integral also is difficult to solve with numerical methods. Note that when $A = 0$ (this condition applies to the calculation of the detection probability of a false target in the sidelobe), (3.8c) becomes:

$$P_u(u/H_2, A = 0) = \frac{1}{2\sigma^2} \exp\left(-\frac{u + C^2}{2\sigma^2}\right) I_0\left(\frac{C\sqrt{u}}{\sigma^2}\right) \quad (3.8d)$$

which looks like (3.8b). Similarly, the nth sample of the auxiliary signal downstream from the envelope detection is

$$v(n) = |r(n)|^2 \qquad (3.9)$$

The corresponding pdf values are

$$P_v(v/H_0) = \frac{1}{2\sigma^2} \exp\left(-\frac{v}{2\sigma^2}\right) \qquad (3.10a)$$

$$P_v(v/H_1) = \frac{1}{2\sigma^2} \exp\left(-\frac{v + (\omega A)^2}{2\sigma^2}\right) I_0\left(\frac{\omega A \sqrt{v}}{\sigma^2}\right) \qquad (3.10b)$$

$$P_v(v/H_2) = \frac{1}{2} \int_0^\infty t J_0(\omega A t) J_0(\beta C t) J_0(t\sqrt{v}) \exp\left(-\frac{\sigma^2 t^2}{2}\right) dt \qquad (3.10c)$$

$$P_v(v/H_2, A = 0) = \frac{1}{2\sigma^2} \exp\left(-\frac{v + (\beta C)^2}{2\sigma^2}\right) I_0\left(\frac{\beta C \sqrt{v}}{\sigma^2}\right) \qquad (3.10d)$$

The SLB performance may be expressed in terms of the following probabilities:

1. The probability of false alarm, P_{FA}, which is the probability of associating the measured signals (u, v) with the region D when the true hypothesis is H_0, i.e.,

$$P_{FA} = \text{Prob}\{(u, v) \in D/H_0\} = \text{func}(\alpha_0, F) \qquad (3.11)$$

the P_{FA} depends on the two thresholds, α_0 normalized to noise power $2\sigma^2$, and F.

2. The probability P_D of detecting a target in the main lobe:

$$P_D = \text{Prob}\{(u, v) \in D/H_1\} = \text{func}(P_{FA}, F, \alpha_0, \text{SNR}) \qquad (3.12)$$

3. The probability P_D^* of detecting a target in the main lobe in presence of a jammer in the sidelobes and in the same range cell:

$$P_D^* = \text{Prob}\{(u, v) \in D/H_2\} = \text{func}(P_{FA}, F, \omega, \beta, \text{SNR}, \text{JNR}) \qquad (3.13)$$

4. The probability P_{FT} of detecting a false target produced by a jammer entering through the radar antenna sidelobes:

$$P_{FT} = \text{Prob}\{(u, v) \in D/H_2, A = 0\} = \text{func}(\alpha_0, F, \beta, \text{JNR}) \qquad (3.14)$$

5. The probability P_{FB} of false blanking, i.e., the probability of delivering a blanking command when the disturbance is just thermal noise in both the main and auxiliary channels:

$$P_{FB} = \text{Prob}\{(u, v) \in B/H_0\} = \text{func}(F) \quad (3.15)$$

This probability is not relevant to the SLB performance and will not be considered further.

6. The probability P_{TB} of blanking a target received in the main lobe:

$$P_{TB} = \text{Prob}\{(u, v) \in B/H_1\} = \text{func}(F, \omega, \text{SNR}) \quad (3.16)$$

7. The probability P_B of blanking a jammer:

$$P_B = \text{Prob}\{(u, v) \in B/H_2, A = 0\} = \text{func}(F, \beta, \text{JNR}) \quad (3.17)$$

To complete the description of the SLB performance, the last figure to consider is the detection loss, L, on the main lobe target. This can be found by comparing the SNR values required to achieve a specified probability of detection for the radar system, with and without the SLB. In decibels, we have

$$L = (\text{SNR})_{\text{with SLB}} - (\text{SNR})_{\text{without SLB}} = \text{func}(P_D, P_{FA}, F, \omega) \quad (3.18)$$

To calculate the probabilities P_{FA}, P_D, and P_{FT}, we need to solve the following integrals:

$$\text{Prob}\{(u, v) \in D/H_i\} = \iint_D P(u, v/H_i) \, du \, dv, \quad i = 0, 1, 2 \quad (3.19)$$

where D is the domain in the (u, v) plane of Figure 3.13. The joint pdf $P(u, v/H_i)$ may be evaluated as the product of the two marginal pdf values because u and v are assumed to be independent as they are provided by two separated channels. Equation (3.19) is rewritten as

$$\text{Prob}\{(u, v) \in D/H_i\} = \int_{\alpha_0}^{\infty} P_u(u/H_i) \int_0^{Fu} P_v(v/H_i) \, dv \, du$$

$$= \int_{\alpha_0}^{\infty} P_u(u/H_i) \left[1 - \int_{Fu}^{\infty} P_v(v/H_i) \, dv \right] du \quad (3.20)$$

The following equations are obtained (see also Maisel (1968)):

$$P_{FA} = \exp(-\alpha_0) - \frac{1}{F+1}\exp[-\alpha_0(F+1)] \qquad (3.21)$$

$$P_D = Q\left(\sqrt{2\mathrm{SNR}}, \sqrt{2\alpha_0}\right)$$
$$-\frac{\exp[-\mathrm{SNR}\cdot F/(F+1)]}{(F+1)}Q\left[\sqrt{\frac{2\mathrm{SNR}}{(F+1)}}, \sqrt{2\alpha_0(F+1)}\right] \qquad (3.22)$$

$$P_{FT} = Q\left(\sqrt{2\mathrm{JNR}}, \sqrt{2\alpha_0}\right) - \frac{F}{F+1}\left[1 - Q\left(\sqrt{\frac{2\mathrm{JNR}F}{F+1}}, \beta\sqrt{\frac{2\mathrm{JNR}}{F+1}}\right)\right]$$
$$+\frac{1}{F+1}Q\left(\beta\sqrt{\frac{2\mathrm{JNR}}{F+1}}, \sqrt{\frac{2\mathrm{JNR}F}{F+1}}\right)$$
$$+e^{-\mathrm{JNR}}\int_0^{\alpha_0}\exp(-u)I_0\left(\sqrt{4\mathrm{JNR}u}\right)Q\left(\beta\sqrt{2\mathrm{JNR}}, \sqrt{2Fu}\right)du \qquad (3.23)$$

where $Q(\cdot)$ is the Marcum function.

We note that, when $F \to \infty$ (3.21) reduces to the well-known equation, valid in the absence of the SLB:

$$P_{FA} = \exp(-\alpha_0)$$

where the threshold α_0 has been normalized to $(2\sigma^2)$. In the same way, we can evaluate the probabilities P_{FB}, P_{TB}, and P_B. To this end, we have the following probabilities:

$$\mathrm{Prob}\{(u, v) \in B/H_i\} = \iint_B P(u, v/H_i)\, du\, dv, \quad i = 0, 1, 2 \qquad (3.24)$$

where B is the domain in the (u, v) plane of Figure 3.13. Assuming, as usual, the independence of u and v, (3.24) becomes

$$\mathrm{Prob}\{(u, v) \in B/H_i\} = \int_0^\infty P_v(v/H_i)\int_0^{v/F} P(u/H_i)\, du\, dv$$
$$= \int_0^\infty P_v(v/H_i)\left[1 - \int_{v/F}^\infty P_u(u/H_i)\, du\right]dv \qquad (3.25)$$

The following equations are found:

$$P_{\text{FB}} = 1 - \frac{F}{F+1} \tag{3.26}$$

$$P_{\text{TB}} = 1 - \frac{1}{F+1}\left[1 - Q\left(\omega\sqrt{\frac{2\text{SNR}}{F+1}}, \sqrt{\frac{2\text{SNR}F}{F+1}}\right)\right]$$
$$- \frac{F}{F+1}Q\left(\sqrt{\frac{2\text{SNR}F}{F+1}}, \omega\sqrt{\frac{2\text{SNR}}{F+1}}\right) \tag{3.27a}$$

$$P_B = 1 - \frac{1}{F+1}\left[1 - Q\left(\beta\sqrt{\frac{2\text{JNR}}{F+1}}, \sqrt{\frac{2\text{JNR}F}{F+1}}\right)\right]$$
$$- \frac{F}{F+1}Q\left(\sqrt{\frac{2\text{JNR}F}{F+1}}, \beta\sqrt{\frac{2\text{JNR}}{F+1}}\right) \tag{3.27b}$$

The last probability P_D^* has been evaluated by means of Monte Carlo simulation.

The above-mentioned probabilities have been represented in a number of tables and curves. Here we show a sample of results.

Table 3.1 shows the threshold α_0 normalized to $(2\sigma^2)$ *versus* the P_{FA} values and the blanking threshold, F. The P_{FA} values are those pertaining to a single pulse, and are in the usual region of interest when an "*m* out of *n*" (Skolnik, 1990, p. 8.9) or other similar noncoherent digital integration schemes are employed downstream of the detection threshold, α_0 as indicated in Figure 3.6. Note that several pairs of (α_0, F) allow us to obtain a certain P_{FA} value. This means that a specific pair should be selected on the basis of other SLB requirements.

Figure 3.15 illustrates the P_D values *versus* the SNR per pulse having the P_{FA} as parameter and $F \geq 0$ dB. We see that as SNR tends toward $-\infty$ the curves reach asymptotic values that are different for different values of P_{FA}. These asymptotes are independent of the values of F. Note that P_D depends on ω through the parameter F. Here, we prefer to show only the case of $\omega^2 = -15$ dB, which gives the worst-case values of P_D.

Figure 3.16 shows the probability P_{FT} of detecting a false target *versus* the JNR per pulse and for different values of P_{FA} and F. More specifically, three figures are depicted for three values of the auxiliary gain margin β^2. Figure 3.16(a) refers to $\beta^2 = 2$ dB, Figure 3.16(b) is related to $\beta^2 = 5$ dB, and the case of $\beta^2 = 8$ dB is considered in Figure 3.16(c). Note that, instead of indicating the dependence of P_{FT} on α_0, we indicate the dependence on P_{FA}. P_{FT} reduces as P_{FA} and F decrease. Also

Table 3.1
Normalized Threshold α_0 versus P_{FA} and the Blanking Threshold

F (dB)	P_{FA}			
	10^{-1}	10^{-2}	10^{-3}	10^{-4}
−10	—	5.56	7.95	9.41
−9	—	5.81	8.06	9.48
−8	—	6.01	8.16	9.53
−7	1.00	6.18	8.23	9.57
−6	1.79	6.32	8.29	9.60
−5	2.36	6.43	8.34	9.62
−4	2.77	6.51	8.36	9.63
−3	3.07	6.56	8.38	9.64
−2	3.28	6.60	8.39	9.64
−1	3.43	6.62	8.39	9.64
0	3.52	6.63	8.39	9.64
1	3.57	6.63	8.39	9.64
2	3.60	6.63	8.39	9.64
3	3.62	6.63	8.39	9.64
∞	3.62	6.63	8.39	9.64

shown is that P_{FT} decreases as β^2, the gain margin of the auxiliary antenna with respect to the radar sidelobes, increases in value.

From Figures 3.15 and 3.16, we note that the probability P_D of detection in the main lobe is monotonic with SNR. However, in the sidelobes, for a given ratio of sidelobe gain β^2, the probability of detection P_{FT} is no longer monotonic with SNR, but possesses a single maximum, which increases in value with increasing values of F. The maximum also increases with decreasing values of β^2. Because we wish to minimize the maximum value of P_{FT}, this can be accomplished either by lowering threshold F, thereby reducing the main-lobe probability of detection P_D, or increasing β^2, thereby requiring a larger auxiliary antenna (Maisel, 1968).

The probability P_{TB} of blanking a target is shown in Figure 3.17 as a function of the SNR per pulse and having the blanking threshold F as parameter and the auxiliary gain $\omega^2 = -15$ dB. The target blanking probability is low for high values of SNR and for high values of the blanking threshold, F. The value of -15 dB for ω^2 has been taken to present a typical situation of radar sidelobes of $\delta^2 = -25$ dB (with respect to the antenna peak) and a gain margin $\beta^2 = 10$ dB.

The probability P_B of blanking a jammer is represented in Figure 3.18. The P_B is drawn as a function of the blanking threshold, F, and having JNR as parameter. Specifically, three figures are depicted for three values of the auxiliary gain margin

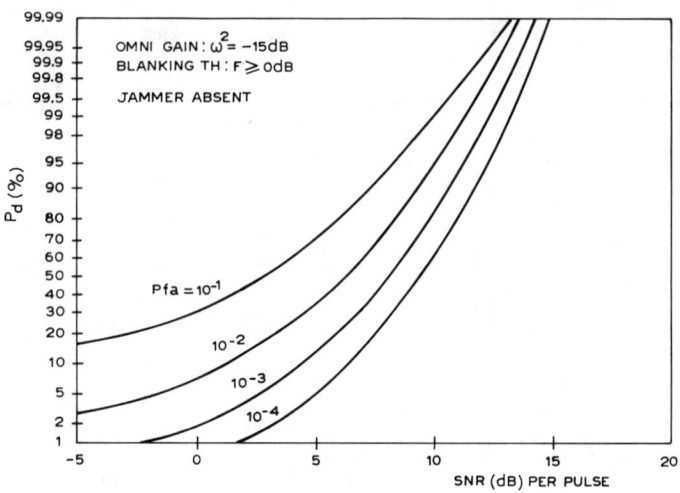

Figure 3.15 Detection probability P_D versus signal-to-noise ratio per pulse with the probability of false alarm P_{FA} as parameter and with the normalized radar sidelobes $\omega^2 = -15$ dB.

(a)

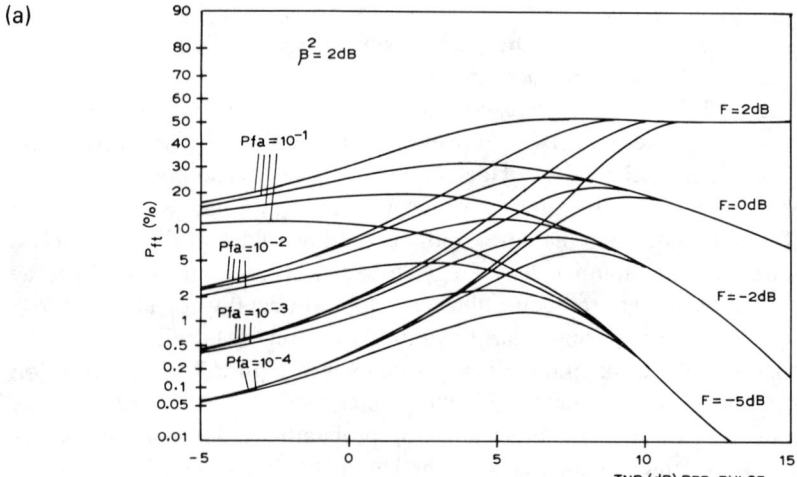

Figure 3.16 Probability P_{FT} of detecting a false target as a function of the jammer-to-noise ratio and having as parameters P_{FA}, the blanking threshold F, and (a) the auxiliary gain margin $\beta^2 = 2$ dB; (b) the auxiliary gain margin $\beta^2 = 5$ dB; (c) the auxiliary gain margin $\beta^2 = 8$ dB.

(b)

(c)

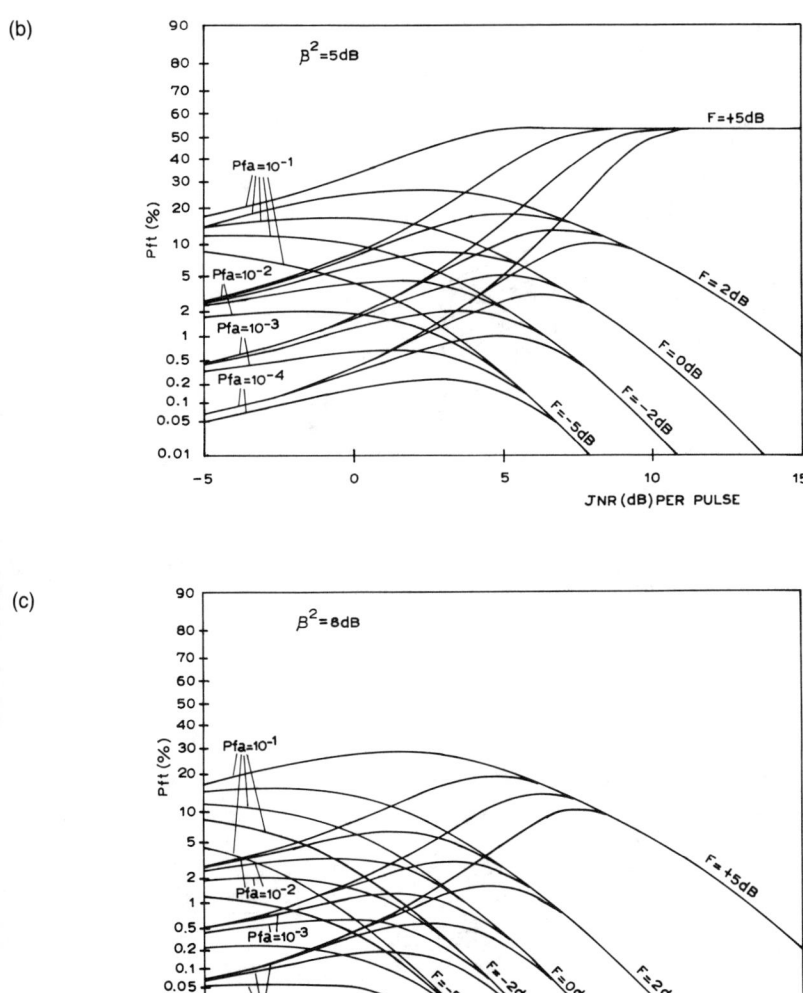

Figure 3.16 Continued

β^2. Figure 3.18(a) refers to $\beta^2 = 2$ dB, Figure 3.18(b) is related to $\beta^2 = 5$ dB, while the case of $\beta^2 = 8$ dB is considered in Figure 3.18(c). Note that good values of P_B are obtained for low values of the blanking threshold F, for high values of the gain margin β, and for large JNR values. Also note that blanking low-power jammers is more difficult, i.e., large gain margin β should be available.

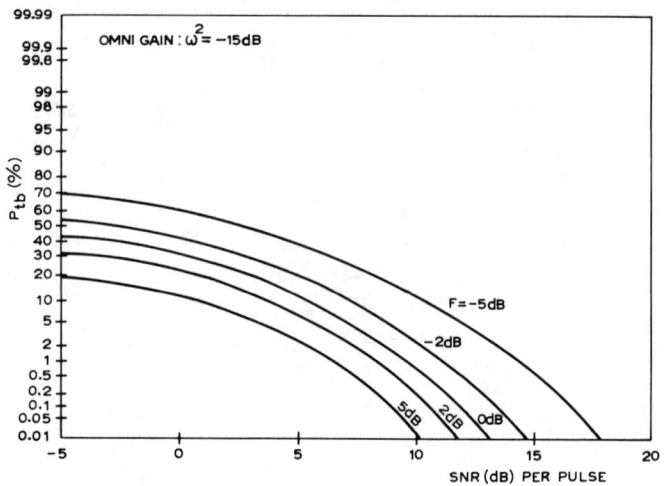

Figure 3.17 Probability P_{TB} of target blanking *versus* the signal-to-noise ratio per pulse and having the blanking threshold F as parameter and the auxiliary gain $\omega^2 = -15$ dB.

(a)

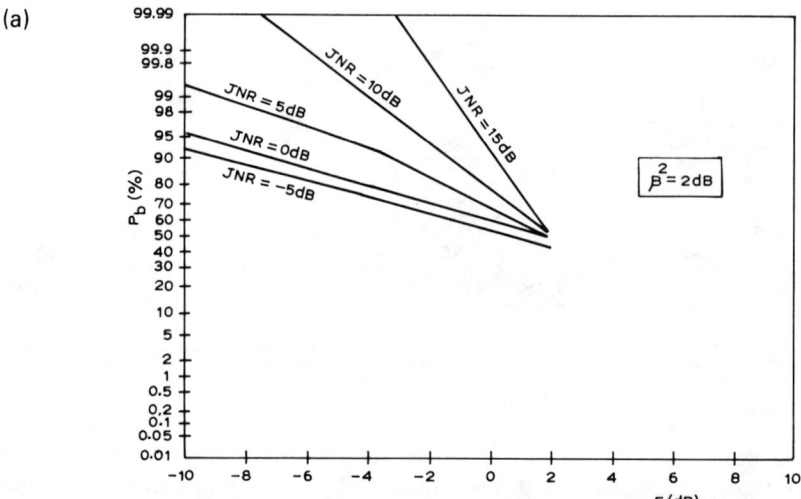

Figure 3.18 Probability P_B of jammer blanking *versus* the blanking threshold F, for different values of the jammer-to-noise ratio, and (a) the auxiliary gain margin $\beta^2 = 2$ dB; (b) the auxiliary gain margin $\beta^2 = 5$ dB; (c) the auxiliary gain margin $\beta^2 = 8$ dB.

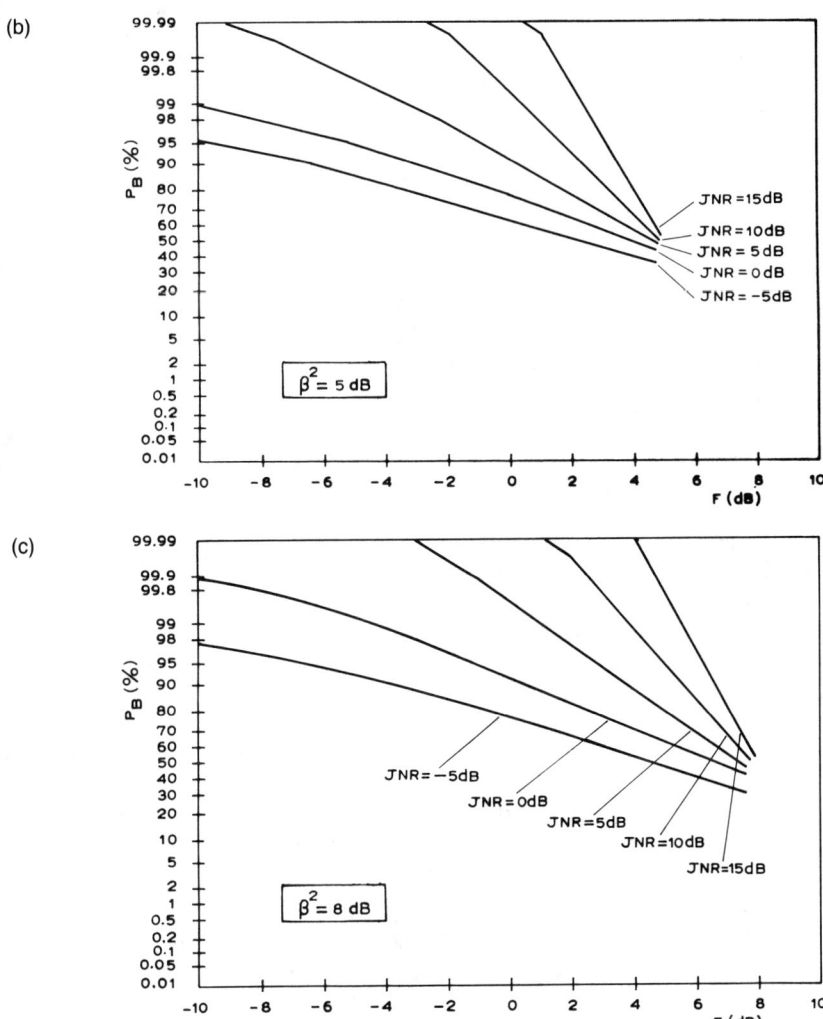

Figure 3.18 Continued

Figure 3.19 illustrates the detection probability P_D^* as a function of the SNR per pulse and having ω, β, F, and JNR as parameters. The curves have been drawn for $\omega^2 = -15$ dB, which gives the worst values of P_D^*. We see that P_D^* is not very dependent on β and F. However, the trend of P_D^* is toward a reduction if β increases and it is toward an increase if the threshold F increases as well. Moreover, as JNR increases, the detection performance deteriorates.

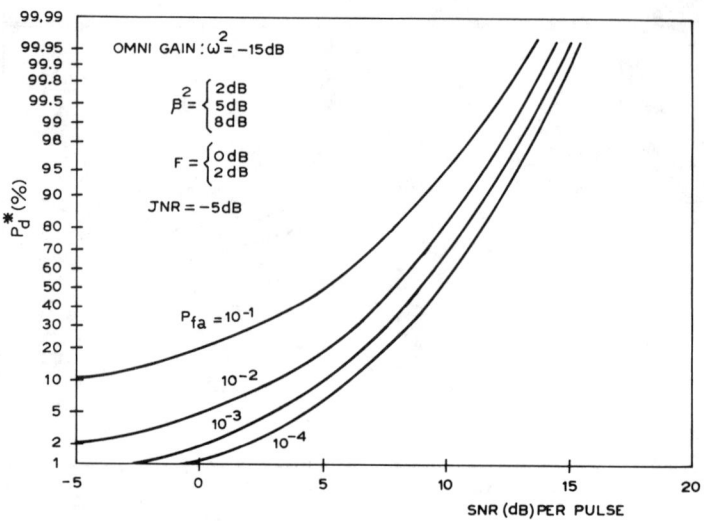

Figure 3.19 Probability P_D^* of detecting a target *versus* the signal-to-noise ratio per pulse, having as parameters the probability P_{FA} of false alarm, the gain $\omega^2 = -15$ dB of the auxiliary antenna, the auxiliary gain margin β^2, the blanking threshold F, and the jammer-to-noise ratio, JNR $= -5$ dB.

Tables 3.2(a) and 3.2(b) illustrate the detection loss in dB, when the jammer is absent, as a function of P_D, P_{FA}, and F. The values of P_D considered in these tables refer to the case of $\omega^2 = -15$ dB. The losses are higher for high values of P_{FA} and P_D and for low values of the blanking threshold, F. Two values of F are shown. $F = -5$ dB and $F = -2$ dB; for $F = 0$ dB, the losses are negligible. The losses are independent of β.

The design curves given in Figures 3.15, 3.17, and 3.19 are for the case where $\omega^2 = -15$ dB, which is an admittedly severe case from the radar sidelobe point of view. In fact, antennas are currently designed with sidelobe levels on the order of -40 to -50 dB. Therefore, to include a few more curves for a better case of $\omega^2 = -40$ dB, for example, seems useful. This is necessary, as the calculations required are difficult. Figure 3.20 illustrates the detection probability P_D showing sometimes better detection values with respect to the companion curves of Figure 3.15. The probability P_{TB} of target blanking is depicted in Figure 3.21 for $\omega^2 = -40$ dB. By comparison with the companion curves of Figure 3.17, we note some difference for high values of the SNR. Of course, P_{TB} is lower when $\omega^2 = -40$ dB. Finally, the detection probability P_D^* is given in Figure 3.22 for the case of $\omega^2 = -40$ dB. Again, better performance is obtained in this case with respect to Figure 3.19.

Table 3.2(a)
Detection Loss L for the Case of Jammer Absent ($F = -5$ dB)

	P_{FA}			
P_D	10^{-1}	10^{-2}	10^{-3}	10^{-4}
.50	1.50	0.46	0.21	0.11
.80	2.59	0.86	0.39	0.21
.95	3.80	1.86	0.91	0.57

Table 3.2(b)
Detection Loss L for the Case of Jammer Absent ($F = -2$ dB)

	P_{FA}			
P_D	10^{-1}	10^{-2}	10^{-3}	10^{-4}
.50	0.37	0.06	0.02	0.01
.80	0.59	0.17	0.03	0.02
.95	1.01	0.22	0.07	0.03

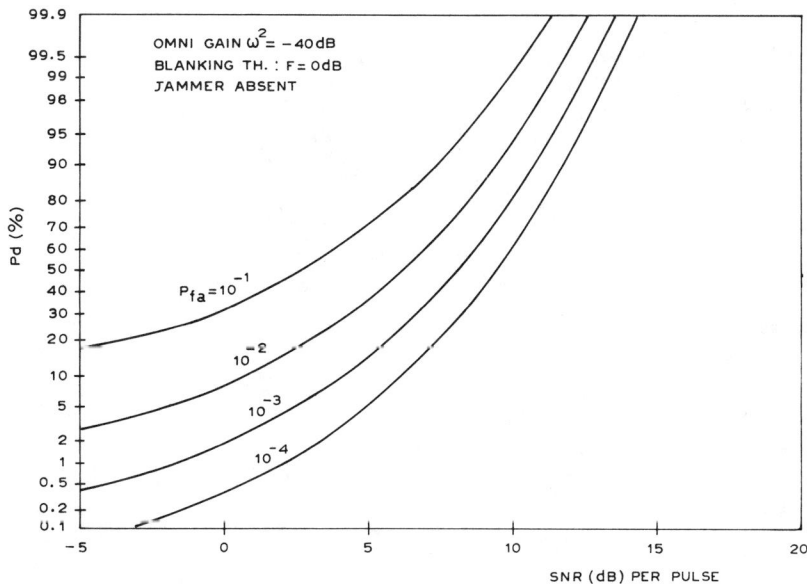

Figure 3.20 Detection probability P_D versus the signal-to-noise ratio per pulse with parameter the probability of false alarm P_{FA} and with the normalized radar sidelobes $\omega^2 = -40$ dB.

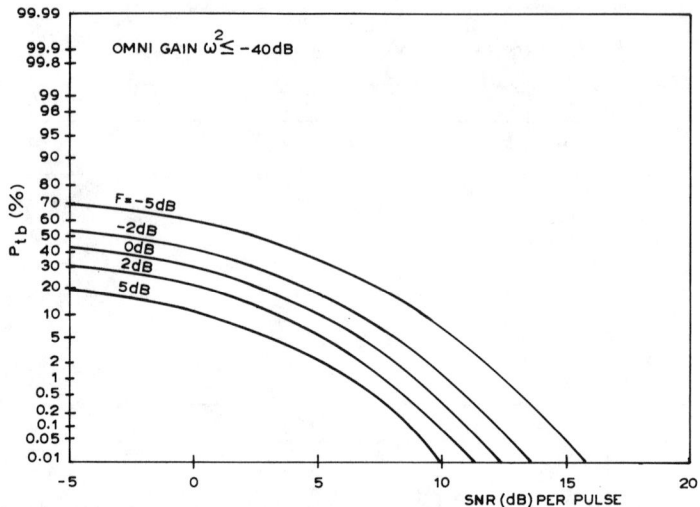

Figure 3.21 Probability P_{TB} of target blanking *versus* the signal-to-noise ratio per pulse and having as parameter the blanking threshold F and with the gain of the auxiliary antenna $\omega^2 = -40$ dB.

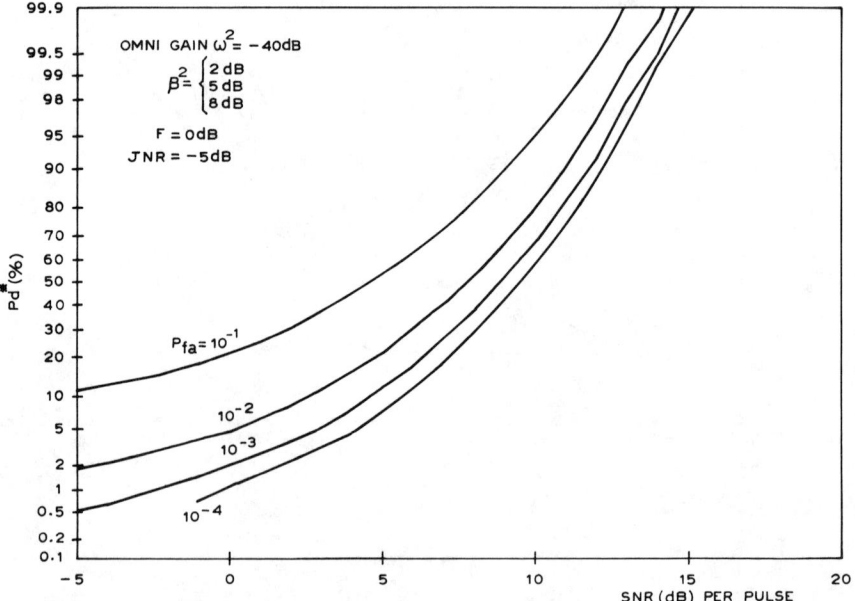

Figure 3.22 Probability P_D^* of detecting a target *versus* the signal-to-noise ratio per pulse, having as parameters the probability P_{FA} of false alarm, the gain $\omega^2 = -40$ dB of the auxiliary antenna, the auxiliary gain margin β^2, the blanking threshold $F = 0$ dB, and the jammer-to-noise ratio, JNR $= -5$ dB.

3.4 DESIGN METHODOLOGY

So far, the curves to quantify the SLB performance have been obtained. Let us now consider the way in which they can be exploited to design an SLB and to evaluate the corresponding performance. The main goal of the SLB design is to select suitable values for the following parameters:

- The gain margin β of the auxiliary antenna with respect to the sidelobes of the radar antenna, and then the normalized gain ω of the auxiliary antenna;
- The blanking threshold F;
- The normalized detection threshold α_0.

The *a priori* known parameters are:

- The normalized sidelobe level δ of the main antenna;
- The SNR and JNR values.

A possible way to proceed to the system design is to start from the requirement on the probability of blanking a jammer. From the JNR and the P_B values several pairs of β and F are found from Figure 3.18. We may select in a first instance the lower value of β (which means mild requirements on the auxiliary antenna gain), but other selections are also possible. Then, the value of F is derived accordingly. From the prescribed value of P_{FA}, the normalized detection threshold α_0 is also derived from Table 3.1. Having selected the parameters α_0, F, and β on the basis of P_B and P_{FA}, it is now possible to check if the other probabilities: P_D, P_{FT}, P_{TB}, and P_D^* and the detection loss L have acceptable values. Otherwise, we need to reconsider Figure 3.18 and to increase the β value.

As an application exercise, consider the following numerical example. The radar antenna has the sidelobe level 20 dB down from the peak, which means $\delta^2 = -20$ dB. Assume an impulse sidelobe jamming producing a JNR = 10 dB upstream from the square-law detector of the radar channel (see Figure 3.12). This means an equivalent value of JNR at the antenna input of 30 dB. By resorting to Figure 3.18, we find the following pairs of F and β values to achieve a probability of blanking the jammer equal to 90%: ($F = -1.5$ dB, $\beta^2 = 2$ dB), ($F = 2.5$ dB, $\beta^2 = 5$ dB), ($F = 5.5$ dB, $\beta^2 = 8$ dB). We select the second pair, corresponding to an intermediate value of β^2, which gives a numerical value $\omega^2 = -15$ dB for the normalized gain of the auxiliary antenna. Table 3.1 provides the normalized detection threshold $\alpha_0 = 6.63$ to set a P_{FA} value of 10^{-2} per pulse. We now need to check whether the values of the other performance parameters are acceptable. We start with the probability P_{FT} of false target (see Figure 3.16(b)), which results to be approximately 10%. To evaluate the detection probability P_D, we need to set a value of the SNR per pulse. A value of 6 dB is deemed to be good, considering that the final radar detection is performed by a second threshold downstream from a noncoherent pulse train integration (implemented, for instance, with a moving window (Skolnik,

1990, p. 8.4; Picardi, 1988, pp. 346–359) as indicated in Figure 3.6. In fact, with SNR = 6 dB, the P_D value on the single pulse is approximately 50% (see Figure 3.15) and, integrating seven pulses in the moving window, we obtain an overall P_D value of 80%. The overall probability of false alarm at the second detection threshold results to be 10^{-6} for a $P_{FA} = 10^{-2}$ on a single pulse and integrating seven pulses (Picardi, 1988, p. 351). The probability P_{TB} of target blanking is less than 5% (see Figure 3.17). The probability P_D^* of detecting a main-beam target when it is on the same range cell of a sidelobe jammer is very low, even when JNR = -5 dB, as shown in Figure 3.19. This unlucky case can be efficiently contrasted by varying the PRI of the radar on a pulse-to-pulse basis (see Section 1.3.3). Finally, the detection losses due to the presence of the SLB are negligible (see Table 3.2).

3.5 CONCLUDING REMARKS

In this chapter, attention has been focused on the sidelobe blanking technique to suppress impulse disturbances, strong targets, and strong pointlike clutter echoes entering the radar through the antenna sidelobes. The working principle of the SLB, performance evaluation, and design criteria have been described in the chapter. The compatibility of the SLB with the MTI and SLC techniques has also been considered. We can state that the SLB is a well tested and valuable technique to suppress low-duty-cycle sidelobe jammers. The SLB should be seriously considered for installation in surveillance radars when they are expected to operate against such disturbances.

The implementation of the SLB does not require special advanced technologies. Care must be paid to the problem of matching the auxiliary and the radar receiving channels with respect to the signal amplitudes. With a simplified reasoning, we may show that an amplitude mismatching produces worse blanking performance. In fact, the blanking rule $v \geq Fu$ of (3.1) would be modified as follows; v' and u' are the new signal amplitudes to compare. Assuming that the amplitude mismatching δ_a is attributable only to the auxiliary channel and that the mismatching can be modeled as multiplicative, i.e.,

$$u = u'$$
$$v = v'\delta_a \tag{3.28}$$

the blanking rule (3.1) becomes

$$v' \geq (F/\delta_a)u' \tag{3.29}$$

which means that the system operates with a new blanking threshold:

$$F' = F/\delta_a \tag{3.30}$$

Assuming that $0 < \delta_a < 1$, F' is larger than F thus resulting in a reduction of the probability P_B of blanking sidelobe jammers; in fact, the region B of Figure 3.13 is reduced in extension. A calibration procedure should be envisaged to compensate for the mismatching of the two channels. To this purpose, an RF calibration tone is fed at the input of the two receivers. The calibration procedure is based on the collection of a number N_C of samples of $u(i)$ and $v(i)$ ($i = 1, 2, \ldots, N_C$). The following average values are calculated:

$$\bar{u} = (1/N_C) \sum_{i=1}^{N_C} u(i)$$
$$\bar{v} = (1/N_C) \sum_{i=1}^{N_C} v(i)$$
(3.31)

and the following corrective multiplication factor:

$$f = \bar{v}/\bar{u} \tag{3.32}$$

is properly used in the current operation of the system.

Another factor affecting the SLB performance, which should be adequately considered in the system design and fabrication, is the maximum dynamic range of the receivers of the two channels. In fact, the SLB cannot work if saturation occurs. Consequently, care should be put in the design of the receivers to guarantee that the input signals, u and v, to the blanking logic are within the corresponding dynamic ranges.

The auxiliary antenna, which is a key component of the SLB system, will be designed and fabricated to provide a receiving pattern with an adequate gain in the azimuthal plane and a proper elevation sector. These performances will be maintained in the whole frequency band of the radar. Indoor and field measurements of the receiving patterns of the radar and auxiliary antennas will check the achievement of the desired patterns.

We have already anticipated that high-duty-cycle sidelobe jammers cannot be countered by an SLB system. A completely different system, the sidelobe canceler, has been conceived for this purpose, and is the theme of Chapter 4.

Chapter 4
Sidelobe Canceler (SLC) System

4.1 INTRODUCTION AND WORKING PRINCIPLE OF SLC

This chapter is devoted to a description of the sidelobe canceler, which has been conceived to suppress directional jammers received through the sidelobes of either a reflector type of antenna or a phased-array antenna (Howells, 1965; Applebaum, 1976). Section 4.1 introduces the basic concepts of the SLC system. Several implementation schemes are presented and discussed at length in Section 4.2, and the assessment of cancellation performance is illustrated in Section 4.3.

Because the adaptive array system, to be discussed in Chapter 5, is a generalization of the SLC technique, there would be a great deal of similarity in the formulation of the fully adaptive array algorithms with those used for sidelobe cancellation. To avoid problems of redundancy, we describe the algorithms for the estimation of the optimum weights for jammer cancellation purpose in this chapter; hence, the generalization to the fully adaptive system is straightforward. The algorithms for calculating the weights to determine the jammer *direction-of-arrival* (DOA) will be presented in Chapter 5. In fact, the problem of estimating the jammer DOA is more pertinent to fully adaptive array systems.

Concerning the mathematical tools, Chapter 5 makes more use of linear algebra, where the concepts of eigenanalysis are extensively recalled and exploited. The theory of stochastic processes having a Gaussian probability density function is largely used in both Chapters 4 and 5. For a complex vector quantity $\mathbf{X}$ having zero mean value, the corresponding covariance matrix definition is $\mathbf{M} \triangleq E\{\mathbf{X}^*\mathbf{X}^T\}$, where the operator $E\{\cdot\}$ denotes the expected value (the overbar $\overline{(\cdot)}$ is also used with the same meaning), the asterisk means complex conjugate, and T stands for transposition operation. An alternative covariance matrix definition is $\mathbf{M} \triangleq E\{\mathbf{X}\mathbf{X}^H\}$, where the Hermitian operator, H, denotes complex conjugate and transpose. The second definition yields a matrix, which is simply the complex conjugate of the first definition. So long as we adhere consistently to either one definition or the other, the results obtained (in terms of selected performance measure) will be the same; so, it is immaterial which definition is adopted for use (Monzingo and Miller, 1980, p. 86).

A search radar operation can be disturbed by high-duty-cycle interference the final goal of which is threefold:

1. To prevent the proper operation via saturation;
2. To reduce drastically the operating range;
3. To overload the search radar by detection of nonexistent targets.

Due to the sidelobes of the antenna radiation pattern, search radars undergo the negative effects of jammers, regardless of the effective antenna azimuth angle. To reduce the jamming effect when the disturbing signals enter the radar via radiation pattern sidelobes, an adaptive system can be used whereby the antenna pattern is modified to present nulls in the jammers' directions. This purpose can be achieved by installing in close proximity to the radar antenna, referred to in the following as the *main*, an adaptive array of auxiliary antennas. A typical sidelobe cancellation system is depicted in Figure 4.1(a). The auxiliary antenna gains are designed to approximate the average sidelobe level of the main antenna gain pattern. The shape of the synthesized array pattern is determined by the weights $W_i\{i = 1, 2, \ldots, N\}$; they are set up in an automatic fashion after a certain learning time. The amount of desired target signal received by the auxiliaries is assumed to be negligible compared to the target signal in the main channel. The target signal time duration is also assumed to be much smaller than the SLC adaptation time. Then, the target signals will pass unchanged through the SLC, while the jammer, which is continuous in time, will be reduced by the adaptation process operated by the canceler.

The purpose of the auxiliaries is to provide replicas of jamming signals in the sidelobes of the main pattern for cancellation. The auxiliary antennas are placed sufficiently close to the phase center of the radar antenna to ensure that the samples of the interference that they obtain may be correlated with the interference received in the radar antenna sidelobes. This requires that the separation, d, between the radar antenna's phase center and the auxiliary antenna's phase center divided by the velocity of light be much less than the reciprocal of the smaller of the radar bandwidth, B, or the interference bandwidth. Also note that the number of auxiliary antennas must at least equal the jamming signals to be suppressed. In fact, N auxiliary radiation patterns controlled by the adaptive weights $W_i\{i = 1, 2, \ldots, N\}$ are needed to force the main antenna radiation pattern in N given directions to zero. The auxiliary antennas may be separate antennas or groups of receiving elements of a phased-array antenna.

The SLC can be implemented with a reflector type of antenna, as depicted in Figure 4.1(b), where the auxiliary antennas are dipoles on the periphery of the reflector. With a phased-array antenna, the auxiliaries can be integrated into the main array as depicted in schematic fashion in Figure 4.1(c) (see also White, 1983). Electromagnetic coupling between these antennas should be minimized to maintain the

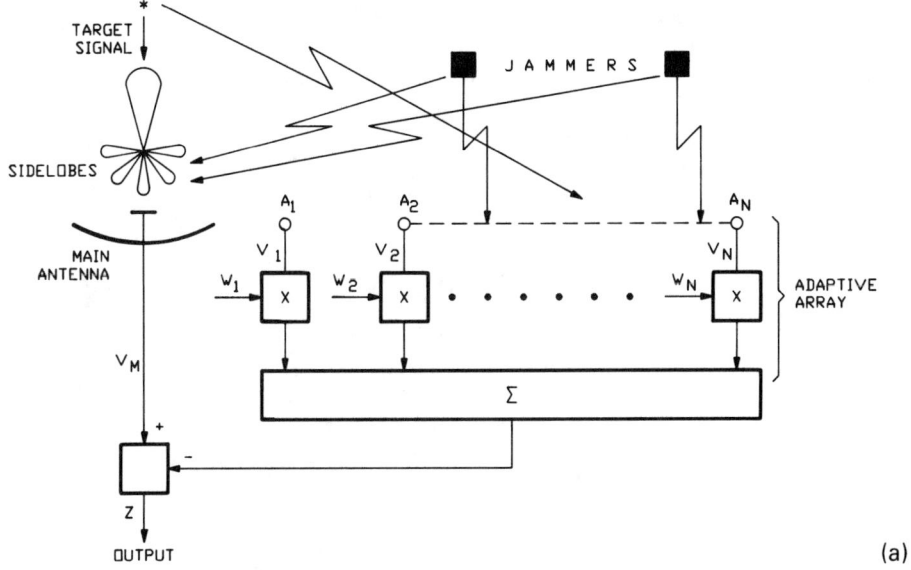

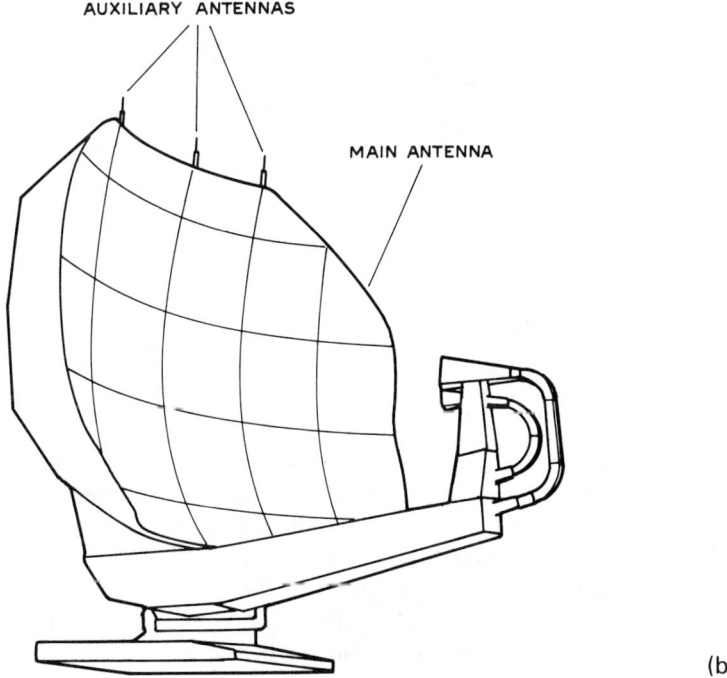

Figure 4.1 (a) Principle of operation of the SLC; (b) SLC with a reflector antenna; (c) SLC with a phased-array antenna.

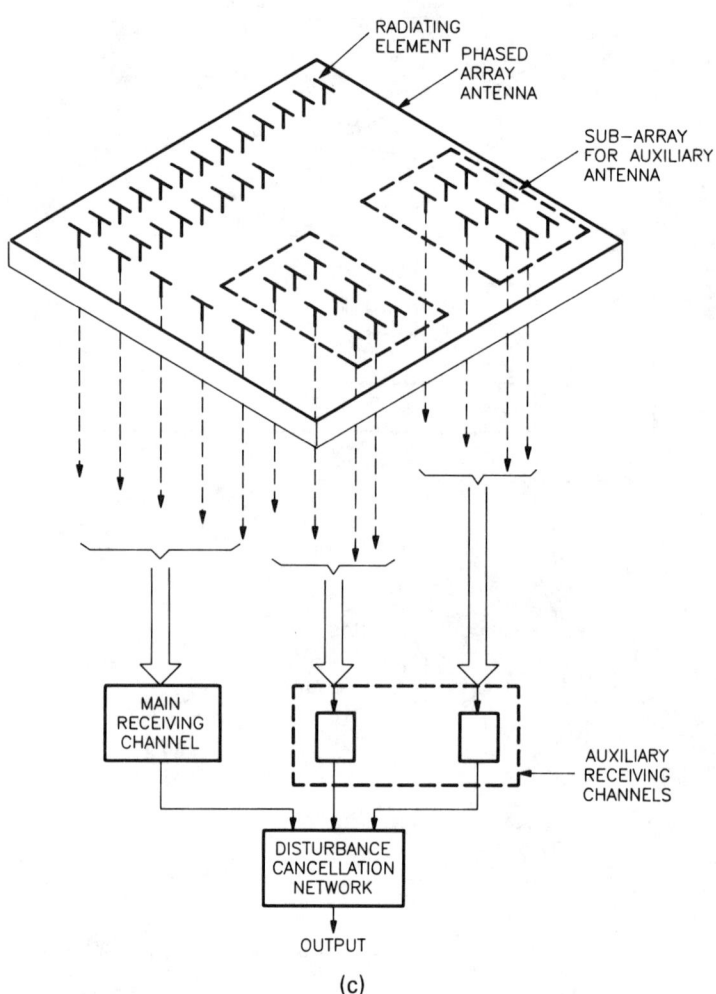

Figure 4.1 Continued

low quiescent sidelobe level. To have the auxiliary antennas within the main antenna is beneficial because the reduced distance between their phase centers mitigates the decorrelation of the jamming signals in the main and auxiliary channels (Farina, 1986; see also Subsection 4.3.2.1). An additional reason is that the 3 dB beamwidth of the auxiliary array should be comparable to the null-to-null spacing of the sidelobes in the radar antenna. This suggests the placement of the auxiliary elements around the periphery of the main antenna to match the sidelobe structure of the main pattern. The broadening of some sidelobes, resulting from the presence of random element errors (random sidelobes, see Section 2.6), or from applying a weighting function to the aperture to achieve low sidelobes (see Section 2.3.2), suggests that to have auxiliary elements within, as well as on, the periphery of the main antenna is desirable (Andrews and Gerlach, 1989, p. 464).

Two distinct ways of forming the auxiliary channels are possible with a phased-array antenna. In the first configuration, individual radiating elements or a cluster of elements are sent through an RF receiver, as indicated in Figure 4.1(c). The corresponding low-gain output forms the auxiliary channel to the SLC. The auxiliary elements should be irregularly spaced to avoid the deleterious effects of grating lobes generated by the auxiliary array (White, 1983). In the second configuration, the auxiliary radiating elements are combined to focus beams in the directions of individual jammers. The beamformer outputs are sent to individual RF receivers, and their outputs form the auxiliary channels to the SLC. The second technique has a less perturbing effect on the quiescent pattern because its action is localized around the jammer direction. Conversely, the first technique modifies the entire quiescent pattern of sidelobes, the perturbation being more pronounced as the number of jammers increases. However, to have adequate information on the jammer's angle direction-of-arrival is necessary to focus a beam toward a jammer. A number of techniques are described in Section 5.7, providing high-quality angular spectra of the electromagnetic emissions surrounding the radar.

The gain margin of the auxiliary antennas with respect to the sidelobe gain of the radar antenna in the jammer direction is a relevant parameter. The choice of the gain margin results from a trade-off analysis. A large value of the gain margin in the steady-state of an adaptive SLC would be desirable. In fact, in this case, the weights of the auxiliary channels would be small and the corresponding internal noise power values in the auxiliary channels would be attenuated. However, in the transient state of the SLC, the transient sidelobes are proportional to the gain margin; therefore, to have a low value of the gain margin would be advisable. A compromise value is around 10 for the gain margin (Andrews and Gerlach, 1989, p. 464).

The auxiliary outputs are weighted and summed, and then the resulting signal is subtracted from that in the main channel. The problem is to find a suitable means of controlling the weights so that the maximum possible cancellation is achieved. Owing to the stochastic nature of the jamming signals in the radar and auxiliary channels and the hypothesized linear combination of signals, the techniques of linear prediction theory for stochastic processes are advisable. V_M denotes the signal from the main antenna and $\mathbf{V}$ is the N-dimensional vector containing the set of signals from the N auxiliary antennas:

$$\mathbf{V}^T = [V_1 V_2 \ldots V_N] \qquad (4.1)$$

We have assumed that all signals have bandpass frequency spectra; therefore, the signals are represented by their complex envelopes, which modulate a common carrier frequency that does not appear explicitly. The jamming signals in the channels may be regarded as samples of a stochastic process having zero mean value and a prescribed autocorrelation function (for linear prediction problems, the first- and second-order statistics are sufficient).

From a statistical viewpoint, the set of samples V_M and **V** are completely described by the N-dimensional covariance matrix **M** of **V** and the N-dimensional cross-correlation vector **R** between V_M and **V**:

$$\mathbf{M} = E\{\mathbf{V}^*\mathbf{V}^T\} \tag{4.2a}$$

$$\mathbf{R} = E\{V_M \mathbf{V}^*\} \tag{4.2b}$$

We indicate with the N-dimensional vector **W** the set of N weights:

$$\mathbf{W}^T = [W_1 W_2 \ldots W_N] \tag{4.3}$$

The rejection of the disturbance in the main channel is achieved by subtracting from V_M the estimate of the jamming signal. The estimation is performed through a linear prediction of the jammer in the main channel on the basis of the jammer samples in the auxiliary channels. The optimum weight vector $\hat{\mathbf{W}}$ is determined by minimizing the mean square prediction error, which equals the output residual power:

$$P_z = E\{|Z|^2\} = E\{|V_M - \mathbf{W}^T\mathbf{V}|^2\} \tag{4.4a}$$

where Z is the SLC output. With little mathematics (4.4a) is shown to be a quadratic function of **W** with parameters **M** and **R**:

$$P_z = E\{|V_M|^2\} - \mathbf{W}^H\mathbf{R} - \mathbf{W}^T\mathbf{R}^* + \mathbf{W}^H\mathbf{M}\mathbf{W} \tag{4.4b}$$

By taking the gradient of P_z with respect to **W** (Haykin, 1986, pp. 105–108):

$$\nabla_\mathbf{W} P_z = \left[\frac{\partial P_z}{\partial W_1} \frac{\partial P_z}{\partial W_2} \ldots \frac{\partial P_z}{\partial W_N} \right]^T$$

$$= 2\,[\mathbf{MW} - \mathbf{R}] \tag{4.5a}$$

and equating it to zero, the following equation is obtained (Brennan and Reed, 1973):

$$\hat{\mathbf{W}} = \mu \mathbf{M}^{-1} \mathbf{R} \tag{4.5b}$$

where μ is an arbitrary constant value. As will be shown in Chapter 5, this equation could also be derived by the more general theory of adaptive arrays. A useful and important statistical condition exists between the output signal V_M and the components of the input signal vector $\mathbf{V}$ when $\mathbf{W}$ is equal to the optimum weight $\hat{\mathbf{W}}$. In fact, if we multiply both sides of the equation:

$$Z = V_M - \mathbf{V}^T\mathbf{W} \tag{4.5c}$$

by $\mathbf{V}^*$:

$$Z\mathbf{V}^* = V_M\mathbf{V}^* - \mathbf{V}^*\mathbf{V}^T\mathbf{W} \tag{4.5d}$$

and take the expected value, we obtain

$$E\{Z\mathbf{V}^*\} = \mathbf{R} - \mathbf{MW} \tag{4.5e}$$

Now, we let $\mathbf{W}$ take its optimum value (4.5b) and obtain

$$E\{Z\mathbf{V}^*\}|_{\mathbf{W}=\hat{\mathbf{W}}} = \mathbf{0} \tag{4.5f}$$

This result is well known in the linear prediction theory; it establishes that the prediction error signal (Z, in our case) is uncorrelated with, or orthogonal to, the input signals $\mathbf{V}$.

From (4.4a) and (4.5b), the power of jammer residue under optimum weights condition is

$$\begin{aligned} P_z &= E\{Z(V_M^* - \hat{\mathbf{W}}^H\mathbf{V}^*)\} \\ &= E\{ZV_M^*\} - \hat{\mathbf{W}}^H E\{Z\mathbf{V}^*\} \\ &= E\{ZV_M^*\} = E\{(V_M - \hat{\mathbf{W}}^T\mathbf{V})V_M^*\} \\ &= E\{|V_M|^2\} - \hat{\mathbf{W}}^T E\{\mathbf{V}V_M^*\} \\ &= E\{|V_M|^2\} - \mathbf{R}^H \mathbf{M}^{-1} \mathbf{R} \end{aligned} \tag{4.6}$$

where the third equality follows from the orthogonality established with (4.5f).

The benefit of using the SLC can be measured by introducing the jammer *cancellation ratio* (CR), defined as the ratio of the output noise power without and with the auxiliary array:

$$\text{CR} = \frac{E\{|V_M|^2\}}{E\{|V_M - \hat{\mathbf{W}}^T\mathbf{V}|^2\}} = \frac{E\{|V_M|^2\}}{E\{|V_M|^2\} - \mathbf{R}^H \mathbf{M}^{-1} \mathbf{R}} \tag{4.7}$$

Section 4.3 is entirely devoted to the calculation of the parameter CR for several practical cases.

The following is an application example of the basic relations (4.5b) and (4.7) to the case of one auxiliary antenna. Assume a narrowband jamming signal at an angular frequency, ω. The mathematical model of the free-space jammer is $a(t)\cos[\omega t + \phi(t)]$, where $a(t)$ and $\phi(t)$ respectively represent the amplitude and phase modulation terms. The signal produced in the main channel is $G_{sl}\, a(t)\cos[\omega t + \phi(t)]$, where G_{sl} is the voltage gain of the radar antenna sidelobe in the jammer direction. The signal produced in the auxiliary antenna is $G_A\, a(t)\cos[\omega t + \omega(d/c)\sin\theta + \phi(t)]$, where G_A is the voltage gain of the auxiliary antenna in the jammer direction of arrival, d is the distance between the phase centers of the two antennas, θ is the jammer angular direction-of-arrival measured with respect to the antenna system boresight, and c is the speed of light. The phase term $\omega d \sin\theta/c$ is due to an extra path length, $d \sin\theta$, with respect to the radar antenna phase center, traveled by the jammer to reach the auxiliary antenna. Note that, owing to the narrowband hypothesis, we have assumed $a(t + d \sin\theta/c) = a(t)$ and $\phi(t + d \sin\theta/c) = \phi(t)$. Adopting the complex notation, the signals can be expressed as the real part of complex quantities. By dropping the terms $\exp(i\omega t)$, as is usually done, the following models can be used for the signals received by the main and auxiliary antennas:

$$V_M(t) = G_{sl}\, j(t) + n_M(t) \qquad (4.8a)$$
$$V_A(t) = G_A\, j(t)\exp(i\alpha) + n_A(t) \qquad (4.8b)$$

where $j(t)$ is the free-space jamming signal, $a(t)\exp[i\phi(t)]$, which can be modeled as a zero-mean random variable with power P_J, α is the phase shift between the auxiliary and the main signals due to the extra path ($d \sin\theta$), and $n_M(t)$, $n_A(t)$ are zero-mean independent random variables with power P_N, which account for the thermal noise in the two receiving channels.

By application of (4.2a), (4.2b), and (4.5b) to our case, we obtain

$$\hat{W} = \frac{E\{V_M V_A^*\}}{E\{|V_A|^2\}} = \frac{G_{sl} G_A P_J \exp(-i\alpha)}{G_A^2 P_J + P_N} \qquad (4.8c)$$

First, we note that the weight $\hat{W}$ plays the role of the correlation coefficient between the main and auxiliary signals. Consequently, $\hat{W} V_A$ is the best linear estimate of V_M so that the residue $Z = V_M - \hat{W} V_A$ will have minimum power value. Another way to prove the cancellation capability of the SLC is the following. We see that, when $G_{sl} = G_A$ and $P_J \gg P_N$, the weight becomes $\hat{W} = \exp(-i\alpha)$; therefore, the auxiliary signal is phase-compensated and cancels, by subtraction, the jamming signal in the radar channel. When $G_{sl} \neq G_A$, the weight $\hat{W}$ will have an amplitude different from one, so that $\hat{W} V_A$ matches V_M in amplitude and phase. When P_J is not large with respect to P_N, the weight value will be influenced by the presence of thermal noise. To be more precise, let us calculate the residue power of the SLC. From (4.6), we obtain

$$P_z = E\{|Z|^2\} = E\{|V_M|^2\} - \frac{|E\{V_M V_A^*\}|^2}{E\{|V_A|^2\}}$$

$$= (G_{sl}^2 P_J + P_N) - \frac{G_A^2 G_{sl}^2 P_J^2}{G_A^2 P_J + P_N} \quad (4.8d)$$

which can be rewritten as

$$\frac{P_z}{P_N} = \frac{\text{JNR}\,(G_{sl}^2 + G_A^2) + 1}{\text{JNR}\,G_A^2 + 1} \quad (4.8e)$$

where JNR is the jammer-to-noise power ratio value P_J/P_N. For JNR $\gg$ 1 and $G_A = G_{sl}$, P_z is $2P_N$. In fact, the jammer is perfectly cancelled, and what remains as residual power is the noise from the two channels. For JNR $\gg$ 1 and unequal antenna gain value, we have

$$P_z = \left(\frac{G_{sl}^2}{G_A^2} + 1\right) P_N \quad (4.8f)$$

which means that the more gain the auxiliary channel has, the smaller is its noise contribution to the output (because it reduces the scale required of the weight $\hat{W}$). Equation (4.8f) shows the effect of the auxiliary gain margin with respect to the radar sidelobes on the residual power. The cancellation ratio (4.7) is as follows:

$$\text{CR} = \frac{E\{|V_M|^2\}}{E\{|Z|^2\}} = \frac{(G_{sl}^2\,\text{JNR} + 1)(G_A^2\,\text{JNR} + 1)}{(G_{sl}^2 + G_A^2)\,\text{JNR} + 1} \quad (4.8g)$$

which, for JNR $\gg$ 1 and $G_A = G_{sl} = g$, becomes

$$\text{CR} = \frac{g^2\,\text{JNR}}{2} = \frac{g^2 P_J}{2P_N} \quad (4.8h)$$

We conclude that the front-end system's thermal-noise power and the maximum expected interference power in the circuit always provide a bound on the cancellation ratio, CR.

4.2 IMPLEMENTATION SCHEMES

In this section, the main techniques that have been devised for implementing the SLC are reviewed. The methods may be divided into two main categories: (1) *closed-loop* or *feedback control* techniques and (2) *direct solution* methods (often referred to as *open-loop*). Broadly speaking, closed-loop methods are simpler and less costly to implement than direct solution methods. By virtue of their self-correcting nature, closed-loop methods do not require components that have a wide dynamic range or high degree of linearity, and so they are well suited to analog implementation. However, closed-loop methods suffer from the fundamental limitation that their speed of convergence must be restricted to achieve stable operation. Direct solution methods, conversely, do not suffer from problems of convergence, but, in general, require components of such high accuracy and wide dynamic range that they can only be realized by digital means. Of course, closed-loop methods can also be implemented by using digital circuitry, in which case the constraints on numerical accuracy are greatly relaxed, and the total number of arithmetic operations is much reduced by comparison with direct solution methods.

A third group of methods that will be described are the so-called *data-domain* or *voltage-domain* methods. They are similar to the open-loop techniques, but do not require estimating and inverting the covariance matrix for data; it will be shown that they offer several advantages from the implementation viewpoint.

4.2.1 Closed-Loop Methods

In this subsection, the main closed-loop methods, both analog and digital, are reviewed.

4.2.1.1 The Howells-Applebaum Closed-Loop Approach

The conventional analog adaptive processor loop is shown in Figure 4.2 for a single auxiliary antenna case. Let us check the correspondence between the circuit and theory previously mentioned. The basic principle of operation is that a weighting signal, W, is developed, in a closed-loop fashion, which causes the power of residual signal, Z, to be a minimum.

The following equations describe the circuit:

$$\tau \dot{u} + u = A \tag{4.9a}$$

$$u = +W/G \tag{4.9b}$$

$$A = ZV_A^* = V_M V_A^* - W|V_A|^2 \tag{4.9c}$$

$$\dot{W} + W[(1 + G|V_A|^2)/\tau] = +(G/\tau) V_M V_A^* \tag{4.9d}$$

$$Z = V_M - WV_A \tag{4.9e}$$

where τ is the integration time constant of the low-pass filter, G is the amplifier

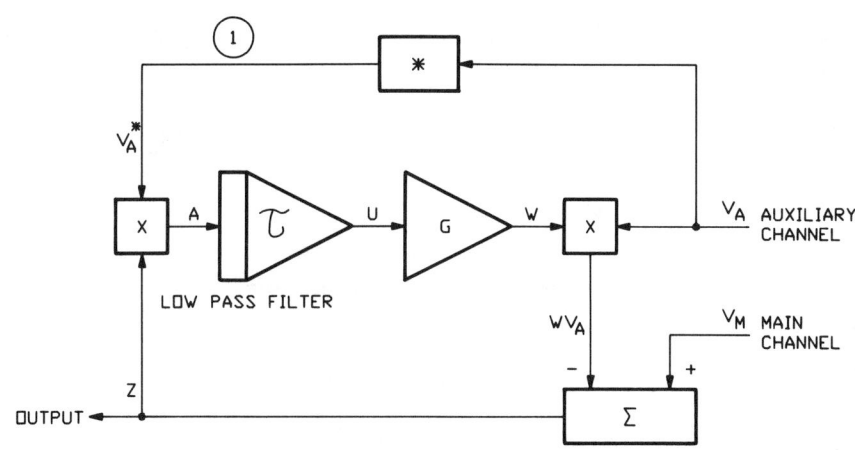

Figure 4.2 Functional diagram of Howells-Applebaum canceler.

gain, and $G\overline{|V_A|^2}$ is the closed-loop gain. To evaluate the SLC performance correctly, it would be necessary to solve the stochastic nonlinear differential equation (4.9d). The stochastic nature of this equation is due to the presence of the stochastic processes, V_A and V_M, which makes the weight to be a stochastic process itself. This problem has been considered in some papers (Farina, 1977; Farina and Studer, 1982; Gerlack and Lang, 1986; Nitzberg, 1981) and accurate expressions for the mean $\overline{W}$, variance σ_w^2, and even the pdf of $W(t)$ have been obtained. This complex analysis allows us to explain the *loop noise* phenomenon (see Subsection 4.3.2.5), which limits the performance of the SLC.

We shall develop an elementary theory, which, however, correctly explains the steady-state performance of the circuit. Making the assumptions of a slow convergence loop and ergodicity, the average value of the steady-state weight approaches (see (4.9d)):

$$\overline{W}_{t,\infty} = \overline{V_M V_A^*} G/(1 + G \overline{|V_A|^2}) \tag{4.9f}$$

The transient time constant with which we reach this value is

$$\tau_{\text{SLC}} = \tau/(1 + G \overline{|V_A|^2}) \tag{4.10}$$

Because τ_{SLC} is always positive, the system is unconditionally stable. Nevertheless, a careful setup of the τ_{SLC} is highly recommended, as will be shown later. From (4.10), the system time constant thus depends on the external conditions (i.e., $\overline{|V_A|^2}$). Inserting a coherent limiter, at point 1 of Figure 4.2, is likely to reduce such a dependence (Brennan and Reed, 1971).

Now we can readily see that the steady-state weight (4.9f) can be expressed as follows:

$$\overline{W}_{t \to \infty} = \rho \left\{ \sqrt{\frac{\overline{|V_M|^2}}{\overline{|V_A|^2}}} \right\} k \qquad (4.11)$$

where ρ is the normalized correlation coefficient between the main and auxiliary signals:

$$\rho = \overline{V_M V_A^*} / [(\overline{|V_M|^2})^{1/2} (\overline{|V_A|^2})^{1/2}] \qquad (4.12a)$$

and the parameter k is

$$k = (G\overline{|V_A|^2})/(1 + G\overline{|V_A|^2}) \qquad (4.12b)$$

Assuming that the jamming power values in the main and auxiliary channels are the same and that $G\overline{|V_A|^2}$ is much greater than unity, the steady-state weight (4.9f) becomes equal to the correlation coefficient, ρ. The same result is obtained from the general expression (4.5b).

The cancellation ratio, CR, of the adaptive loop is next derived in terms of the parameter ρ. From (4.7) and (4.9f), with little mathematics, we find (Lewis et al., 1986; Kretschmer and Lewis, 1978a):

$$\text{CR} = (1 - 2|\rho|^2 k + |\rho|^2 k^2)^{-1} \qquad (4.13)$$

For the case where $|\rho|^2 = 1$, corresponding to perfectly correlated signals, we obtain from (4.12b) and (4.13):

$$\text{CR} = (1 + G\overline{|V_A|^2})^2 \qquad (4.14)$$

For the case of $G\overline{|V_A|^2} \gg 1$ and any value of ρ, (4.13) gives

$$\text{CR} = \frac{1}{1 - |\rho|^2} \qquad (4.15)$$

From the previous equations, high CR values are thus obtained for values of ρ close to unit and for large values of $G\overline{|V_A|^2}$.

Increasing $G\overline{|V_A|^2}$ is not always a good strategy. This is so because the loop time constant (4.10) would be reduced, or, in other words, the loop bandwidth would be increased, thus increasing the amount of jamming power transferred to the output. A short loop-time constant would allow tracking nonstationary jammer situations. As a matter of fact, the mean value of the weight would quickly reach the steady-state condition. However, we can show (Farina and Studer, 1982; Nitzberg, 1981) that the variance of the weight in the steady-state condition increases with $G |V_A|^2$, thus reducing the efficacy of the jammer cancellation (see also Section 4.3.2). This is the major limitation of the Howells-Applebaum loop.

The original Howells-Applebaum loop was implemented at the IF stage by using analog devices. However, the SLC can be implemented today at baseband with digital techniques. Both techniques are now briefly reviewed.

4.2.1.2 Analog Implementation at IF Stage of the Howells-Applebaum Loop

The auxiliary antenna feeds a receiver which is assumed to be identical to that of the radar, and the two receivers share the same local oscillator as illustrated in Figure 4.3 (see Lewis *et al.*, 1986). The difference between the time-of-arrival of the interference at the radar and at the auxiliary antenna, separated by a distance d, is given by

$$T_a = (d \sin\theta)/c \qquad (4.16)$$

where θ is the angle of incidence of the interference, and c is the speed of light. For proper sidelobe canceler operation, T_a must be limited so that its amplitude is much less than $1/B$ to maintain correlation between the main and auxiliary interference signals.

The output of the radar and auxiliary receivers at IF feed an analog loop. The interference voltages V_M and V_A are assumed to be functions of time, $A(t)$, modulating an IF carrier ($\omega_1 = 2\pi f_1$, rad/s) that is much higher in frequency than the bandwidth B on the modulating function, $A(t)$. In addition, V_A is assumed to be correlated with V_M (i.e., V_A only differs from V_M by a constant amplitude and constant phase shift of the carrier frequency). This is assumed to result from the difference in gain and location of the phase centers of the radar and auxiliary antennas, as illustrated in Figure 4.3. The Howells adaptive loop employs a local oscillator (LO) with an angular frequency, ω_2, to translate the radar receiver's output from IF to $(\omega_1 - \omega_2)$. The resulting signal, V_M, enters the positive input of a subtractor, the output of which is both the adaptive-loop residue, Z, and the input to a multiplier, where it translates the auxiliary signal down to a center frequency ω_2 with phase ϕ. The resulting signal, which passes through a narrowband filter centered on frequency ω_2 and is amplified by a gain, G, translates the auxiliary signal to $(\omega_1 - \omega_2)$ and zero phase in a multiplier and bandpass filter centered on $(\omega_1 - \omega_2)$. The resulting zero-phase auxiliary signal is then subtracted from the frequency-translated zero-phase radar signal to form the

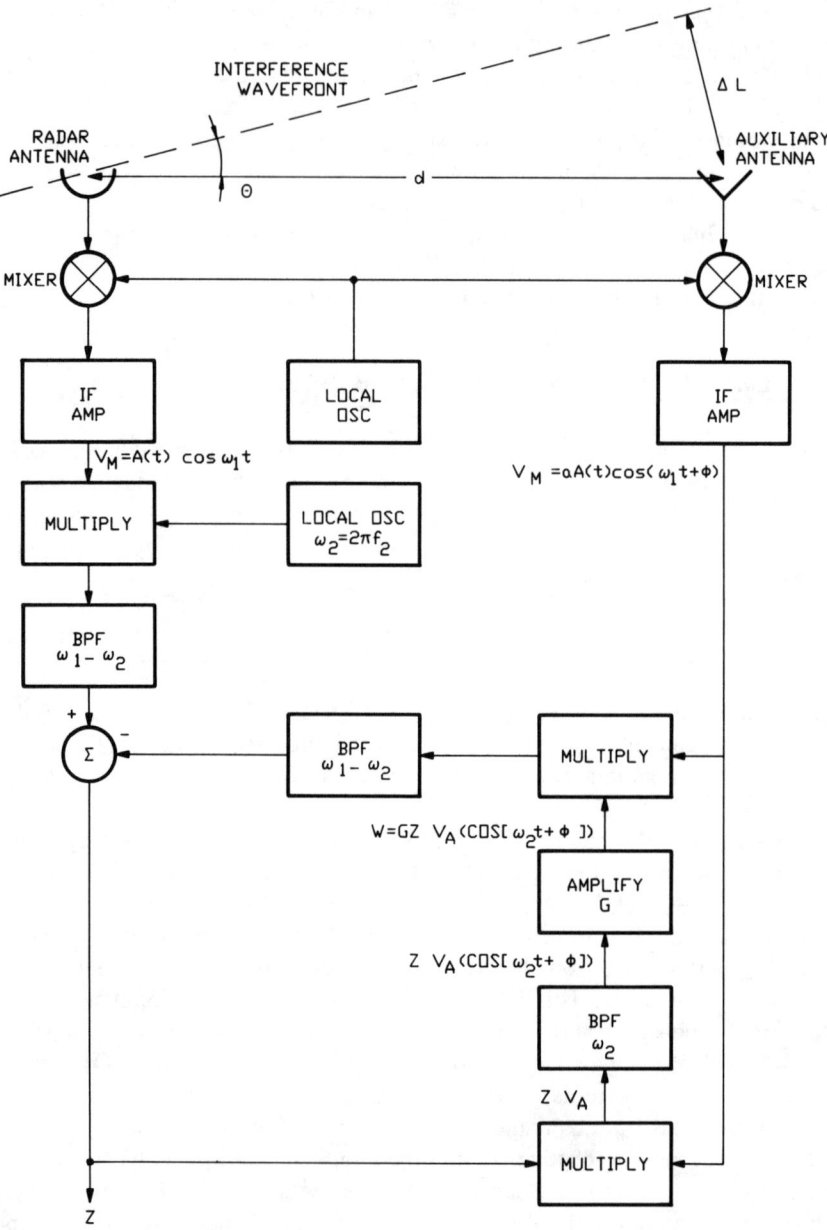

Figure 4.3 Howells analog adaptive loop. (From Lewis, Kretschmer, and Shelton, *Aspects of Radar Signal Processing,* Artech House, Norwood, MA, 1986.)

loop residue, Z. The output of the amplifier (gain G) in the loop is the signal that controls the amplitude and phase of V_A so that, when the weighted V_A is subtracted from V_M, the residue Z approaches zero.

The SLC system with only one auxiliary antenna is unable to cancel more than one jammer. In fact, a single loop represents one amplitude and phase change on V_A; more interference sources from different directions require that different weights be used. The problem of cancelling interference from multiple independent jammers at different angular locations can be approached by the system of Figure 4.4, which shows a correlation loop attached to each auxiliary antenna. To demonstrate that this scheme implements the optimum weights of (4.5b), note that the residue power P_z of (4.4b) is a bowl-shaped quadratic function of $\mathbf{W}$. The weight corresponding to the point at the bottom of the bowl individuates the optimum set $\hat{\mathbf{W}}$ of the weights. $\hat{\mathbf{W}}$ can be calculated by P_z using the gradient algorithm. It is a recursive algorithm

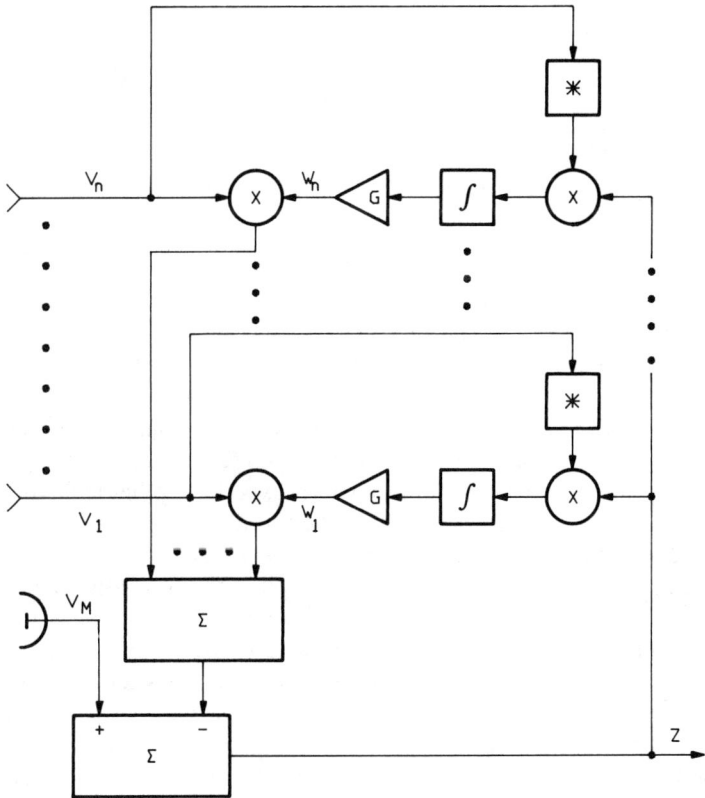

Figure 4.4 Howells-Applebaum implementation of the SLC.

in the sense that it modifies the guess of the weight at each step by forcing it in the direction of the negative gradient of P_z and with a step amplitude that is proportional to the residue power. The algorithm is

$$\mathbf{W}(t + \Delta t) = \mathbf{W}(t) - k\nabla_w E[|Z(t)|^2] \quad (4.17)$$

By approximating the expected value $E[|Z(t)|^2]$ to the measured value $|Z(t)|^2$, we thus have

$$\mathbf{W}(t + \Delta t) = \mathbf{W}(t) - k\nabla_w |Z(t)|^2 \quad (4.18)$$

$$\nabla_w |Z(t)|^2 = 2 Z(t)\nabla_w Z^*(t) = 2 Z(t)\nabla_w [V_M^* - (\mathbf{W}^T\mathbf{V})^*] = -2Z(t)\mathbf{V}^* \quad (4.19)$$

$$\mathbf{W}(t + \Delta t) = \mathbf{W}(t) + 2 kZ(t)\mathbf{V}^*(t) \quad (4.20)$$

In the continuous time case, (4.20) becomes

$$\frac{d\mathbf{W}}{dt} = GZ(t)\mathbf{V}^*(t) \quad (4.21)$$

which together with

$$Z = V_M - \mathbf{W}^T\mathbf{V} \quad (4.22)$$

corresponds to the scheme of Figure 4.4.

An illustrative implementation for two auxiliary antennas is shown in Figure 4.5, which is a generalization of the scheme of Figure 4.3.

The *multiple SLC* (sometimes referred to as MSLC) can be analyzed by studying the set of N-coupled nonlinear stochastic differential equations (4.21) and (4.22). A common procedure to handle this problem uses a decoupling technique based on the eigenvalue and eigenvector analysis applied to the set of equations, which consequently becomes uncoupled. Examples of this technique will be given afterwards. More often, the set of equations are simulated on a digital computer to study their transient and steady-state behavior (see, for example, Klemm, 1977; Bucciarelli *et al.*, 1982). Another procedure has been suggested by Lewis and Hansen (see Lewis *et al.*, 1986, pp. 137–152) and applies directly to schemes similar to that of Figure 4.5. This approximate approach, which is referred to as the *vector technique*, allows an analyst to determine the residue of multiple loops under any given geometry of jammers and auxiliary antennas through a complete 360° scan of the radar antenna if desired. The analyst then may determine the effect of repositioning the auxiliary antennas or see that more antennas and loops must be added to obtain the desired

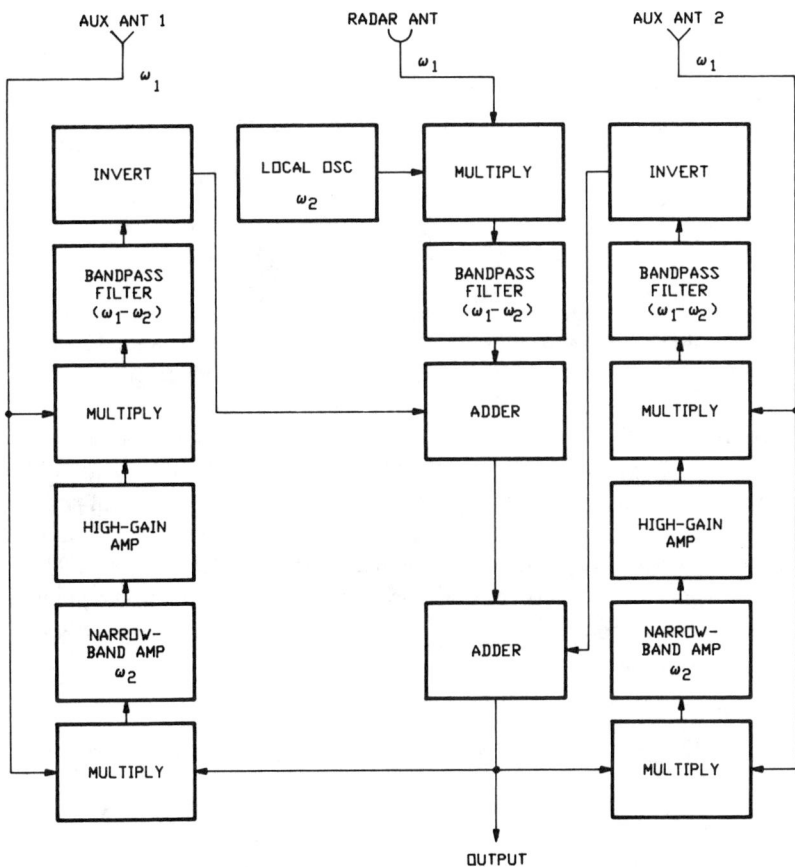

Figure 4.5 Functional block diagram of a two-degree-of-freedom coherent sidelobe canceler. (From Lewis, Kretschmer, and Shelton, *Aspects of Radar Signal Processing,* Artech House, Norwood, MA, 1986.)

performance. The technique is based on the fact that the operation of any loop is independent of the operation of all other loops to a first approximation. This allows an analyst to pick any loop to start the analysis and to ignore the operation of all others while completing the analysis of the first loop. In this way, he can analyze one loop after another, with each successive loop operating on the residue of those already analyzed and thus determine an initial approximation of the final residue produced by the group operating together. The approximation can be made much more accurate by repeating the process, using this initial result as the radar input to the cancelers.

4.2.1.3 Least-Mean-Square (LMS) and Modified LMS (MLMS) Algorithms

The Widrow-Hoff LMS is a closed-loop digital algorithm described by the following equations (Widrow, 1966; Kretschmer and Lewis, 1978b):

$$Z(j) = V_M(j) - \mathbf{V}^T(j)\mathbf{W}(j) \qquad (4.23a)$$

$$\mathbf{W}(j+1) = \mathbf{W}(j) + 2\mu Z(j)\mathbf{V}^*(j) \qquad (4.23b)$$

where $\mathbf{V}(j)$, $V_M(j)$, $Z(j)$, and $\mathbf{W}(j)$ denote the jth discrete time samples of the complex modulation functions respectively corresponding to the vector of auxiliary signals, main signal, output from the canceler, and weight vector. Discrete data are normally sampled at the Nyquist rate. The parameter μ is a gain factor that controls the stability and convergence rate of the MSLC. The algorithm is particularly efficient, requiring only approximately $4N$ real multiplications and $4N$ real additions per sample time to update the weight vector. An attractive feature of this algorithm is that it is suitable for computer analysis as will be shown in Section 4.3.2. The scheme of the LMS algorithms is shown in Figure 4.6 for the ith auxiliary antenna. The LMS algorithm comprises N loops operating in parallel with a common gain factor,

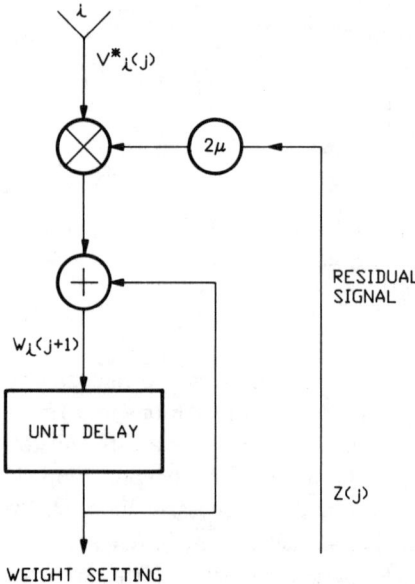

Figure 4.6 Digital cancellation loop.

μ. Widrow (1966) has shown that if the weight vector is updated according to (4.23b), then

$$\lim_{j\to\infty} E\{W(j)\} = M^{-1} R\mu \qquad (4.24)$$

i.e., the weight vector tends in the mean to the optimum value given by (4.5b) provided that the gain constant μ satisfies the condition of stability:

$$0 < \mu < \frac{1}{\lambda_{max}} \qquad (4.25)$$

where λ_{max} denotes the largest eigenvalue of the covariance matrix M.

The rate of convergence depends on the value of μ. For small values of μ, the algorithm converges slowly and the weight vector does not fluctuate much around the mean. For large values of μ, the algorithm converges more rapidly, but the weight vector is subject to larger fluctuations due to the fact that the integration time of the loop and hence the statistical accuracy in the determination of W are reduced. These fluctuations, in turn, lead to *misadjustment noise*, which causes the output power to increase above its optimum level. Misadjustment is defined as follows (Widrow *et al.*, 1976): misadjustment $\triangleq$ excess mean squared error, (mse) divided by mse obtained with optimum weight setting. The time constant of adaptation for the kth normal component of the weight vector (i.e., the component associated with the kth eigenvector of M) is given by

$$t_k = \frac{1}{2\mu\lambda_k} \qquad (4.26a)$$

where λ_k is the kth eigenvalue. If the spread of eigenvalues is large, the smaller components must suffer very slow convergence if the stability condition (4.25) is to be satisfied. This fundamental limit to the overall rate of convergence is common to all closed-loop techniques for adaptive nulling. Considering that the maximum transient time constant, τ_{max}, is related to λ_{min}:

$$\tau_{max} = \frac{1}{2\mu\lambda_{min}} \qquad (4.26b)$$

by combination of (4.26b) with (4.25), we thus have

$$\tau_{max} > \lambda_{max}/2\lambda_{min} \qquad (4.26c)$$

In practice, the parameter μ must be very small compared with $1/\lambda_{max}$ to keep the misadjustment within an acceptable limit. Therefore, τ_{max} is much greater than is indicated by the lower bound, which is the right-hand side of (4.26c). Nevertheless, the ratio ($\lambda_{max}/\lambda_{min}$) gives a good measure of the maximum speed of convergence.

Let us now consider in some detail the LMS algorithm for the case of one auxiliary antenna (Kretschmer and Lewis, 1978b). Substituting (4.23a) into (4.23b) results in

$$W(j+1) = W(j)(1 - 2\mu|V_A(j)|^2) + 2\mu V_M(j)V_A^*(j) \qquad (4.27)$$

where $V_A(\cdot)$ is the auxiliary antenna signal. The solution of this equation for a deterministic step input of V_M and V_A is readily determined by iterating (4.27), using the initial condition $W(1) = 0$, and by determining the sum of j terms of the resulting geometric series. The results are

$$W(j) = B[1 - A^{j-1}]/(1 - A), j \geq 2 \qquad (4.28)$$

where $W(1) = 0$, $A = 1 - 2\mu|V_A|^2$, and $B = 2\mu V_M V_A^*$. Stability requires that

$$|A| < 1 \qquad (4.29)$$

which reduces to the requirement:

$$0 < \mu < 1/|V_A|^2 \qquad (4.30)$$

For the general N-dimensional or N-channel case, conventional analysis reduces the problem to an orthogonal coordinate system, where the covariance matrix of the input signal is diagonalized. The result is that each weight can be expressed independently of the other weights by equations having the same form. The stability criterion is then determined as in (4.30), except the maximum eigenvalue is used instead of $|V_A|^2$. In other words, the stability condition is expressed by (4.25). In our one-dimensional analysis, $|V_A|^2$ corresponds to the unique eigenvalue.

The mean value of the steady-state weight $\overline{W}_\infty$, for any stochastic input, is obtained from averaging (4.27) and making the assumption that

$$\overline{W(j)V_A(j)} = \overline{W(j)}\,\overline{V_A(j)} \qquad (4.31)$$

The result is

$$\overline{W}_\infty = \overline{V_M V_A^*}/|V_A|^2 \qquad (4.32)$$

which corresponds to (4.5b) for $N = 1$.

Next we investigate the time constant of the LMS algorithm for a step input of V_M and V_A. The time constant is defined as the duration for the weight to reach $[1 - \exp(-1)]$ of the steady-state value. From (4.28), we obtain

$$1 - A^{j-1} = 1 - \exp(-1) \tag{4.33}$$

which results in the time constant, normalized by the sampling interval:

$$T = -1/\ln|A| + 1 \tag{4.34}$$

The magnitude sign is necessary because A may become negative, which corresponds to a damped oscillation of $W(j)$, provided that $|A| < 1$. From (4.34), the minimum time constant T_{min} occurs for

$$1 - 2\mu|V_A|^2 = 0 \tag{4.35a}$$

in which case, we have

$$T_{min} = 1 \tag{4.35b}$$

Now we investigate the modified LMS or MLMS algorithm, which is obtained by modifying the indexing of the LMS algorithm of (4.23b). The MLMS algorithm is determined from the feedback control-loop equation (Kretschmer and Lewis, 1978b):

$$W(j) = W(j-1) + 2\mu Z(j) V_A^*(j) \tag{4.36a}$$

where

$$Z(j) = V_M(j) - W(j) V_A(j) \tag{4.36b}$$

Here the present weight $W(j)$ is obtained from the present values of V_A and Z in contrast to the LMS algorithm, which generates the next weight from the present V_A and Z. The difference equation for $W(j)$ is

$$W(j) = [W(j-1) + 2\mu V_M(j) V_A^*(j)]/[1 + 2\mu|V_A(j)|^2] \tag{4.37}$$

This equation is of the form $W(j) = CW(j-1) + D$, which, for an input step of V_M and V_A, has the solution:

$$W(j) = [D(1 - C^j)]/(1 - C), \quad j \geq 1 \tag{4.38}$$

where $C = 1/(1 + 2\mu|V_A|^2)$ and $D = 2\mu V_M V_A^*/[1 + 2\mu|V_A|^2]$. For stability, $|C| < 1$ is required, or

$$1/(1 + 2\mu|V_A|^2) < 1 \tag{4.39}$$

which is always satisfied when μ is positive and $|V_A|^2$ is nonzero. The mean steady-state weight is obtained from (4.37) as

$$\overline{W_\infty} = \overline{W}_{opt} = \overline{V_M V_A^*}/|V_A|^2 \tag{4.40}$$

so that the weight of the MLMS algorithm is also unbiased. The time constant for the MLMS algorithm for a step input of V_M and V_A is

$$T = 1/\ln[1 + 2\mu|V_A|^2] \tag{4.41}$$

The minimum time constant for the LMS criterion requires from (4.35a) that

$$2\mu|V_A|^2 = 1 \tag{4.42}$$

Substituting this value in (4.41), the MLMS algorithm time constant is

$$T = 1/\ln 2 = 1.44 \tag{4.43}$$

Hence, for the same parameters the MLMS algorithm is slightly slower. However, this is somewhat deceiving because the time constant of the MLMS algorithm can be reduced by simply increasing the gain term without becoming unstable.

Next consider the case of a nonideal or "leaky" integrator, such as that found in a low-pass filter. The scheme of system based on the LMS algorithm is shown in Figure 4.7, where T_s indicates the sampling period of the signals. The integrator is represented by means of one-pole low-pass filter, where β is the pole value. The loop equation is

$$W(j + 1) = W(j)[\beta - G(1 - \beta)|V_A(j)|^2] + G(1 - \beta)V_M(j)V_A^*(j) \tag{4.44}$$

The corresponding stability condition is

$$|\beta - G(1 - \beta)|V_A|^2| < 1 \tag{4.45a}$$

bounding the parameter β as follows:

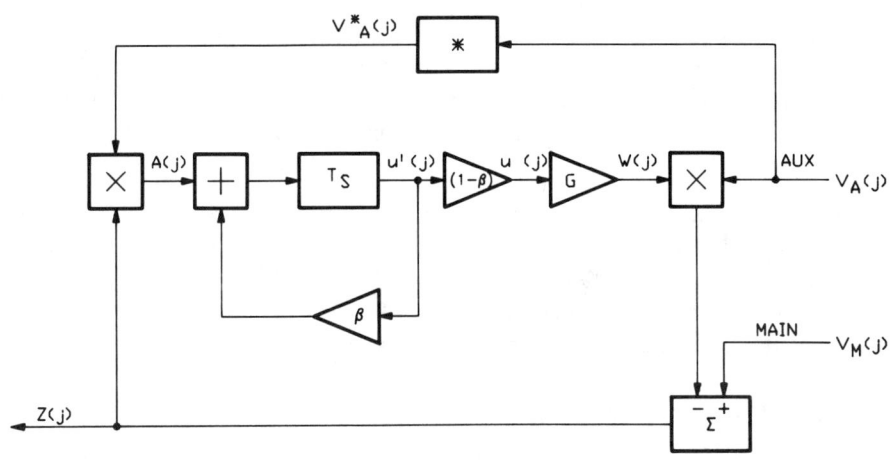

Figure 4.7 LMS algorithm with leaky integrator.

$$\frac{G|V_A|^2 - 1}{G|V_A|^2 + 1} < \beta < 1 \qquad (4.45b)$$

Taking the following expression for β:

$$\beta = \frac{\tau}{T_S + \tau} \qquad (4.45c)$$

where τ is the desired time constant of the low-pass filter and T_S is the sampling period, the stability condition becomes

$$\frac{|\tau - GT_S|V_A|^2|}{\tau + T_S} < 1 \qquad (4.45d)$$

which is equivalent to

$$0 < T_S < \frac{2\tau}{G|V_A|^2 + 1} = 2\tau_{\text{SLC}}$$

where τ_{SLC} has been defined in (4.10). The above equation indicates that the sampling period should be less than the loop constant time (Farina, 1977). The steady-state solution of (4.44) is again given by (4.32).

This scheme has been widely analyzed from the viewpoint of performance assessment (see Section 4.3.2). Here, we should mention the results of a simulation study undertaken to evaluate the number of bits required in the digital implementation. An SLC system with one auxiliary antenna has been considered. The system parameters are

- Receiver bandwidth, $B = 1$ MHz;
- Sampling time interval, $T_S = 0.5$ μs;
- Amplifier gain, $G = 9000$;
- Parameter of low-pass filter, $\beta = 0.99999$;
- Equivalent constant time of the filter, $\tau = 50$ ms;
- Input jamming power, $\sigma^2 = 1$;
- Loop time constant, $\tau_{SLC} = 5.55$ μs.

Figure 4.8 shows the cancellation ratio, CR, as defined by (4.7), *versus* the number of bits used to represent the signals, the parameters β and G of the circuit, and the rounding errors of the complex multiplications. The following curves are shown:

- Curve (a), illustrating the effects of the quantization of V_M and V_A;
- Curve (b), illustrating the effects of the quantization of the parameter β;
- Curve (c), illustrating the effects of the rounding errors in the two multiplications;
- Curve (d), illustrating the effects of the quantization of V_M and V_A plus the rounding errors, by using the same number of bits and β represented by 16 bits;
- Curve (e), the same as curve (d), but with β represented by 14 bits.

By comparison with the ideal cancellation ratio value of 27 dB (corresponding to a correlation coefficient of $\rho = 0.999$), we can see from a quantization viewpoint, β is a critical parameter. In fact, if less than 13 bits are used, the loop becomes unstable. The other signals may be represented by a reduced number of bits (e.g., seven). With the reduction of ρ, the number of bits required to represent the input signals and the parameter β are reduced as well. The parameter G is not critical, as it requires a number of bits equal to four. When considering an SLC system with two auxiliary antennas, the number of bits required to represent the input signals V_M and V_A may increase to 12 bits, especially when the correlation coefficient ρ has a high value (e.g., $\rho = 0.999$).

Before closing this subsection, additional references are worthy of mention: Bock (1978) and Klemm (1977). An experimental SLC with six auxiliary antennas is described by Bock, where the Howells-Applebaum loop is implemented with digital technology. The influence of the number of bits on the CR is discussed at length. An eight-bit arithmetic accuracy is a good compromise between interference suppression rate and implementation complexity. The same conclusions are shared by Klemm.

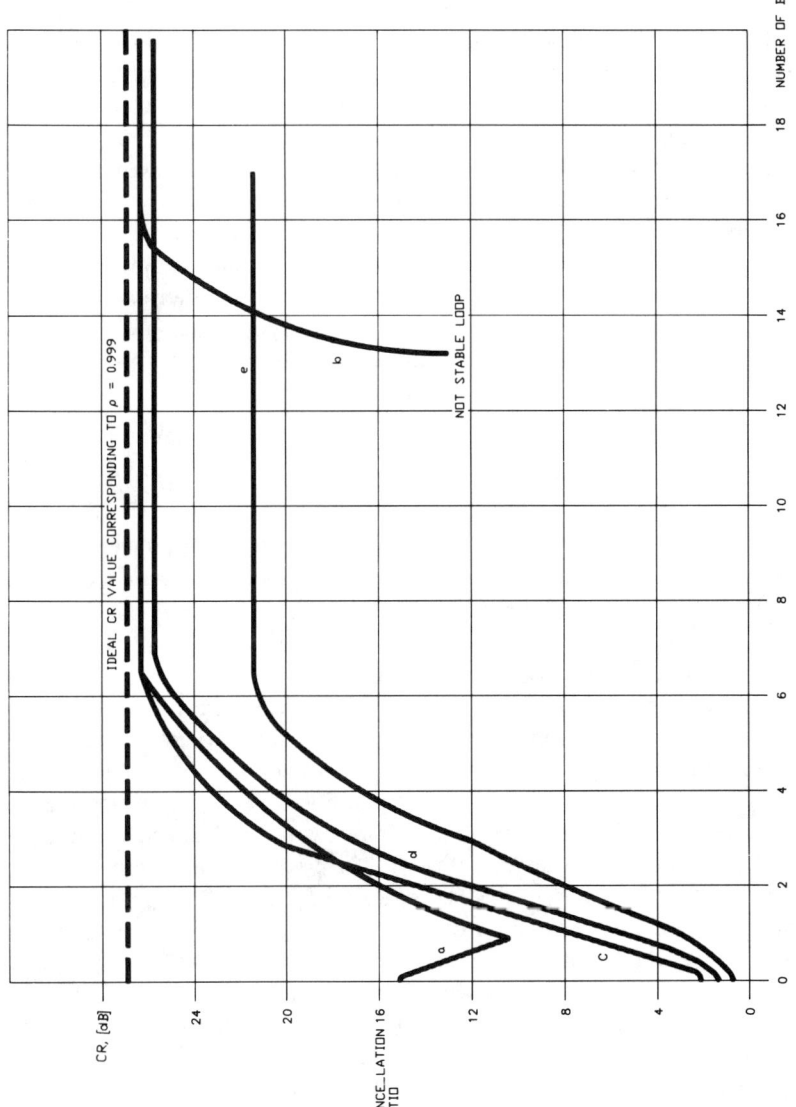

Figure 4.8 Effects of quantization in the SLC cancellation ratio.

4.2.1.4 Hybrid Analog-Digital Techniques

So far in this section, closed-loop processing methods have been described, which are either entirely analog or entirely digital. However, a number of hybrid analog-digital methods have also been proposed in which the weight is computed digitally, but the weighting itself is performed by analog means (Powell *et al.*, 1982). This combines the wide bandwidth and signal-handling capability of analog circuits with the flexibility and high dynamic range of digital circuitry. Figure 4.9 illustrates schematically an adaptive sidelobe canceler based on hybrid loop (Masenten, 1979). In the figure, ADC denotes analog-to-digital conversion and DAC means digital-to-analog conversion. In this case, the multiplication of the residual signal by each of the auxiliary signals is also done by an analog multiplier, and the product is passed through a low-pass filter before being digitized. The weight vector can then be computed in a flexible way, using proper algorithms. The hybrid analog-digital approach is not only relevant to closed-loop techniques, but also to the direct solution methods, which are discussed in the section to follow.

4.2.2 Direct Solution Methods

Very fast adaptive nulling is considered to be essential in a number of important military applications. For example, the radar may have a fast rotating antenna, have to cope with rapidly switched jammers, or require changing frequency rapidly. In such situations, direct solution methods are essential for adaptive nulling, and a number of these will now be discussed.

In the following, a number of technical approaches are presented under the heading of *direct solution methods* because they are characterized by the direct calculation of the interference covariance matrix, thus avoiding the problem of eigenvalue spread, which limits the convergence capabilities of the closed-loop algorithms mentioned in Section 4.2.1. Specifically, we start with the Gram-Schmidt approach (Section 4.2.2.1), which can be considered as a preprocessor to other SLC algorithms or an SLC algorithm by itself. The rationale for this approach is to transform the original problem into one which does not have an eigenvalue spread. Subsequently, we consider the simple open-loop adaptive processor (Section 4.2.2.2), which is very attractive for an SLC system with only one or very few auxiliary antennas. Section 4.2.2.3 is dedicated to the basic direct matrix inversion (DMI) algorithm, where the covariance matrix is estimated at first by means of a sufficient number of data "snapshots" and then inverted with a proper numerical algorithm. Finally, the Kalman prediction algorithm is the subject of Section 4.2.2.4. This algorithm is particularly recommended to handle nonstationary problems. We should remember that the basic expression (4.5b) gives the optimum weights for the stationary case. For the nonstationary case, the Wiener filter, characterized by the weights (4.5b),

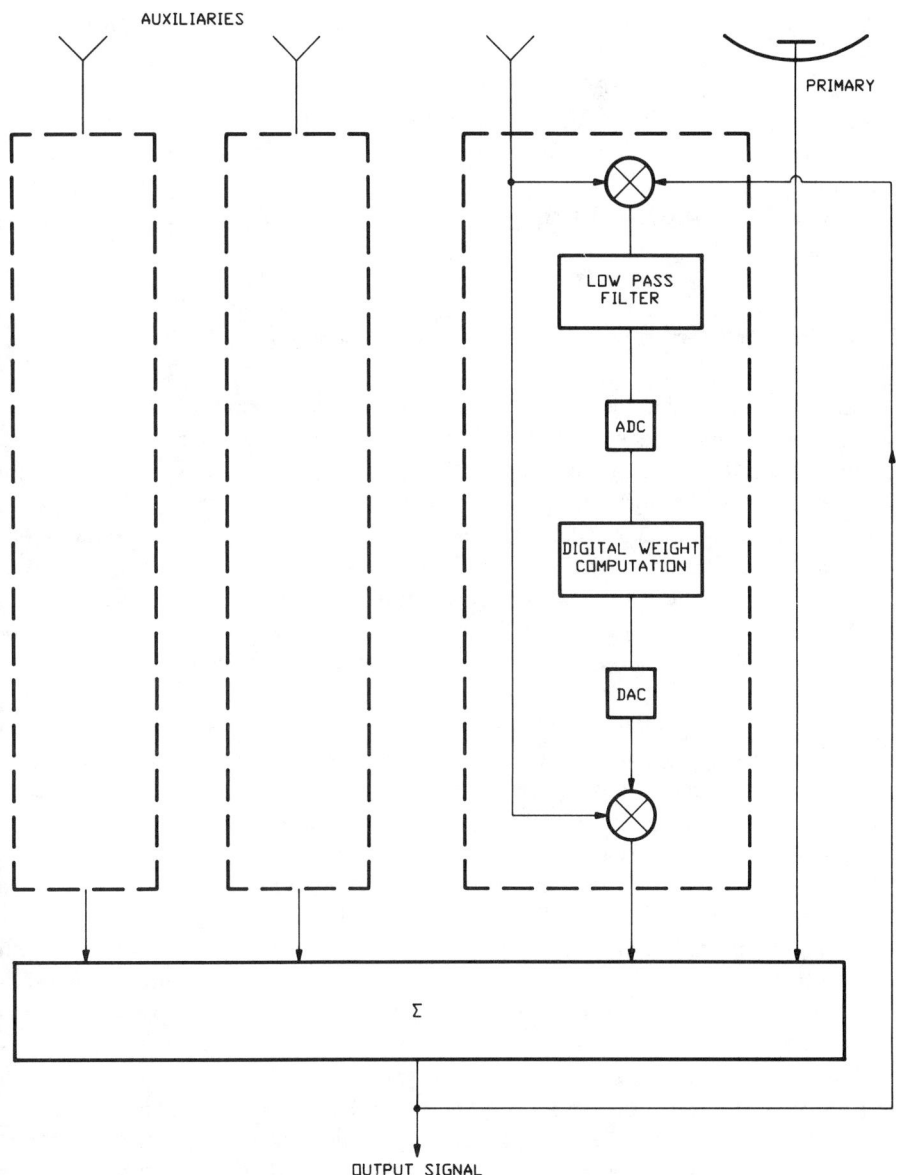

Figure 4.9 Use of hybrid analog-digital loops for SLC.

should be substituted with the Kalman filter, which is optimum for nonstationary situations, provided that an adequate model of the dynamic evolution of the interfering environment is available. The Kalman filter is recommended to deal with SLC on-board a moving platform with well-behaved dynamics. The Kalman filter requires the inclusion of more *a priori* information on the signal environment and opens the possibility of constructing processors that integrate adaptive array control with other system functions, such as position location and navigation (Monzingo and Miller, 1980, p. 326).

4.2.2.1 Gram-Schmidt Approach or Sequential Decorrelator

The potentially slow convergence speeds of the Howells parallel loops have encouraged researchers to seek a better solution to multiple degree-of-freedom systems. The problem of slow convergence using the LMS algorithms arises whenever there is a wide spread in the eigenvalues of the input signal correlation matrix. The condition of wide spread of eigenvalues occurs if the signal environment includes a very strong source of interference together with other weaker interferers. Another operational situation of interest occurs if two or more very strong interference sources arrive at the array from closely spaced but not identical directions. An approach for obtaining rapid convergence is to rescale the vector space in which the minimization of SLC output power is occurring by appropriately transforming the input signal samples so that no eigenvalue spread is present in the rescaled space. The function of this preprocessing may be regarded as one of resolving the input signals into their eigenvector components. By equalizing the resolved signals with *automatic gain control* (AGC) amplifiers, the eigenvalue spread may be considerably reduced, thereby simplifying the task of any gradient type of algorithm.

An orthogonal signal set (i.e., the signal eigenvectors) can be obtained by a transformation based on the Gram-Schmidt orthogonalization procedure. In the Gram-Schmidt approach, the signals are preprocessed to be mutually orthogonal to each other, similar in principle to the mathematical method of Gram-Schmidt for orthogonalization of dependent vectors. While a Gram-Schmidt preprocessor may be followed by a Howells-Applebaum adaptive processor to realize accelerated convergence, as illustrated in Figure 4.10, we shall show that it is also simply possible to utilize the preprocessor alone in an SLC configuration to achieve interference cancellation. We shall see that the Gram-Schmidt (GS) processor has excellent transient response characteristics. The use of GS in the SLC and adaptive array systems has been proposed and described by many authors (Monzingo and Miller, 1980; Compton, 1988; Giraudon, 1975; Farina and Studer, 1984; Yuen, 1989, 1991; Lewis and Kretschmer, 1984; Farina and Studer, 1988; Hara *et al.*, 1989, Gerlach and Kretschmer, 1990, 1991; Ling *et al.*, 1986). Here the working principles, basic schemes, and main advantages and drawbacks will be reviewed.

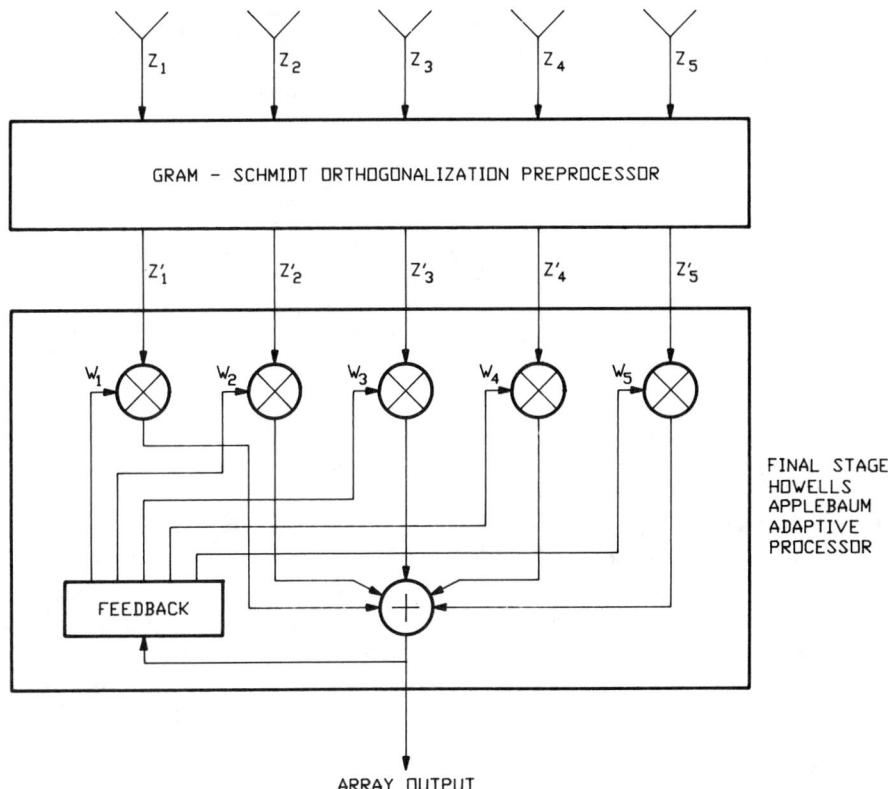

Figure 4.10 Gram-Schmidt orthogonalization network with Howells-Applebaum adaptive processor for accelerated convergence.

Let us explain the working principle of the GS orthogonalization procedure. Consider a set $\mathbf{Z}^T = \{Z_1, Z_2, \ldots, Z_N\}$ of samples of a stochastic sequence having zero mean value and a covariance matrix $\mathbf{M}$. We are to find a transformation $\mathbf{D}$ such that the new set of samples:

$$\mathbf{Z}' = \mathbf{DZ} \qquad (4.46)$$

has the covariance matrix:

$$\text{cov}\{\mathbf{Z}'\} = \mathbf{D}^*\mathbf{MD}^T \qquad (4.47)$$

which is diagonal, (i.e., $\mathbf{Z}'$ is a set of mutually uncorrelated samples). To this end, the GS orthogonalization procedure is applied. This algorithm is well known in functional analysis theory and is used to obtain an orthogonal basis in a vector space, starting from a linearly independent set of vectors. When applied to stochastic processes, the GS algorithm gives the following equations:

$$Z'_1 = Z_1$$
$$Z'_i = Z_i - \sum_{m=1}^{i-1} W_{mi} Z'_m; \quad i = 2, \ldots, N \quad (4.48)$$

with

$$W_{mi} = E\{Z_i Z'^*_m\}/E\{|Z'_m|^2\}$$

The meaning of such equations is
- The first sample Z'_1 is merely a replica of Z_1;
- For any successive sample $Z'_i (i = 2, \ldots, N)$, the first step of the processing requires the estimation of the correlation between Z_i, and all the previous Z'_m already calculated. Therefore, the quantity $W_{mi} Z'_m$ represents the component of Z_i, which is correlated with Z'_m. Hence, the new sample Z'_i is evaluated by progressively cancelling from Z_i these correlated components. As a consequence, the output $\mathbf{Z}'$ has a diagonal covariance matrix. Thus, the GS procedure is equivalent to multiplying the incoming $\mathbf{Z}$ by $\mathbf{D}$ (4.46).

In the specific case of $N = 5$, by application of (4.48), the transformation matrix $\mathbf{D}$ appears to have the following lower triangular expression:

$$\mathbf{D} = \begin{bmatrix} 1 & 0 & 0 & 0 & 0 \\ -W_{12} & 1 & 0 & 0 & 0 \\ -W_{13} & -W_{23} & 1 & 0 & 0 \\ -W_{14} & -W_{24} & -W_{34} & 1 & 0 \\ -W_{15} & -W_{25} & -W_{35} & -W_{45} & 1 \end{bmatrix} \quad (4.49)$$

We can show that the algorithm is conveniently implemented by means of the scheme of Figure 4.11, which refers to the case of $N = 5$ samples. The orthogonalization is obtained through $(N - 1)$ steps; at each step, one sample is taken as a reference and all the successive samples are decorrelated from it. This is obtained by means of a set of blocks, A (see Figure 4.11), each producing an output orthogonal to the reference input of the corresponding step. A modular processing architecture is obtained consisting of $N(N - 1)/2$ blocks for N samples. The need for

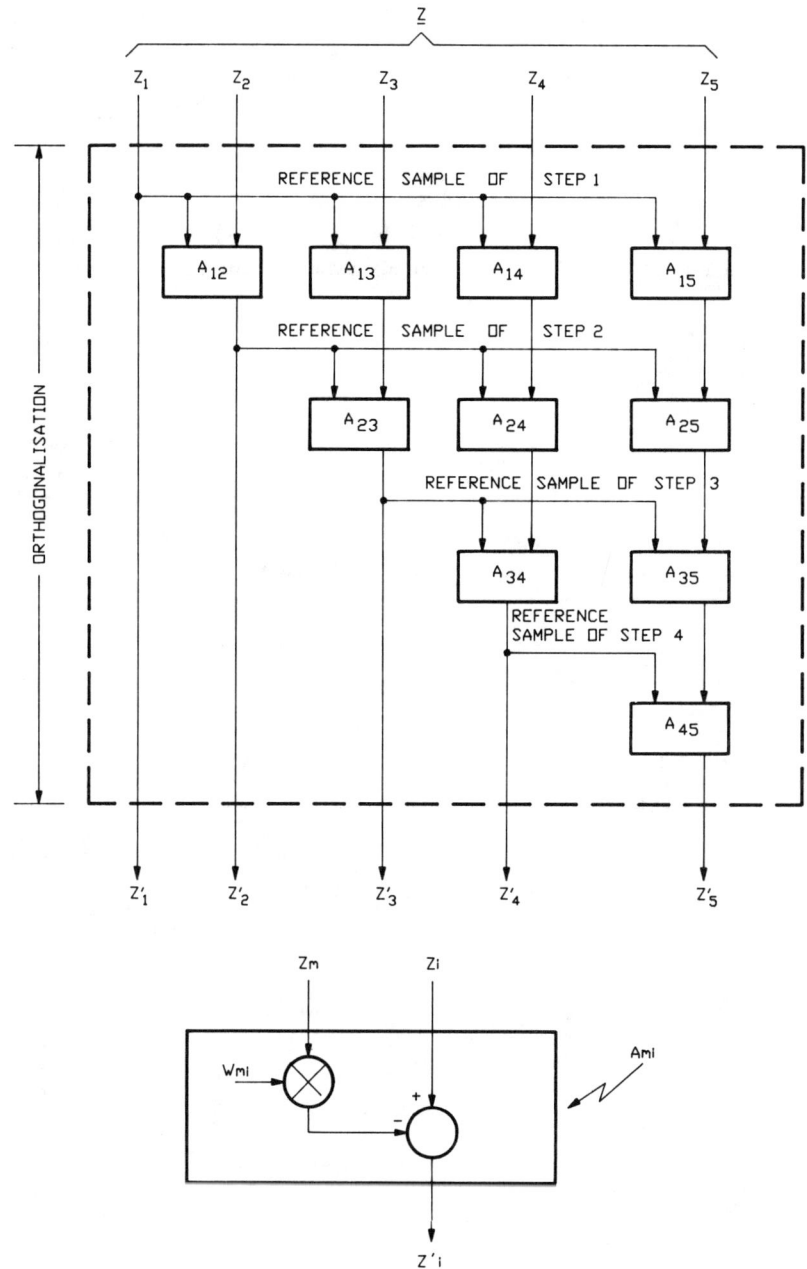

Figure 4.11 Scheme of the Gram-Schmidt processor.

more hardware is a disadvantage of GS with respect to the Howells-Applebaum processor. For a digital implementation, the decorrelation blocks operate sequentially, row by row, this means that systolic architectures can be efficiently adopted.

A possible implementation of each decorrelation block A_{mi} is represented by the standard SLC with a single auxiliary signal and an adaptation weight W_{mi} (see Figure 4.2) because, in the steady-state condition, the output signal is uncorrelated with the reference signal; thus, from (4.5f), $E\{Z_i'Z_m^*\} = 0$.

Another implementation is shown in Figure 4.12, which is now briefly discussed (Farina and Studer, 1984). In the steady-state condition, the output expression is

$$Z_i' = Z_i E\{Z_m Z_m^*\} - Z_m E\{Z_i Z_m^*\} \qquad (4.50\text{a})$$

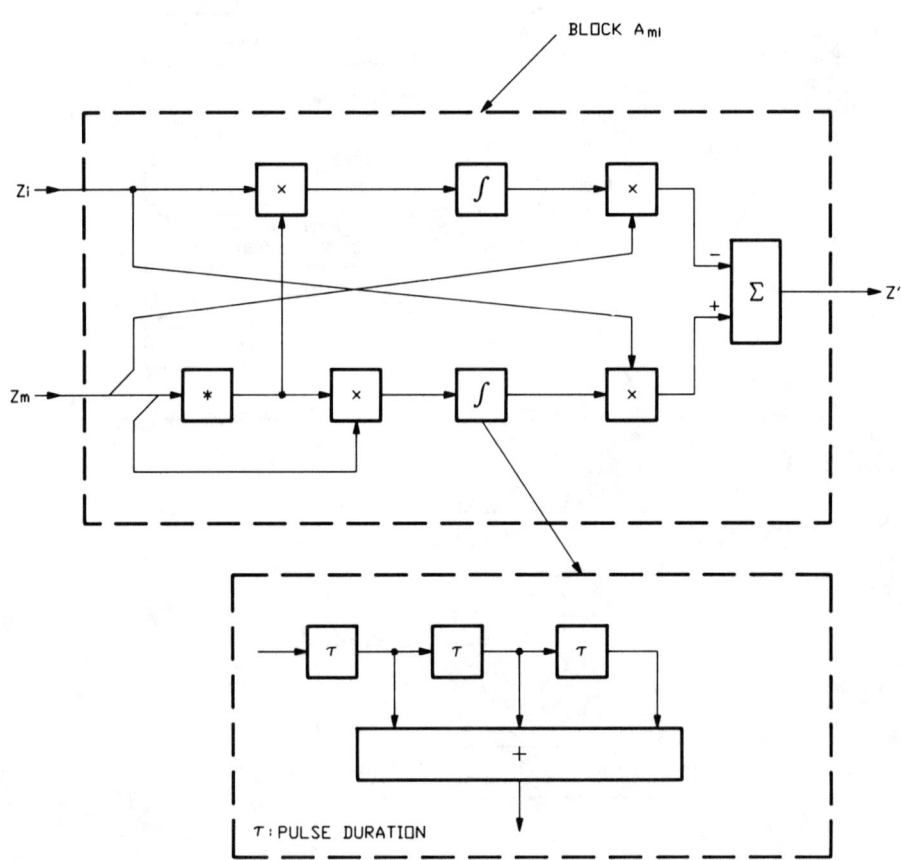

Figure 4.12 Detailed scheme of block A. (From Farina and Studer, IEE Proc., Pt. F, Vol. 131, No. 2, April 1984, pp. 139–145.)

The expectations are estimated as time averages of independent samples taken from consecutive range cells. That the following condition holds is easy to verify:

$$E\{Z_i'Z_m^*\} = 0 \tag{4.50b}$$

which corresponds to the required orthogonalization of the output sample with the reference input.

From the previous description, we expect that the proposed architecture has a fast adaptation with respect to the classic Howells-Applebaum approach. This is mainly due to the absence of any feedback loop, which would make the processing of the samples strongly dependent on each other. In fact, the weight at each block A is determined only by the input signals at that block. Moreover, the stability problem is completely avoided. This item is widely confirmed by the performance analyses reported (Farina and Studer, 1984; Bucciarelli et al., 1982; see also Subsection 4.3.2.5).

We should note that the transformation of the input signal vector $\mathbf{Z}$ into a set of uncorrelated output signals $\mathbf{Z}'$ is not unique. The GS transformation can adopt any component of $\mathbf{Z}$ as the first output component, and the remaining components of $\mathbf{Z}$ may be taken in any order as the signals to be decorrelated, using the procedure illustrated in Figure 4.11. The problem of conceiving a numerically efficient and stable algorithm for adaptively filtering multiple input channels into desired multiple output channels has been tackled in (Gerlach, 1986; Yuen, et al., 1988). This approach allows solving the problem of decorrelating each input channel with all the other input channels without necessarily implementing the GS decomposition N times.

The fact that the transformation network of Figure 4.11 yields a set of uncorrelated output signals suggests that this network may be capable of functioning in the manner of an SLC system having an output signal (in the steady-state) that is uncorrelated with each of the auxiliary channel input signals. By selecting Z_5 of Figure 4.11 as the main-beam signal V_M and Z_5' as the output signal Z of the SLC, the cascade preprocessor converges in the steady-state to $Z = V_M - \hat{\mathbf{W}}^T\mathbf{V}$, where now $\mathbf{V}$ is the set of the remaining inputs $\{Z_1, Z_2, Z_3, Z_4\}$. This is because Z_5' has been obtained by adding to Z_5 a linear combination of the other input samples. Furthermore, as Z_5' is orthogonal to the input samples $\{Z_1, Z_2, Z_3, Z_4\}$, it must, in fact, be the residual signal for which $P_Z = E\{|Z|^2\}$ is minimized. A scheme based on the GS approach for an SLC with two auxiliaries is shown in Figure 4.13.

A more detailed analysis can also be done (Monzingo and Miller, 1980, pp. 364–382; Bucciarelli et al., 1982; Compton, 1988, pp. 304–317) to show that the conventional SLC and GS systems have the same steady-state cancellation values. Concerning stability considerations, we may readily see that, because no loop is in another's feedback path, the loops of each block A do not interact. Then, the stability of the GS system is related to the block A, which can be unconditionally stable if implemented according to the scheme of Figure 4.12. This is a great advantage of the GS with respect to the conventional SLC schemes.

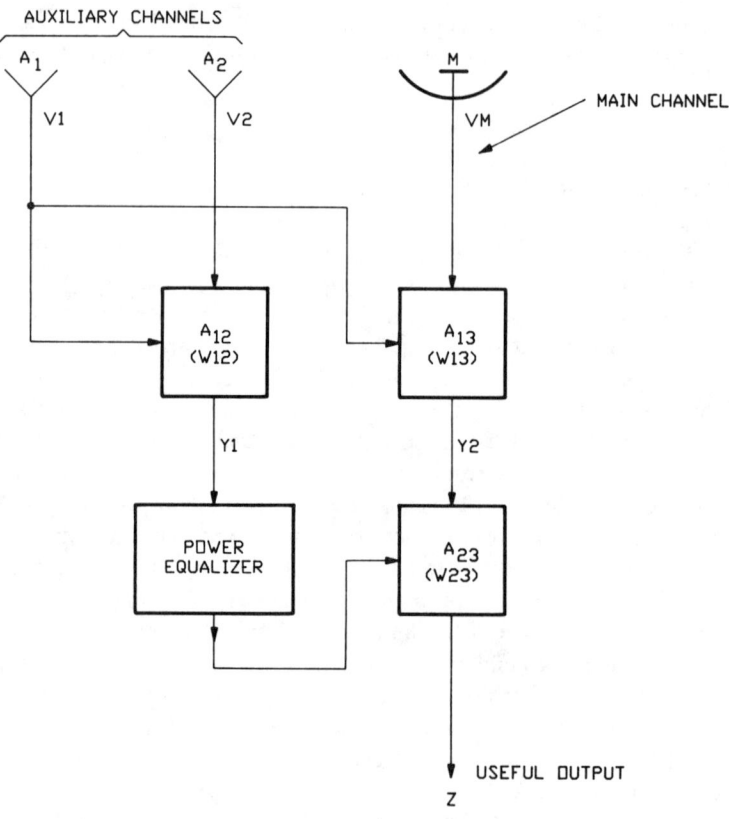

Figure 4.13 SLC with two auxiliary antennas implemented with the Gram-Schmidt algorithm.

Another advantage of the GS system is its transient time behavior. This feature will be deeply analyzed in Subsection 4.3.2.5 in connection with the so-called "loop-noise" problem. The GS has excellent transient response characteristics. If the same transient time would be required for the conventional (i.e., Howells-Applebaum) SLC, a reduction of steady-state cancellation would occur due to the increase of loop noise. This is because the GS is less prone to the loop-noise effect with respect to the Howells-Applebaum architecture. As a final remark concerning the convergence time of the GS, note that it is strongly related to the settling time of each block A as determined by its auxiliary input. Now, moving downward through the levels of the GS processor, the auxiliary inputs have reduced power because of the cancellation of the previous processing levels. This would increase the settling time of the decorrelation blocks pertaining to the processing level closer to the useful output. A way to equalize the settling time of all the blocks A is to use the power equalizer, as shown in the Figure 4.13. Consequently, the reaction time of this SLC is independent of the interference strength and the eigenvalue spread in the original interference covariance matrix, **M**.

4.2.2.2 Digital Open-Loop Adaptive Processor

This approach has been widely studied by Kretschmer and Lewis (1978a). Note that the feedback loop is important in analog circuitry to compensate for component drifts. In digital implementation, however, this problem is not present, and the optimum weight can be determined directly in an open-loop configuration from the input signals without any feedback loop.

In digital applications, the ensemble average indicated in (4.9f) is approximated by a sample average over a finite time window. Thus, the jth sampled residue is computed as

$$Z(j) = V_M(j) - \hat{W}(j)V_A(j) \tag{4.51a}$$

where, for an n-sample window size, we have

$$\hat{W}(j) = \sum_{k=j-n+1}^{j} V_M(k)V_A^*(k) \bigg/ \sum_{k=j-n+1}^{j} |V_A(k)|^2 \tag{4.51b}$$

We can see from (4.51b) that, in the case of a single sample, W becomes V_M/V_A. Substitution of this into (4.51a) results in perfect cancellation, irrespective of V_M or V_A. This suggests that more samples are averaged to prevent cancellation of a valid target signal by an uncorrelated auxiliary signal. An illustrative implementation is shown in Figure 4.14 for a two-sample average; the delay time shown in the figure corresponds to one sampling interval. Note that the stability considerations are completely eliminated because the loop is open. For a thorough discussion of the performance of the simulated processor, the reader should refer to Kretschmer and Lewis (1978a).

A generalization of this method to more than one auxiliary antenna is discussed in Subsection 4.2.2.3. The calculation of weights requires the estimation, by means of sample averages, of the elements of the covariance matrix, **M**, and of the correlation vector, **R**. Subsequently, the matrix **M** must be inverted and the result multiplied by **R**. Matrix inversion can be done with the Gaussian elimination method or other standard techniques. These operations can be implemented by using digital signal-processing units.

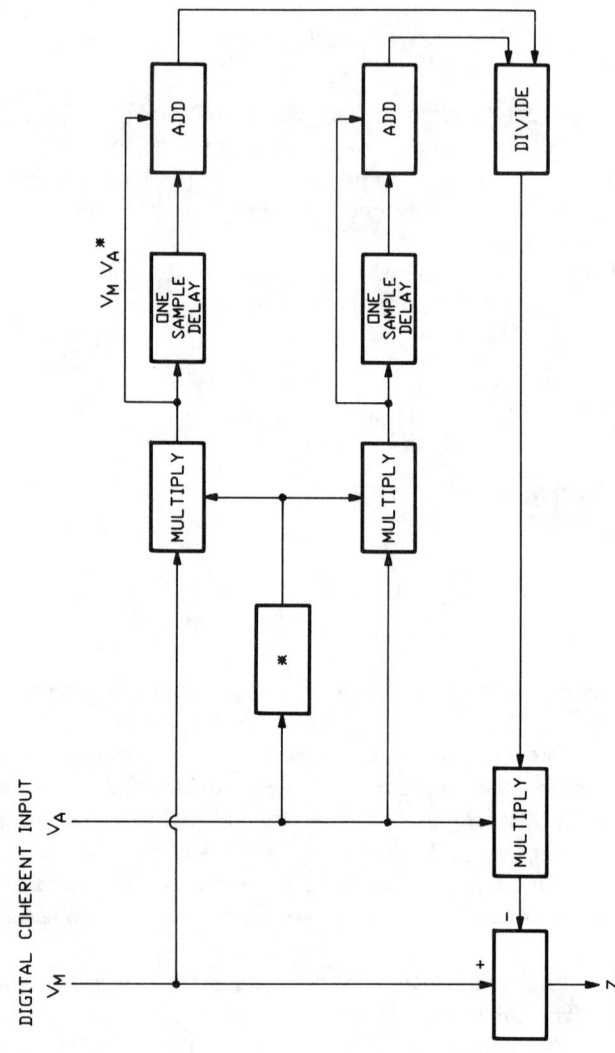

Figure 4.14 Digital open-loop adaptive processor. (From Kretschmer and Lewis, IEEE Trans. on Aerospace and Electronic Systems, Vol. AES-14, No. 1, January 1978, pp. 165–171, ©1978 IEEE.)

4.2.2.3 Direct Inversion of Estimated Covariance Matrix

One of first methods conceived to speed the convergence of optimum weight estimates and circumvent the convergence rate dependence on eigenvalue distribution is the so-called direct matrix inversion (DMI) (Reed et al., 1974; Monzingo and Miller, 1980). It is essentially based on the maximum likelihood estimate of the interference covariance matrix and the numerical inversion of the matrix to find the optimum weights. In mathematical terms, the problem is stated as follows. Indicate with $\mathbf{Z}(i)$ the collection of $N + 1$ samples from the main and auxiliary antennas at the ith time instant:

$$\mathbf{Z}^T(i) = [V_M(i) V_1(i) \ldots V_N(i)] \tag{4.52}$$

The maximum likelihood estimate of the interference covariance matrix is given by

$$\mathbf{M}_{ML} = \frac{1}{k} \sum_{i=1}^{k} \mathbf{Z}^*(i) \mathbf{Z}^T(i) \tag{4.53a}$$

where k is the number of independent samples over which the estimation is performed. Once the covariance matrix has been estimated, a numerical method for matrix inversion is applied. A very efficient technique is the Cholesky factorization, which expresses the matrix $\mathbf{M}_{ML}$ as the product of lower triangular matrices and allows performance of the matrix inversion by resorting to backsubstitution algorithm (Golub and VanLoan, 1983, p. 53). The weight vector $\hat{\mathbf{W}}$ is obtained as follows:

$$\mathbf{M}_{ML} \hat{\mathbf{W}} = \mathbf{E}_1 \tag{4.53b}$$

where $\mathbf{E}_1^T = [1, 0, \ldots, 0]$ indicates that the weight of the main channel has to be clamped to one.

As has been shown, the convergence rate of the algorithm, i.e., the number k of snapshots, to have an accurate estimate of the interference covariance matrix, is independent of both the amount of interference and the eigenvalue spread of the covariance matrix. For an $N + 1$ element array, Reed et al. (1974) and Monzingo and Miller (1980) found that $k = 2(N + 1)$ snapshots are sufficient to provide a resulting SNR at the SLC output that is within 3 dB of the optimum value (obtained after an infinitely long averaging window). A considerably increased number of data snapshots may, however, need to be used to estimate the covariance matrix of the interference when the measured data snapshots contain samples of the wanted signal (Monzingo and Miller, 1980, p. 298). A tighter criterion of convergence has also been proposed (Nitzberg, 1984).

Under the hypothesis that the matrix inversion is computed exactly, the eigenvalue spread has no effect on the convergence because the estimation error is the main cause of the adaptation convergence. The eigenvalue spread instead plays a role for the accuracy of the steady-state solution. Suppose that the number k of independent samples is large enough to ensure that $\mathbf{M}_{ML}$ is very close to the exact value of $\mathbf{M}$. In this case, the weight would not be optimum only if the covariance matrix inversion is not accomplished with sufficient accuracy due to matrix ill conditioning, as measured by the eigenvalue spread (Golub and VanLoan, 1983) and (Haykin, 1989). Of course, as the number of bits available in the computer to perform the matrix inversion increases, the degree of matrix ill conditioning that can be handled by the matrix inversion algorithm increases as well. Therefore, there is a link between the number of bits available in the computer to accomplish the matrix inversion and the allowable eigenvalue spread of the covariance matrix (Monzingo and Miller, 1980, pp. 312–313). This is a key point, which will be reconsidered in detail in Section 4.2.3. In fact, the methods based on the covariance matrix estimation and inversion, such as the DMI, require a lot of bits (e.g., 24 bits for an eigenvalue spread of 70 dB), which would impair the real-time applications of these methods.

The major price to be paid for very fast adaptation against adverse interference environments, entailing a wide spread of jamming power levels, is processing complexity. In addition to the large number of bits needed to perform covariance matrix inversion, another issue is the number of multiplication and addition operations. The estimation of the covariance matrix requires a number of multiplications proportional to kN^2. The inversion of the covariance matrix requires a number of multiplications and divisions proportional to N^3, which reduces to N^2 if the matrix is Toeplitz and the inversion algorithm is the Levinson-Durbin (Haykin, 1986, p. 137). Finally, the weights computation requires N^2 multiplications.

A modification to the original DMI method has been suggested by B.D. Carlson (1988) and Carlson *et al.* (1990). The technique, named *diagonal loading,* artificially increases the diagonal elements of the estimated covariance matrix $\mathbf{M}_{ML}$, producing a new matrix $\mathbf{M}' = \mathbf{M}_{ML} + \beta\mathbf{I}$ to invert. Typically, β is chosen to be 3 to 10 dB greater than the thermal-noise power on each array element. The advantage gained in using $\mathbf{M}'$ in lieu of $\mathbf{M}_{ML}$ is mainly related to reduced distortion of the main beam and sidelobe structure in the adapted pattern. The adaptive nulling introduces relevant distortions in the quiescent pattern that can be mitigated by the diagonal loading technique. This is because the modified matrix $\mathbf{M}'$ enforces, by means of the term $\beta\mathbf{I}$, the features of the quiescent weights (i.e., those related to only white-noise interference). One undesired effect of this averaging of the quiescent and adapted patterns is a reduction in the null depth. However, the null is still deep enough so that the interference has a negligible effect. This technique has found a practical implementation in an array operating at UHF and having 14 degrees of freedom in elevation (Carlson *et al.*, 1990; Teitelbaum, 1991).

Additional interesting modifications of the basic DMI algorithm are available in the literature. Among others, we mention a recursive algorithm to form the inverse of the covariance matrix directly, starting from the data snapshots $\mathbf{Z}(i)(i = 1, 2, \ldots, k)$. The algorithm is described in Monzingo and Miller (1980, pp. 324–326) and specifically applied also to adaptive clutter cancellation as shown in Farina (1987a, pp. 23–28). The rate of convergence is comparable with that of the original DMI algorithm. The number of multiplications required for each updating cycle of the recursive algorithm is on the order of N^2.

4.2.2.4 The Kalman Algorithm Applied to the SLC Implementation

In the 1960s, the *Kalman filter* (KF) was introduced for signal processing. Its application requires signal modeling by means of a dynamic state equation and the assumption that the probability density of the stochastic processes involved is Gaussian. The theory provides the optimum estimation filter (which is linear) in a recursive form (well suited for computer implementation) rather than a closed form for the impulse response as in the case of Wiener filtering. Kalman filtering theory is suitable for processing nonstationary sequences or data window of finite length, as opposed to the Wiener filter, requiring a window of infinite length or samples from a stationary process. Despite its generality, KF has found limited application to the radar case, essentially because it needs a precise modeling of the processes to be estimated. In the field of optimum radar signal processing, the KF can be used for on-line estimates of the coefficients of an autoregressive (AR) process, either to model the temporal correlation of a clutter sequence or the angular spectrum of a jammer (Krucker, 1981; 1983; Monzingo and Miller, 1980; Haykin, 1986). The advantage in building the Kalman filter based on this model is that the inversion of the interference covariance matrix is not required. Another advantage is the capability of tracking the interference changes owing to specific jamming operation or antenna motion.

In this subsection, we consider the sidelobe canceler as a *prediction error filter* (PEF), which "whitens" the angular interference spectrum, modeled by an all-pole AR process, thereby nulling the interference. The task hence is to estimate the coefficients of the PEF. A recursive estimation procedure based on the Kalman algorithm can be applied, which leads to an iterative solution of the optimum weight (4.5b). The following material is essentially from Krucker (1983).

Let us start with the derivation of the model for the jamming signal received by the main and auxiliary antennas. We assume that the measured signal samples, representing the angular spectrum of an interference, are generated by a filter of order N (equal to the number of auxiliary antennas) as follows:

$$y(i) = \mathbf{H}^T(i)\mathbf{X}(i) + v(i) \qquad (4.54)$$

where

- $y(i)$ = the ith scalar jamming sample measured at the radar antenna (i.e., our V_M signal in the SLC terminology);
- $\mathbf{H}(i)$ = an N-dimensional vector containing the past samples of the jamming process; $\mathbf{H}^T(i) = [y(i-1), y(i-2), \ldots, y(i-N)]$. This vector corresponds to the vector $\mathbf{V}$ of auxiliary signals in our standard terminology. The interference covariance matrix $\mathbf{M}$ is $E[\mathbf{H}^*(i)\mathbf{H}^T(i)]$. $\mathbf{H}^T(i)$ is the transformation matrix of the measurement equation in the KF terminology;
- $\mathbf{X}(i)$ = an N-dimensional vector representing, in the KF terminology, the state of the dynamic system to be estimated. $\mathbf{X}^T(i) = [x_1(i), x_2(i), \ldots, x_N(i)]$ corresponds to the weight vector $\mathbf{W}$ in the SLC terminology;
- $v(i)$ = a zero-mean white Gaussian scalar sequence with variance σ_v^2. It corresponds to the measurement noise in the KF terminology and to the thermal noise in the auxiliary channels in the SLC terminology.

The model (4.54) represents an autoregressive (AR) stochastic sequence. The corresponding transfer function in the complex plane is given by the z-transform of (4.54):

$$A(z) = \frac{y(z)}{v(z)} = \left(1 - \sum_{\nu=1}^{N} x_\nu z^{-\nu}\right)^{-1} \quad (4.55)$$

with $z = \exp(j2\pi(d/\lambda)\sin\theta)$, where d is the spatial sample distance, λ is the wavelength, and θ is the angle of incidence. To whiten the colored angular spectrum described by (4.54), we need to estimate the state vector $\mathbf{X}$. To this end, the dynamic evolution of the state vector $\mathbf{X}$ is assumed to be described by the following linear stochastic difference equation:

$$\mathbf{X}(i+1) = \boldsymbol{\phi}(i+1, i)\mathbf{X}(i) + \boldsymbol{\Gamma}(i)\mathbf{W}(i) \quad (4.56)$$

where

- $\boldsymbol{\phi}(i+1, i)$ = an $N \times N$ transition matrix;
- $\boldsymbol{\Gamma}(i)$ = an $N \times N$ transformation matrix;
- $\mathbf{W}(i)$ = an N vector of random modeling errors (model noise in the KF terminology), represented by a zero-mean white-noise process with covariance matrix $\sigma_w^2 \mathbf{I}$, where $\mathbf{I}$ is the identity matrix.

Applying the discrete Kalman filter algorithm, the state estimate is updated via

$$\hat{\mathbf{X}}(i) = \boldsymbol{\phi}(i, i-1)\hat{\mathbf{X}}(i-1) + \mathbf{K}(i)[y(i) - \mathbf{H}^T(i)\boldsymbol{\phi}(i, i-1)\hat{\mathbf{X}}(i-1)] \quad (4.57a)$$

where the gain vector $\mathbf{K}(i)$ and error covariance $\mathbf{P}(i,i) = E\{[\mathbf{X}(i) - \hat{\mathbf{X}}(i)]*[\cdot]^T\}$ are calculated recursively by the following formulas:

$$\mathbf{K}(i) = \mathbf{P}(i, i-1)\mathbf{H}^*(i)[\mathbf{H}^T(i)\mathbf{P}(i, i-1)\mathbf{H}^*(i) + \sigma_v^2]^{-1} \quad (4.57\mathrm{b})$$

$$\mathbf{P}(i+1, i) = \boldsymbol{\phi}(i+1, i)\{[\mathbf{I} - \mathbf{K}(i)\mathbf{H}^T(i)]\mathbf{P}(i, i-1)\}\boldsymbol{\phi}^T(i+1, i) + \boldsymbol{\Gamma}(i)\boldsymbol{\Gamma}^T(i)\sigma_w^2 \quad (4.57\mathrm{c})$$

where the notation $\mathbf{P}(i+1, i)$ implies the predicted or *a priori* error covariance at time instant $(i+1)$ given the estimation $\hat{\mathbf{X}}(i)$ and observation $y(i)$ at the ith time instant.

The filter is initialized by an estimate $\hat{\mathbf{X}}(0)$, a random variable with known statistics, and a corresponding $\mathbf{P}(1,0)$, the latter reflecting the confidence that we have in the initial estimate. Note that it is only necessary to invert a scalar value in (4.57b).

An interpretation of (4.57a) in terms of SLC signal processing shows that $y(i)$ is the main channel signal from which the weighted auxiliary array signals $\mathbf{H}^T\hat{\mathbf{X}}$ are subtracted. The residual error signal is weighted by the gain vector $\mathbf{K}(i)$, reflecting the confidence the PEF has in the actual observation. This residue is used to update the weights $\hat{\mathbf{X}}(i-1)$ in the new set of weights $\hat{\mathbf{X}}(i)$. The residual error signal will tend to be white at the steady state.

Now let us apply the general theory thus described to the case in which the jamming angular spectrum is stationary. The dynamic state equations (4.56) reduce to

$$\mathbf{X}(i+1) = \mathbf{X}(i) \quad (4.58)$$

corresponding to the assumptions of $\boldsymbol{\phi} = \mathbf{I}$ and $\sigma_w^2 = 0$. The solution of (4.57a) is shown to be

$$\lim_{i \to \infty} E\{\hat{\mathbf{X}}(i)\} = \mathbf{M}^{-1}\mathbf{R} \quad \text{with} \quad \mathbf{R} \triangleq E\{\mathbf{H}^*(i)y(i)\} \quad (4.59)$$

which is the optimum SLC weight vector (4.5b). The influence of the initial guess $\hat{\mathbf{X}}(0)$ vanishes with an increasing number of iterations (i.e., the estimation is asymptotically unbiased). The convergence rate can be calculated from the SLC residue power $E\{|\varepsilon(i)|^2\}$, where:

$$\varepsilon(i) = y(i) - \hat{\mathbf{X}}^T(i)\mathbf{H}(i) \quad (4.60)$$

The residue signal $\varepsilon(i)$ corresponds to the output signal $Z(i)$ in our standard SLC terminology. We find that

$$E\{|\varepsilon(i)|^2\} \approx \sigma_v^2 \left(1 + \frac{N}{i}\right) \qquad (4.61)$$

Further, with a definition of the cancellation ratio as $CR = 1/E(|\varepsilon|^2)$, we have

$$CR = CR_{max} \frac{1}{1 + (N/i)} \qquad (4.62)$$

where $CR_{max} = 1/\sigma_v^2$. Expressed as 1 dB loss relative to CR_{max}, approximately $i = 4N$ iterations are needed. This describes, as a rule of thumb, the requirement on the number of data samples which agrees with results obtained by simulation (Krucker, 1983). A diagram indicating the processor structure of an SLC with a stationary Kalman filter is shown in Figure 4.15.

Let us now consider in some detail the nonstationary interference case. A nonstationary interference environment, caused by the scanning antenna or a moving interference source, must be taken into account by the state dynamic model (4.56); otherwise, divergence may occur owing to discrepancies between the model and real process. This can be done in several ways, resulting in different filter performance at different numerical expense (Krucker, 1983). Here we make reference to a model that involves the introduction of a "velocity" term $\dot{\mathbf{X}} = d\mathbf{X}/dt$. The basic idea is to make use of the known scan pattern of the antenna, continuously scanning in the mechanical case, where a change in the PEF state is assumed to be caused predominantly by the scanning antenna. A further assumption is that the initial angles of incidence of the interference sources are not known.

Consider a linear approximation of the state vector (weight) temporal variation:

$$\mathbf{X}(i+1) = \mathbf{X}(i) + \Delta t \dot{\mathbf{X}}(i) \qquad (4.63)$$

with the νth component

$$X_\nu(i) = \exp(j2\pi(d/\lambda)\nu \sin\theta_s(i)) \qquad (4.64)$$

where

$$\theta_s(i) = \alpha - \dot{\theta} i \Delta t \qquad (4.65)$$

describes the scan. The scan rate $\dot{\theta}$ is known, the initial angle α is assumed to be uniformly distributed over $[0, 2\pi]$. With (4.64) and (4.65), we obtain from (4.63):

$$\mathbf{X}(i+1) = \boldsymbol{\phi}(i+1, i)\mathbf{X}(i) \qquad (4.66)$$

where the transition matrix is

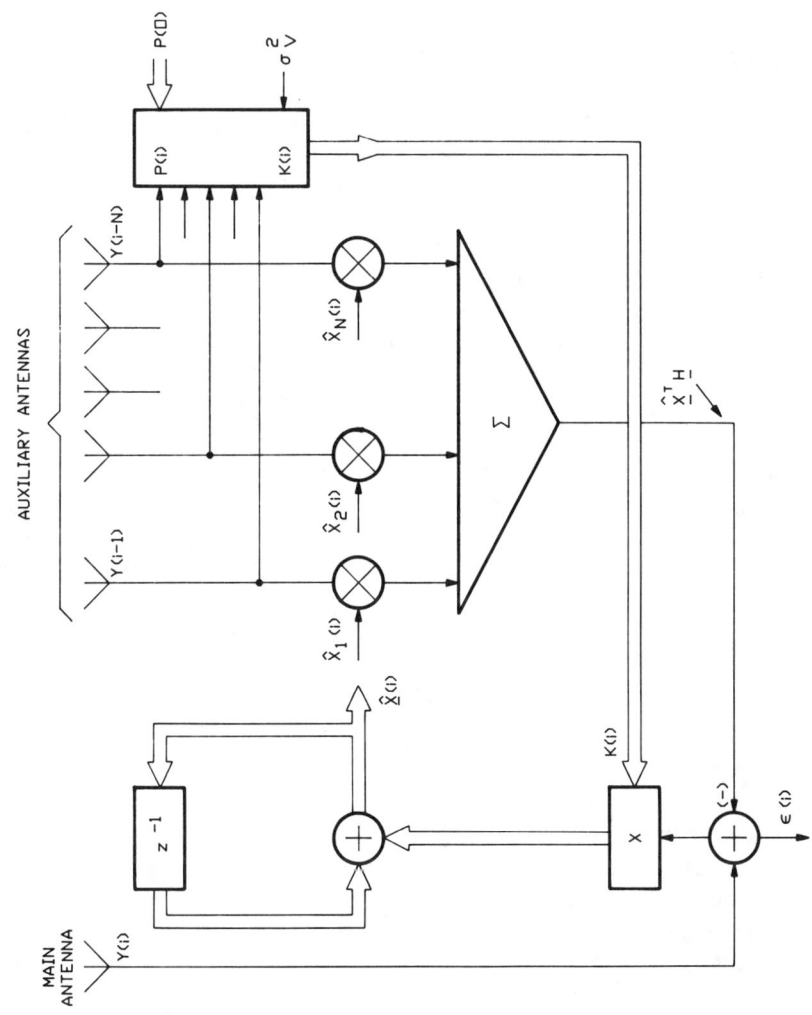

Figure 4.15 Block diagram of an SLC with a one-step Kalman predictor. (From Krucker, IEE Proc., Vol. 130, Pts F and H, No. 1, February 1983, pp. 36–40.)

$$\varphi(i+1,i) = \mathbf{I} - j2\pi(d/\lambda)\Delta t[\cos\theta_s(i)]\dot\theta \begin{bmatrix} 1 & 0 & 0 & \dots & 0 \\ 0 & 2 & 0 & \dots & 0 \\ 0 & 0 & 0 & \dots & N \end{bmatrix} \quad (4.67)$$

Simulation exercises have shown the excellent tracking capability of this system concept with respect to the classical Howells-Applebaum approach (Krucker, 1983). In the same reference, an evaluation of the required bit number for numerical operations was also done. The most critical point is the error covariance update (4.57c) because it may lose its positive-definiteness owing to small word-length and round-off errors. This effect increases with increasing number of auxiliary elements. To overcome this drawback, a small amount of model noise σ_w^2 can be introduced. This technique has proved to be useful in reducing the number of bits required for quantization. In the one-element SLC, 12 bits are required, the three-element SLC needs 14 bits, and 20 bits are necessary for the six-element SLC for stable operation. Finally, the number of complex multiplications are evaluated to provide an idea of the complexity and cost of the processor. The simple Kalman predictor of Figure 4.15 needs approximately $(3N^2 + 3N)$ multiplications, while the scan adapted Kalman filter needs $(6N^2 + 3N)$ multiplications. In comparison, the well known Howells-Applebaum method requires only $2N$ multiplications, but this simple implementation results in very poor steady-state cancellation performance and poor tracking capabilities against a nonstationary jamming environment.

Before closing the section, we mention recent advances of the research in the field of the systolic implementation of the *square root covariance Kalman filter*. The square root covariance Kalman filter is a modification of the original Kalman algorithm with better numerical properties. Parallel processing and, in particular, systolic arrays are seen as solutions to the problem of real-time implementation of the Kalman filter. An interesting reference on this new research topic is (Gaston et al., 1990).

4.2.3 Data-Domain Methods

Closed-loop methods, the subject of Section 4.2.1, trade off the simplicity of implementation (the number of multiplications per iteration is on the order of N, where N is the number of auxiliary antennas) for their poor convergence time. These methods become ineffective for more than a few jammers.

The direct solution methods of Section 4.2.2 offer fast transient response at the expense of an increase in the number of multiplications per iteration, which is on the order of N^2. Unfortunately, as we shall show, they have a disadvantage related to their poor numerical stability in the inversion of the interference covariance matrix. They need extremely high arithmetic precision during digital calculations. However, the so-called data-domain algorithms, the theme of this section, are currently making great strides in fast computational capability (comparable to that of direct solution methods) and, at the same time, they do not suffer from numerical instabilities. This is because they operate on the data (voltage values) received by the array, thus avoiding the need to calculate the sample covariance matrix (power values), which doubles the dynamic range of digital words. Additionally, these algorithms seem to be well suited to mapping onto highly regular computing structures such as the systolic array computers. Systolic arrays, parallel processors offering a high data throughput, can be implemented with VLSI and WSI technologies. The design of an SLC system (even with tens of auxiliary antennas) according the principle of the data domain algorithms and the corresponding implementation with VLSI and WSI technologies should lead to highly compact, low-power-consumption processors, offering high data throughput values.

The data-domain algorithms have been studied in the United Kingdom and United States. Notably, in the United Kingdom, a group of researchers, from the Royal Signals and Radar Establishment (RSRE) Electronics Division-Defense Research Agency, and BNR Europe Ltd have performed theoretical and experimental work, as documented in the literature: (McWhirter, 1983a, 1983b, 1989; Hargrave and Ward, 1985; Hargrave et al., 1988; Hargrave, 1990; Ward et al., 1984, 1985, 1986). In the United States, among the others, an experimental SLC with 64 auxiliary antennas was implemented at MIT Lincoln Laboratory (Rader et al., 1990). We also note the research studies by A.O. Steinhardt (1988), Rader with Steinhardt (1986), and on the use of systolic arrays in data-domain problems by H.T. Kung (1981); and S.Y. Kung (1985). A reference book on data-domain algorithms and their implementation by systolic array processors is authored by S. Haykin (1986). Another book for in-depth mathematical studies is Golub and VanLoan (1983). Additional references of interest are Schreiber and Kuekes (1985) and VanVeen (1989).

The section is organized as follows. We start (Subsection 4.2.3.1) with the analysis of numerical sensitivity of the power-domain methods, i.e., those that need to calculate and invert the covariance matrix of the interference. Having recognized the poor numerical stability of these methods, we introduce in Subsection 4.2.3.2 the data-domain formulation of the SLC problem. Next, we introduce the key topic of QR decomposition (Subsection 4.2.3.3), which is the technical means to handle the data-domain SLC problem. Then, we illustrate in Subsection 4.2.3.4 the implementation of QR decomposition by means of the Givens rotations. The mapping of QR decomposition onto a systolic array processor is the topic of Subsection 4.2.3.5. Some simulation results concerning an explanatory example of an SLC implemented by a triangular systolic array processor are described in Subsection 4.2.3.6. Finally, we introduce the method of Householder reflections as an alternative data-domain technique to solve the SLC problem (Subsection 4.2.3.7).

4.2.3.1 Numerical Sensitivity of Power-Domain Methods

The direct solution techniques described in Section 4.2.2 are essentially power-domain methods because they need to estimate and invert the interference covariance matrix. This section evaluates the numerical stability issues related to these algorithms.

From (4.5b), we note that small perturbations in the covariance matrix **M** or in the crosscorrelation vector **R** can induce relatively large changes in the weight vector $\hat{\mathbf{W}}$. The perturbations may result from limited computational accuracy or parametric changes due to a nonstationary environment. To obtain a quantitative measure of sensitivity, let the covariance matrix assume a slightly perturbed value, $\mathbf{M} + \delta\mathbf{M}$. Similarly, let the crosscorrelation vector assume a slightly perturbed value, $\mathbf{R} + \delta\mathbf{R}$. Correspondingly, denote $\hat{\mathbf{W}} + \delta\mathbf{W}$ as the weight vector of the slightly perturbed sidelobe canceler, for which the following equation applies:

$$\hat{\mathbf{W}} + \delta\mathbf{W} = (\mathbf{M} + \delta\mathbf{M})^{-1}(\mathbf{R} + \delta\mathbf{R}) \tag{4.68}$$

After some numerical manipulations, the following equation defines the perturbation $\delta\mathbf{W}$ in the optimum weights (Haykin, 1989, pp. 275–281):

$$\delta\mathbf{W} \approx \mathbf{M}^{-1}(-\delta\mathbf{M}\,\hat{\mathbf{W}} + \delta\mathbf{R}) \tag{4.69}$$

To translate the perturbation vector $\delta\mathbf{W}$ into a single number that may be used to describe the condition of a vector and a matrix, we use *vector norm* and *matrix norm*. We remember that the Euclidean norm of a complex valued vector **X** is defined as follows:

$$\|\mathbf{X}\| = (\mathbf{X}^T\mathbf{X}*)^{1/2} \tag{4.70}$$

The vector norm defines the length of the vector. Similarly, the matrix norm enables us to assess the size of matrices. The Euclidean norm of a matrix **A** is defined by

$$\|\mathbf{A}\| = \max_{\mathbf{X} \neq 0} \frac{\|\mathbf{A}\mathbf{X}\|}{\|\mathbf{X}\|} \tag{4.71}$$

and measures the largest amount by which the length of any vector **X** is amplified as a result of matrix multiplication by **A**.

Having recalled the concepts of vector and matrix norms, we are able to evaluate the norm of the perturbation vector $\delta\mathbf{W}$. From (4.69), we obtain the following inequality in terms of norms (Haykin, 1989):

$$\|\delta\mathbf{W}\| \leq \|\mathbf{M}^{-1}\|(\|\delta\mathbf{M}\|\,\|\hat{\mathbf{W}}\| + \|\delta\mathbf{R}\|) \tag{4.72a}$$

or, equivalently,

$$\rho_W \triangleq \frac{\|\delta\mathbf{W}\|}{\|\hat{\mathbf{W}}\|} \leq \|\mathbf{M}^{-1}\| \left(\|\delta\mathbf{M}\| + \frac{\|\delta\mathbf{R}\|}{\|\hat{\mathbf{W}}\|} \right) \tag{4.72b}$$

Define the condition number of matrix **M** as

$$\chi(\mathbf{M}) \triangleq \|\mathbf{M}\| \, \|\mathbf{M}^{-1}\| \tag{4.73}$$

and use this definition to rewrite the previous equation as (Haykin, 1989):

$$\rho_W \leq \chi(\mathbf{M})(\rho_M + \rho_R) \tag{4.74a}$$

where

$$\rho_M = \frac{\|\delta \mathbf{M}\|}{\|\mathbf{M}\|} \tag{4.74b}$$

$$\rho_R = \frac{\|\delta \mathbf{R}\|}{\|\mathbf{R}\|} \tag{4.74c}$$

The parameter, ρ_M, defines the relative error in the covariance matrix $\mathbf{M}$. Similarly, the parameter, ρ_R, denotes the relative error in the correlation vector, $\mathbf{R}$. Equation (4.74a) shows that the relative error ρ_W in the optimum weight $\hat{\mathbf{W}}$ can be $\chi(\mathbf{M})$ times the sum of the relative errors, ρ_M and ρ_R. In this sense, the condition number $\chi(\mathbf{M})$ quantifies the sensitivity of the SLC filter to perturbations in $\mathbf{M}$ and $\mathbf{R}$. If we bear in mind that the norm of a matrix is a measure of its size, the following two expressions are relatively easy to understand (Haykin, 1989):

$$\|\mathbf{M}\| = \lambda_{\text{MAX}} \tag{4.75a}$$

$$\|\mathbf{M}^{-1}\| = \frac{1}{\lambda_{\text{MIN}}} \tag{4.75b}$$

where λ_{MAX} and λ_{MIN} are the maximum and minimum eigenvalues of the covariance matrix $\mathbf{M}$. Consequently, the condition number of the matrix $\mathbf{M}$ is the eigenvalue spread:

$$\chi(\mathbf{M}) = \frac{\lambda_{\text{MAX}}}{\lambda_{\text{MIN}}} \tag{4.76}$$

The lower bound of the eigenvalue spread is one, which happens, for instance, when $\mathbf{M} = \sigma^2 \mathbf{I}$. In the more general case of correlated noise, the eigenvalue spread will be greater than one, which causes trouble in computing the optimum weight vector. In particular, small perturbations on the correlation matrix $\mathbf{M}$ or on the correlation vector $\mathbf{R}$ will be amplified by the condition number, thereby producing a large change in the optimum weight vector, $\hat{\mathbf{W}}$. In this condition, we customarily say that the set of linear equations providing the optimum weights is poorly conditioned. Therefore, there is great interest in finding numerical algorithms that are better conditioned than those which need the direct solution of the (4.5b). This is done in the following subsection by introducing the QR decomposition, which is a numerically stable algorithm to derive the optimum weights.

4.2.3.2 Data-Domain Formulation of SLC Problem

In this section, we reformulate the basic equation of an SLC system by making explicit reference to the matrix of received voltage data snapshots. We also compare the numerical stability of power-domain methods with that expected with data-domain methods (Ward *et al.*, 1986).

With reference to Figure 4.1(a), we denote the SLC output at time instant t_i by

$$z(t_i) = -\mathbf{V}^T(t_i)\mathbf{W} + V_M(t_i) \qquad (4.77)$$

where $\mathbf{V}(t_i)$ is the so-called data snapshot from auxiliary antennas; that is, the set of received signals at time instant t_i and $V_M(t_i)$ is the corresponding sample from the main radar antenna. The optimum weight vector $\hat{\mathbf{W}}$ is obtained by minimizing the residual output power at time t_n as estimated by the quantity:

$$P_Z(n) = \sum_{j=1}^{n} |z(t_j)|^2 \beta^{2(j-1)} \qquad (4.78)$$

where the factor $\beta (0 \leq \beta \leq 1)$ generates an exponential time window in the averaging procedure. A more compact matrix notation is as follows:

$$\mathbf{Z}(n) = \mathbf{B}(n) \begin{bmatrix} z(t_1) \\ z(t_2) \\ \vdots \\ z(t_n) \end{bmatrix} \qquad (4.79a)$$

where

$$\mathbf{B} = \text{diag}[\beta^{n-1} \beta^{n-2} \ldots 1] \qquad (4.79b)$$

Therefore, the vector of SLC residuals may be written in the form:

$$Z(n) = X(n)W + Y(n) \qquad (4.80a)$$

where

$$X(n) = B(n) \begin{bmatrix} V^T(t_1) \\ V^T(t_2) \\ \vdots \\ V^T(t_n) \end{bmatrix} \qquad (4.80b)$$

$$Y(n) = B(n) \begin{bmatrix} V_M(t_1) \\ V_M(t_2) \\ \vdots \\ V_M(t_n) \end{bmatrix} \qquad (4.80c)$$

In summary, the n-dimensional vector $Z(n)$ contains the samples of the SLC output, the $(n \times N)$ data matrix $X(n)$ contains the n data snapshots from the N auxiliary antennas, and the n-dimensional vector $Y(n)$ has the n samples of the received signal from the radar antenna.

The determination of the optimum weight vector $\hat{W}$ is obtained by minimization of $P_Z(n)$. This is done, as usual, by taking the gradient of P_Z with respect to W and equating it to zero. The result of this procedure is (Ward et al., 1986):

$$[X^H(n)X(n)]\hat{W}(n) = -[X^H(n)Y(n)] \qquad (4.81)$$

The term in brackets on the left-hand side is the estimated interference covariance matrix (Haykin, 1989, p. 292); the other term in brackets in the right-hand side of the equation is the estimated crosscorrelation vector between the auxiliary signals and the radar signals. Equation (4.81) is another way to write the basic equation (4.5b). We shall see why we are rewriting the basic equation of the SLC. We note that the filter (4.5b) is optimum in a probabilistic sense. That is, one filter performs well for all realizations of the operational environment. On the other hand, equation (4.81) yields a different filter for each collection of input data.

The calculation of $\hat{W}$ involves the inversion of the estimated covariance matrix. This operation is numerically unstable if the matrix is ill conditioned, i.e., if the condition number $\chi[X^H(n)X(n)]$ is large. At this point, we note the following crucial relationship:

$$\chi[X^H(n)X(n)] = \chi^2[X(n)] \qquad (4.82)$$

which indicates that the condition number of the data matrix $X(n)$ is simply the square-root value of the condition number of the covariance matrix. In other words, any numerical algorithm that avoids the estimation of the interference covariance matrix and operates directly on the data is better conditioned from a numerical point of view. This is the rationale of the so-called data-domain algorithms, which will be described in the following subsections.

Another negative aspect of so-called power-domain algorithms (i.e., those that operate on the covariance matrix) is related to the type of signal processing architecture required to implement and solve (4.5b). Essentially, the architecture comprises a number of distinct components:

1. A memory that stores the data snapshots;
2. A processor unit with high arithmetic precision to form the covariance matrix and the crosscorrelation vector to calculate the weight vector and combine it with the data snapshots;
3. A high-speed communication link between the memory and processor unit;
4. A control unit to deliver the proper sequence of instructions to each component of the processor.

This type of architecture is a conventional one and seems unsuitable to be implemented with the modern technology of VLSI. The trend is toward the use of parallel processors, such as systolic arrays, which are indeed suited to implementation with VLSI technology (Kung and Gentleman, 1981; Kung, 1985). As will be subsequently shown, the data-domain algorithms based on QR decomposition can be mapped onto a systolic array computer to calculate the optimum weights and the canceler output. The convergence speed of the data-domain algorithms is comparable to that of power-domain algorithms (Section 4.2.2), but they place minimal requirements on computer word length and arithmetic precision. We should note that, traditionally, the computational complexity of an algorithm, measured in terms of the number of arithmetic operations per iteration, has been used as a criterion for selecting one algorithm over another. In parallel processing, an issue of concern is that of throughput or data rate rather than computational complexity; throughput is the total rate, in bits per second, at which the data are transmitted from the input to the output of the processor. Accordingly, when considering a parallel processor like the systolic array implementing the QR decomposition, a criterion of interest is the maximum attainable data throughput (Haykin, 1989, p. 247).

4.2.3.3 QR Decomposition

QR decomposition is a numerical technique that is used to solve least-squares problems (Ward et al., 1986). One of the advantages of this technique is that it avoids direct computation and inversion of the interference covariance matrix. Additionally, it operates directly on the matrix of received data rather than on the corresponding covariance matrix. The computation of the covariance matrix involves a process of squaring voltage data, and, in many practical situations, it results in a set of linear equations (4.81) that are poorly conditioned, possibly leading to numerical instability and significantly degraded cancellation performance unless very precise arithmetic is used in the computations. We shall show that the direct use of the data matrix leads to a better conditioned problem; i.e., it leads to greater accuracy in the determination of the weight vector with corresponding improvements in adaptive nulling performance.

An $n \times n$ unitary matrix $\mathbf{Q}$ (i.e., a matrix such that $\mathbf{Q}^H\mathbf{Q} = \mathbf{I}$) is generated that, when applied to the data matrix $\mathbf{X}(n)$:

$$\mathbf{Q}(n)\mathbf{X}(n) = \begin{bmatrix} \mathbf{R}(n) \\ \mathbf{0} \end{bmatrix} \quad (4.83)$$

produces an upper triangular matrix $\mathbf{R}(n)$ having dimensions $(N \times N)$. Applying the same unitary transformation to the vector $\mathbf{Y}(n)$, we obtain

$$\mathbf{Q}(n)\mathbf{Y}(n) = \begin{bmatrix} \mathbf{u}(n) \\ \mathbf{v}(n) \end{bmatrix} \quad (4.84)$$

where $\mathbf{u}$ is an N-dimensional vector and $\mathbf{v}$ is $(n - N)$-dimensional vector. Applying the same transformation $\mathbf{Q}$ to (4.80a), we have

$$\mathbf{Q}(n)\mathbf{Z}(n) = \mathbf{Q}(n)\mathbf{X}(n)\mathbf{W} + \mathbf{Q}(n)\mathbf{Y}(n) \quad (4.85a)$$

which, when using equations (4.83) and (4.84), becomes:

$$\mathbf{Q}(n)\mathbf{Z}(n) = \begin{bmatrix} \mathbf{R}(n) \\ \mathbf{0} \end{bmatrix} \mathbf{W} + \begin{bmatrix} \mathbf{u}(n) \\ \mathbf{v}(n) \end{bmatrix} \quad (4.85b)$$

As the matrix $\mathbf{Q}(n)$ is unitary (i.e., it does not change the residual output power), it is evident that the optimum weight vector should satisfy the new equation:

$$\mathbf{R}(n)\hat{\mathbf{W}} + \mathbf{u}(n) = \mathbf{0} \quad (4.86)$$

As the matrix $\mathbf{R}(n)$ is upper triangular, the weight $\hat{\mathbf{W}}$ is easily solved by backsubstitution as shown in the following simple example for $N = 3$:

$$\begin{bmatrix} R_{11} & R_{12} & R_{13} \\ & R_{22} & R_{23} \\ 0 & & R_{33} \end{bmatrix} \begin{bmatrix} \hat{W}_1 \\ \hat{W}_2 \\ \hat{W}_3 \end{bmatrix} = - \begin{bmatrix} u_1 \\ u_2 \\ u_3 \end{bmatrix} \quad (4.87a)$$

From the third equation, we obtain immediately $\hat{W}_3$:

$$\hat{W}_3 = -u_3/R_{33} \tag{4.87b}$$

which, substituted in the second equation:

$$\hat{W}_2 R_{22} + \hat{W}_3 R_{23} = -u_2$$

gives:

$$\hat{W}_2 = \frac{u_3 R_{23} - u_2 R_{33}}{R_{22} R_{33}} \tag{4.87c}$$

The third and second components of the weights can be substituted in the first equation:

$$\hat{W}_1 R_{11} + \hat{W}_2 R_{12} + \hat{W}_3 R_{13} = -u_1$$

to provide the first component of the weight:

$$\hat{W}_1 = -\frac{u_1}{R_{11}} + u_2 \frac{R_{12}}{R_{11} R_{22}} + u_3 \left(\frac{R_{13} R_{22} - R_{12} R_{23}}{R_{11} R_{22} R_{33}} \right) \tag{4.87d}$$

Additionally, (4.86) is better conditioned than its power-domain counterpart (4.81) because the condition number of $\mathbf{R}(n)$ is given by

$$\chi[\mathbf{R}(n)] = \chi[\mathbf{Q}(n)\mathbf{X}(n)] = \chi[\mathbf{X}(n)] \tag{4.88}$$

The second equality follows directly from the fact that $\mathbf{Q}(n)$ is unitary. By comparison of (4.88) with (4.82), which gives the condition number for the power-domain method, it is evident that the condition number of the data-domain algorithm is the square-root of the corresponding value of the power-domain algorithm.

4.2.3.4 Implementation of QR Decomposition by Givens Rotations

The triangularization process described by (4.85) may be accomplished by Givens rotations, the subject of this subsection, or by Householder reflections, to be described in Subsection 4.2.3.7. The Givens rotation method performs the triangularization of the data matrix in a recursive fashion as each new snapshot of data enters the array computer.

A complex Givens rotation is an elementary transformation of the form (Ward et al., 1984)

$$\begin{bmatrix} c & s^* \\ -s & c \end{bmatrix} \begin{bmatrix} 0 \ldots 0, \beta r_i \ldots \beta r_k \ldots \\ 0 \ldots 0, \ x_i \ \ldots \ x_k \ \ldots \end{bmatrix} = \begin{bmatrix} 0 \ldots 0, r'_i \ldots r'_k \ldots \\ 0 \ldots 0, 0 \ldots x'_k \ldots \end{bmatrix} \quad (4.89)$$

where β is a scaling factor. The rotation coefficients, c and s, are

$$c = \frac{\beta r_i}{[|x_i|^2 + \beta^2 r_i^2]^{1/2}} \quad (4.90a)$$

$$s = \frac{x_i}{[|x_i|^2 + \beta^2 r_i^2]^{1/2}} \quad (4.90b)$$

where r_i and r'_i are real and nonnegative values. The parameters c and s are generalizations of the cosine and sine for an angular rotation in a two-dimensional complex space. One Givens rotation eliminates but one element of a matrix. A sequence of such eliminations may be used to triangulate the data matrix $\mathbf{X}(n)$.

4.2.3.5 Mapping of QR Decomposition and Givens Rotations onto a Systolic Array Processor

The Givens rotation algorithm may be efficiently mapped onto a systolic array computer (Kung, 1985; Ward et al., 1986). A systolic computer is an array of processing cells. Each cell has a local memory and is connected with its neighboring cells in the form of a regular grid. The more common configurations of the systolic array are linear and triangular. The internal cells are all identical, while the boundary cells are different. The operations performed by the systolic array are synchronized by a clock. On each clock cycle, every cell receives data from its neighboring cells and performs operations. The resulting data are stored within the cell and passed to the neighboring cells on the next clock cycle.

An instructive example (Ward et al., 1984) of systolic array is the linear configuration shown in Figure 4.16 to perform the rotation between the following two rows of data:

$$\begin{bmatrix} y_1 & y_2 & y_3 & y_4 \\ x_1 & x_2 & x_3 & x_4 \end{bmatrix} \quad (4.91)$$

We see that, except for the left boundary cell, all cells of the array are equal and perform the same operation. As shown in the figure, the data x_1 to x_4 are stored in the computer, while the data y_1 to y_4 are applied to the array in a time-skewed fashion. The cell computation and the data flow through the array computer are controlled by a clock. At the first clock, the data x_1 and y_1 are processed within the

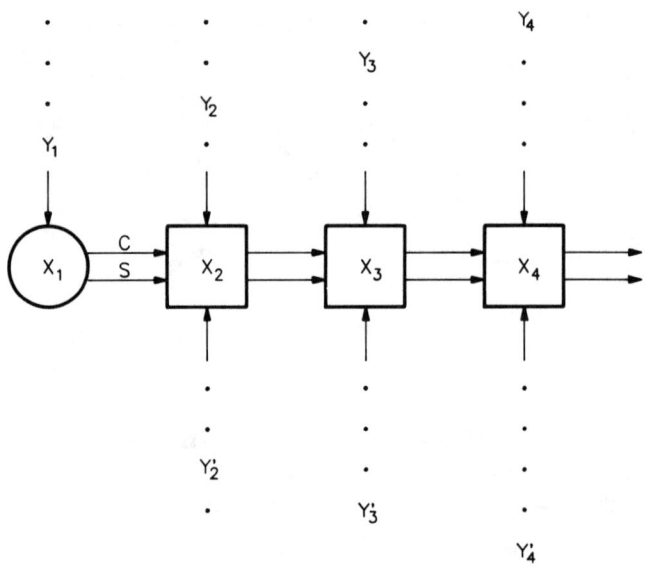

Figure 4.16 Linear systolic array to perform the elementary Givens rotation. (Adapted from Ward *et al.*, IEE Proc. Vol. 131, Pt. F, 1984, pp. 638–645.)

boundary cell (i.e., the round cell), and the rotation coefficients c and s are calculated and prepared to propagate horizontally in the computer. At the second clock cycle, the coefficients c and s together with the data y_2 are received by the first internal cell, and an elementary rotation is performed on the data x_2 and y_2. On the next cycle, the transformed value y_2' leaves the first internal cell, and the rotation parameters c and s are received by the second internal cell, which applies the same rotation to the stored data x_3 and the input data y_3. Proceeding in this way, the data vector $[x_1, x_2, x_3, x_4]$ is updated and the reduced vector $[0, y_2', y_3', y_4']$ leaves the array on a time-skewed fashion.

The implementation of an SLC with four auxiliary antennas by QR decomposition mapped onto a systolic array is shown in Figure 4.17. The computer array is a stack of linear systolic arrays of the type illustrated in the previous Figure 4.16. A continuous stream of data is sent to the computer from the auxiliary and main channels. Specifically, $[x_1(i), x_2(i), x_3(i), x_4(i)]$ represents the snapshot of auxiliary data at the ith time instant, while $y(i)$ is the ith sample from the main antenna. As pointed out by Ward *et al.* (1984; 1986), three distinct sections can be identified in the array, namely:

1. The triangular section, *ABC*, which stores the elements of the matrix $\mathbf{R}(n)$, which is recursively updated;

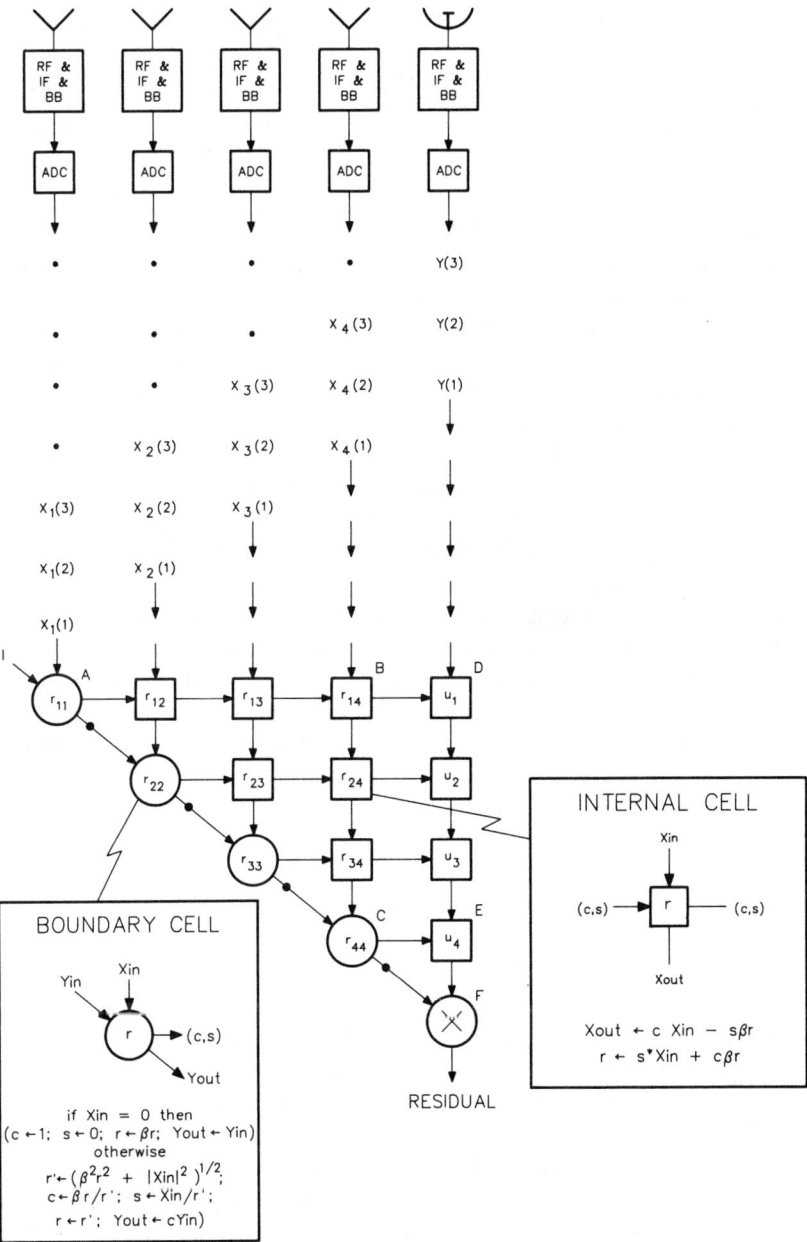

Figure 4.17 Implementation of an SLC with four auxiliary antennas by means of a systolic array computer. (Adapted from Ward *et al.*, IEEE Trans on Antennas and Propagation, Vol. AP–34, No. 3, March 1986, pp. 338–346, ©1986 IEEE.)

2. The right-hand column of computing cells, *DE*, which stores the updated version of the vector **u**(*n*);
3. The final processing cell, *F*, which provides the residual output of the SLC.

The figure also shows the functions performed by the three types of processing cells. Note that the final cell executes a multiplication.

Several modifications have been conceived for this basic processing structure. Among others, the following are interesting:

1. The *square-root-free systolic processor* (Gentleman, 1973), which avoids performing the square root in the round boundary cells;
2. The *systolic processor* to calculate, in addition to the residual output, the adapted weight vector, **Ŵ**. This can be done by using an additional linear systolic array in conjunction with the triangular systolic array described in this section, as explained in Haykin (1986). An alternative, interesting method is the so-called "weight freezing and flushing" mentioned in Ward *et al.* (1986). This method does not need any additional hardware. It freezes the stored values $r_{ij}(i,j = 1, 2, \ldots, N)$ and $u_i(i = 1, \ldots, N)$ in the triangular array and stimulates the array with particular input data to generate the weight components of **Ŵ** at the output at successive instants of time. To understand the so-called flushing operation, we can imagine that the frozen systolic array operates as a linear combiner of the input data x and y, with the combination coefficients represented by the components of **Ŵ**, to provide the output signal (4.80a). To generate the *i*th component of the weight vector **Ŵ** at the output, the array is fed with all signals equal to zero except the *i*th auxiliary signal which is set equal to one. In practice, the signals sent to the systolic array are as follows:

$$\mathbf{Y}(n) = \begin{bmatrix} 0 \\ \vdots \\ 0 \\ 0 \end{bmatrix}, \quad \mathbf{X}(n) = \begin{bmatrix} 0 & & & 1 \\ & & 1 & \\ & \cdot\cdot\cdot & & \\ 1 & & & 0 \end{bmatrix} \qquad (4.92)$$

In fact, the use of an identity for the data matrix **X** produces an orderly sequence of weight components at the processor output.
3. The systolic processor for implementation of the *generalized SLC* (GSLC). The GSLC is a technique for efficiently implementing a constrained minimization problem (Buckley and Griffiths, 1986; Jablon, 1986; VanVeen and Buckley, 1988). To be more specific, let us leave the sidelobe canceler system for a moment and consider the adaptive array system. As will be explained in Chapter 5, the adaptive array problem refers to the calculation of a set of optimum weights to combine linearly a number of omnidirectional antennas with the purpose of receiving a target signal from a desired direction and to cancel

interference impinging on the array from other directions. The adaptive array differs from the SLC system in that it does not have a main radar antenna to gather the desired target echoes. Nevertheless, the adaptive array problem can be recast as an SLC, or, more precisely, as a GSLC problem, by introducing a constraint on the minimization of the residual power $P_Z(n)$ for a maximum adapted pattern along a desired direction. We can show that the constrained minimization problem for the GSLC results in a suitable preprocessor upstream of an SLC system. Figure 4.18 illustrates the scheme of such a system. The signals from the N omnidirectional antennas are processed in parallel into two blocks. The left-hand-side block performs the product of the N signals with a *blocking matrix*, **BM**, with dimensions $(N - 1) \times N$ and delivers a new set of $(N - 1)$ signals:

$$\mathbf{V'} = \mathbf{BM}\ \mathbf{V} \qquad (4.93)$$

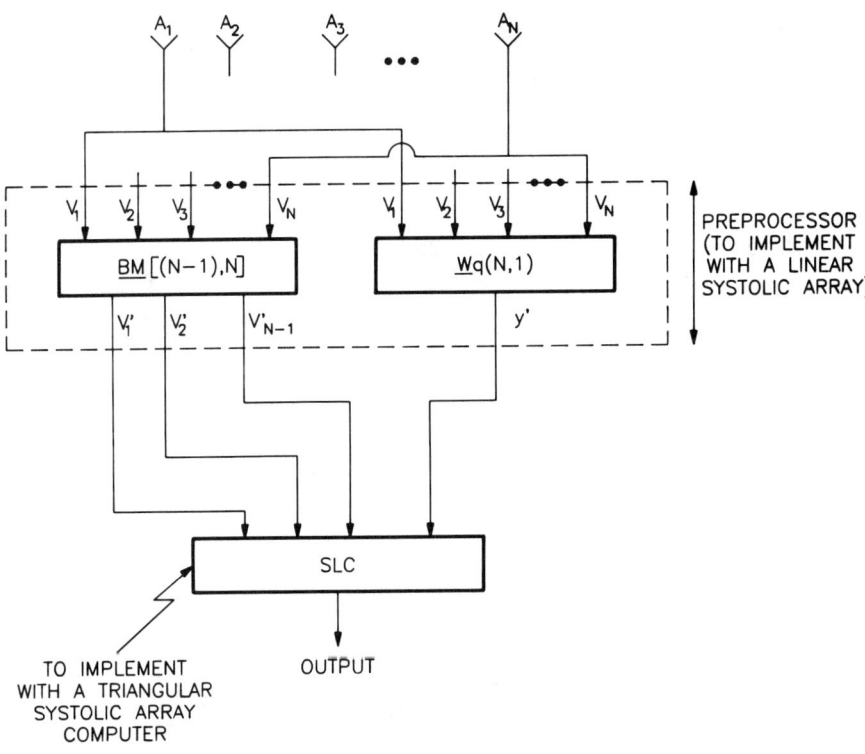

Figure 4.18 Generalized SLC.

where

$$\mathbf{V}' = [V_1' V_2' \ldots V_{N-1}']^T$$
$$\mathbf{V} = [V_1 V_2 \ldots V_N]^T \qquad (4.94)$$

The right-hand block performs the inner product between the signal vector $\mathbf{V}$ and the so-called quiescent N-dimensional vector $\mathbf{W}_q$, thus producing a scalar quantity y':

$$y' = \mathbf{W}_q^T \mathbf{V} \qquad (4.95)$$

Now, the set of signals $\mathbf{V}'$ is interpreted as the auxiliary signals of an SLC system, while y' plays the role of the main signal. The jammer cancellation is operated by an SLC system, which processes $\mathbf{V}'$ and y'. The SLC system can be efficiently implemented by the triangular systolic array already discussed. We can also show that the blocking matrix and combination with the quiescent vector $\mathbf{W}_q$ are efficiently implemented with a linear systolic array (e.g., the top row of a larger triangular systolic array used to do the SLC processing). The quiescent vector $\mathbf{W}_q$ is proportional to a set of phasors to coherently combine the target signals from a wished direction. If we express the constraint of the problem as

$$\mathbf{C}^T \mathbf{W} = \mu \qquad (4.96)$$

where the vector $\mathbf{C}$ indicates the desired "look" direction and μ is the gain of the adapted array along the desired direction, the quiescent vector can be expressed as

$$\mathbf{W}_q = \mu \frac{\mathbf{C}}{\mathbf{C}^T \mathbf{C}^*} \qquad (4.97)$$

In fact, $\mathbf{W}_q$ satisfies the condition:

$$\mathbf{W}_q^T \mathbf{C} = \mu \qquad (4.98)$$

The expression of the blocking matrix $\mathbf{BM}$ is

$$\mathbf{BM} = \begin{bmatrix} -c_2/c_1 & & \\ -c_3/c_1 & & \\ \vdots & \mathbf{I}_{(N-1)\times(N-1)} \\ -c_N/c_1 & & \end{bmatrix} \qquad (4.99)$$

where $c_1, c_2, \ldots, c_N$ are the components (phasors related to the look direction) of the vector $\mathbf{C}$, and $\mathbf{I}$ is the $(N - 1)$-dimensional unit matrix. This blocking matrix satisfies the condition:

$$\mathbf{BM\ C} = \mathbf{0} \tag{4.100}$$

Thus, a signal wavefront impinging on the array from a direction specified by the vector $\mathbf{C}$ is "blocked" or nulled by the matrix $\mathbf{BM}$. In summary, the GSLC produces an output:

$$z = \mathbf{W}_q^T \mathbf{V} + \mathbf{W}_{\text{SLC}}^T \mathbf{BM\ V} \tag{4.101a}$$

where $\mathbf{W}_{\text{SLC}}$ is the set of $(N - 1)$ weights generated in the SLC section. This equation can be rewritten by considering $\mathbf{W}_{\text{SLC}}^T \mathbf{BM}$ as an additional N-dimensional weight vector $\mathbf{W}_0^T$:

$$z = \mathbf{W}_q^T \mathbf{V} + \mathbf{W}_0^T \mathbf{V} \tag{4.101b}$$

and, hence, the optimum weight $\hat{\mathbf{W}}$ of the original adaptive array problem is

$$\hat{\mathbf{W}} = \mathbf{W}_q + \mathbf{W}_0 \tag{4.101c}$$

which satisfies the linear constraint of (4.96). In fact,

$$\hat{\mathbf{W}}^T \mathbf{C} = \mathbf{W}_q^T \mathbf{C} + \mathbf{W}_0^T \mathbf{C} = \mu + \mathbf{W}_{\text{SLC}}^T \mathbf{BM\ C} = \mu \tag{4.101d}$$

The last equality derives from (4.100).

4.2.3.6 Performance Evaluation by Simulation

A Monte Carlo simulation program has been used to evaluate the cancellation performance of the SLC mapped onto a triangular systolic array computer. The simulation exercise refers to 15 auxiliary antennas and one main radar antenna. The main antenna pattern has been synthesized by using an additional set of 15 auxiliary antennas. The interfering environment is constituted by three monochromatic jammers, characterized by a same value of jammer-to-noise power ratio, JNR = 20 dB. The directions of arrival of the jammers with respect to the array boresight are $\theta_1 = -80°$, $\theta_2 = 25°$, and $\theta_3 = 30°$. The corresponding covariance matrix of the disturbance is characterized by a ratio of the maximum to minimum eigenvalue of 34 dB. The maximum cancellation ratio that can be achieved by knowing *a priori* the interference covariance matrix is CR = 24.8 dB. This value has been calculated by using (4.7). Figure 4.19 show a number of adapted antenna patterns, namely:

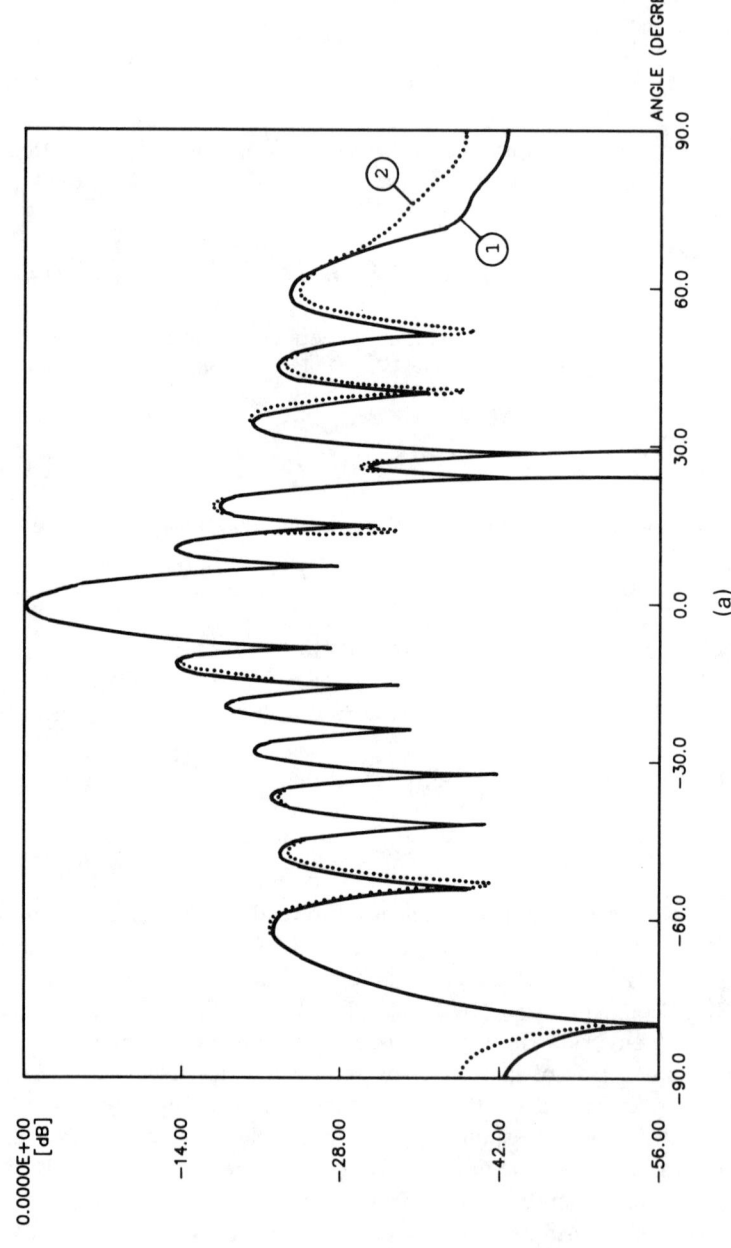

Figure 4.19 (a) Adapted patterns for *a priori* known weights (1) and systolic array at steady state (2); (b) adapted patterns of the systolic array after 31 snapshots (1) and after 100 snapshots (2).

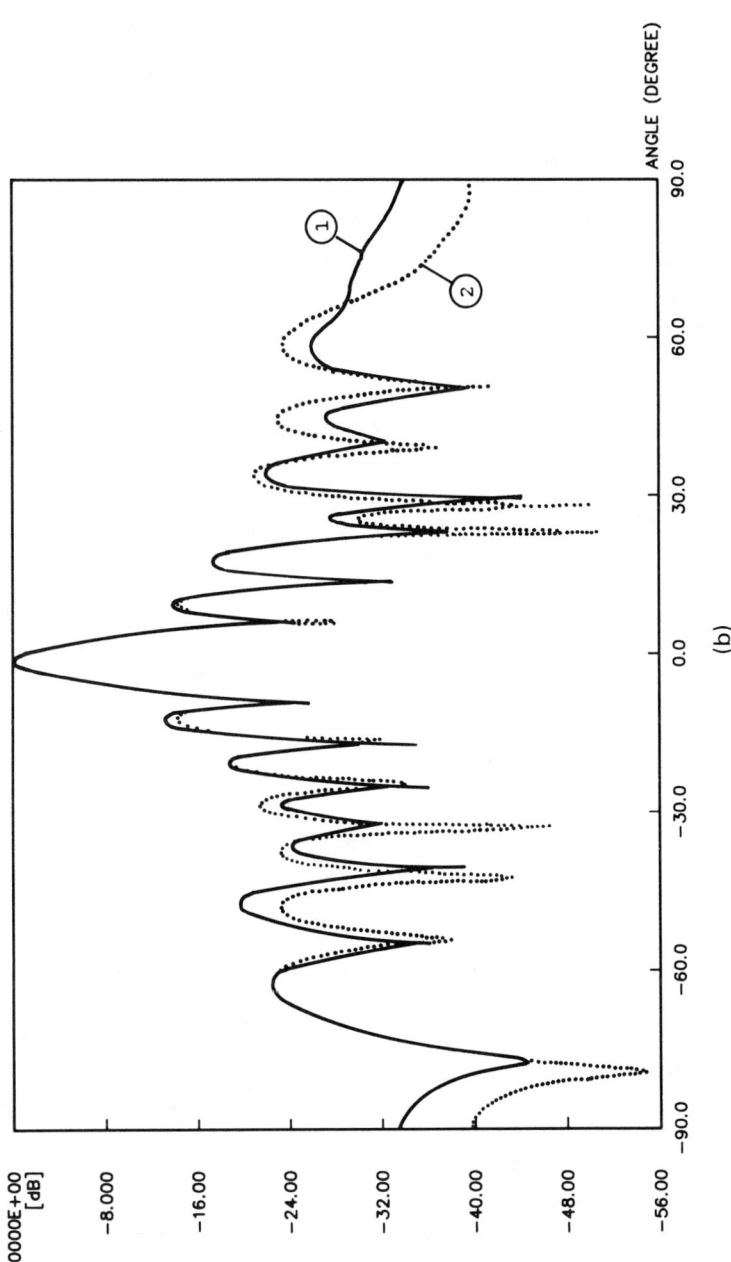

Figure 4.19 Continued

1. Figure 4.19(a) gives the pattern obtained by knowing *a priori* the optimum set of weights, and the pattern (averaged over 20 statistically independent trials) obtained by the systolic array when it has reached the steady-state (i.e., after 150 snapshots);
2. The pattern of the systolic array after 31 clock cycles, or, equivalently, data snapshots (which are necessary to fill and initialize the systolic array) is given in Figure 4.19(b) with the adapted pattern from the systolic array after 100 clock cycles.

The CR values obtained after 31 and 100 snapshots are 14 dB and 23 dB, respectively, while the cancellation ratio when the systolic array has reached the steady-state is 24.5 dB. Apart from the data snapshots needed for initialization the transient time of the systolic array is approximately the same as the very fast DMI method (see Subsection 4.2.2.3). Remember that the DMI method reaches the CR value, within 3 dB of the optimum CR value, in $2N$ data snapshots, where N is the number of auxiliary antennas.

The considerable difference between the data-domain and power-domain methods is related to their sensitivity to the arithmetic precision with which the calculations need to be performed. The problem is to find and compare the minimum word lengths, n_p and n_d, respectively, required to invert the covariance matrix of the power-domain algorithm and to calculate the weights by operating directly on the matrix of data. As has been shown (Rader and Steinhardt, 1986, p. 1593), the two inequalities apply:

$$n_p \geq \log_2(\lambda_{MAX}/\lambda_{MIN}) \tag{4.102a}$$

$$n_d \geq \frac{1}{2}\log_2(\lambda_{MAX}/\lambda_{MIN}) \tag{4.102b}$$

where λ_{MAX} and λ_{MIN} are the maximum and minimum eigenvalues of the interference covariance matrix. The data-domain method enjoys a saving of a factor of two in the required minimum word length by avoiding the formation of the covariance matrix. By application of the previous equations to the aforementioned simulated example, we find the following values, $n_p \geq 12$ and $n_d \geq 6$. We also note that a ratio of $\lambda_{MAX}/\lambda_{MIN} = 34$ dB is too optimistic. In practice, the eigenvalue spread is larger than 70 dB, giving rise to $n_p \geq 24$ and $n_d \geq 12$. Generally the power-domain method requires double precision, which costs four times more than single precision.

To complete the comparison between the power-domain and data-domain methods, we should note that the numbers of multiplications and summations are on the order of N^2 for both methods. The count of operations has been done for the square-root-free Givens rotation and DMI (Subsection 4.2.2.3) algorithms.

4.2.3.7 Implementation by Householder Transformation

Householder transformation (HT) is another data-domain technique, in addition to the Givens transformation, that maps a matrix of data in a low triangular matrix. If the original data matrix pertains to a linear set of equations, the transformed low triangular matrix allows us to solve the system of equations quite easily by back-substitution. The original (dense) and transformed (sparse) data sets have the same covariance matrix because the HT operates by orthonormal matrices. This property permits applying the HT to the calculation of the optimum weights of the SLC system, which are dependent upon the covariance data matrix. The motivation for using the HT is related to its numerical stability.

The HT and its application to the solution of least-squares problems have been studied in Golub and VanLoan (1983), Steinhardt (1988), and Rader and Steinhardt (1986). The material presented in this section is essentially derived from the latter two references. The section is organized as follows. We start with the definition of the HT of a vector and with its geometrical interpretation, as suggested by Steinhardt (1988). Next, we consider the application of the HT to a data matrix. Equipped with these basic concepts, we then introduce the solution of the least-squares problem for the batch case, i.e., the case in which the equations are solved once for a block of data of fixed size.

A simulation study has been done by Rader and Steinhardt (1986) to compare the effect of finite word length on the SLC performance of a power-domain method and the two data-domain methods, based on the HT and Givens rotations. As has been shown, the HT has better numerical accuracy than the Givens rotations. In other words, the HT is closer to the optimal numerical accuracy predicted by (4.102b). However, in favor of the Givens rotations, we should mention its easy and natural way of solving the recursive least-squares problem and the nice mapping of this algorithm onto systolic array computers. A systolic configuration does not seem readily available for the recursive least-squares problem, based on the HT.

We start with the definition of the HT and its geometric interpretation as suggested in Steinhardt (1988). Given two complex-valued N-dimensional vectors, $\mathbf{x}$ and $\mathbf{b}$, we define the projection of $\mathbf{x}$ onto $\mathbf{b}$, denoted by $\mathbf{P_b(x)}$, as follows:

$$\mathbf{P_b(x)} = \frac{\langle \mathbf{x},\mathbf{b} \rangle}{\|\mathbf{b}\|^2} \mathbf{b} \qquad (4.103)$$

which is a new N-dimensional vector. The numerator denotes the inner product of two vectors, which is a scalar quantity:

$$\langle \mathbf{x},\mathbf{b} \rangle = \sum_{i=1}^{N} x_i b_i^* = \mathbf{x}^T \mathbf{b}^* \qquad (4.104)$$

The denominator is the squared value of the norm of a vector:

$$\|\mathbf{x}\| = \langle \mathbf{x},\mathbf{x}\rangle^{1/2} \qquad (4.105)$$

The projection of **x** onto **b** is a vector in the direction of **b** with a length equal to that of the component of **x**, which lies in the **b** direction. If **x** and **b** are orthogonal, the projection $\mathbf{P}_b(\mathbf{x})$ is obviously equal to zero. Now calculate the projection of **x** onto the direction $\mathbf{b}^\perp$ orthogonal to **b**. This projection, indicated by $\mathbf{P}_b^\perp(\mathbf{x})$, is a vector orthogonal to $\mathbf{P}_b$ and is expressed as follows:

$$\mathbf{P}_b^\perp(\mathbf{x}) = \mathbf{x} - \mathbf{P}_b(\mathbf{x}) \qquad (4.106)$$

Having defined the projections of **x** onto **b** and onto $\mathbf{b}^\perp$, to introduce the HT of a vector **x** with respect to another vector **b** is now possible. The transformation, denoted by the

$$\mathbf{Q}_b(\mathbf{x}) = \mathbf{P}_b^\perp(\mathbf{x}) - \mathbf{P}_b(\mathbf{x}) \qquad (4.107)$$

is again an N-dimensional vector. By resorting to a geometric illustration of this transformation (see Figure 4.20 for $N = 2$ and real-valued vectors), we easily note that $\mathbf{Q}_b(\mathbf{x})$ has the same length of **x** and appears as the reflection of **x** about the line perpendicular to **b**. Because of this property, the HT is also named the *Householder reflection*. Another relevant property of the HT is the invariance of the inner product (Steinhardt, 1988):

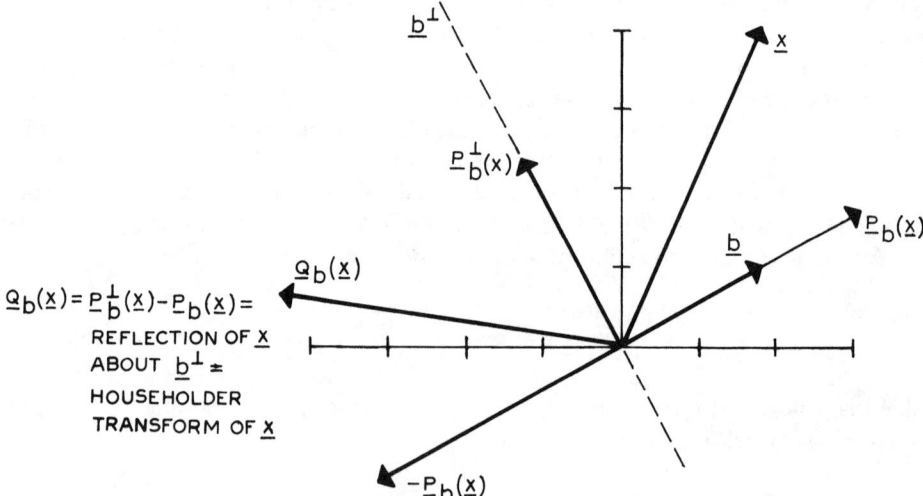

Figure 4.20 The Householder transformation of the vector **x** with respect to **b**. (From Steinhardt, IEEE ASSP Magazine, Vol. 5, No. 3, July 1988, pp. 4–12, ©1988 IEEE.)

$$\langle \mathbf{Q_b(x)}, \mathbf{Q_b(y)} \rangle = \langle \mathbf{x,y} \rangle \tag{4.108}$$

Because the terms of a covariance matrix are obtained as inner products of data vectors, this equation establishes that the HT does not modify the covariance matrix of the original data. As already indicated, this property is fundamental for the application of the HT to the calculation of the SLC optimum weights.

An alternative way to write the HT of a vector $\mathbf{x}$ is

$$\mathbf{Q_b(x)} = \mathbf{P_b^\perp(x)} - \mathbf{P_b(x)}$$

$$= \mathbf{x} - 2\mathbf{P_b(x)} = \mathbf{x} - 2\frac{\mathbf{x^T b^*}}{\|\mathbf{b}\|^2}\mathbf{b}$$

$$= \mathbf{x} - 2\frac{\mathbf{bb^H}}{\mathbf{b^H b}}\mathbf{x} = \mathbf{Q_b x} \tag{4.109}$$

where the N-dimensional square matrix $\mathbf{Q_b}$, named the *Householder matrix*, is

$$\mathbf{Q_b} = \mathbf{I} - 2\frac{\mathbf{bb^H}}{\mathbf{b^H b}} \tag{4.110a}$$

The next point to consider is to show how to select $\mathbf{b}$ given $\mathbf{x}$ to obtain a sparse reflection. Thus, we want to transform a vector $\mathbf{x}$ into a new vector having the same energy, but all concentrated in one element of the new vector. Specifically, we wish to concentrate the vector energy in its kth component. In geometric terms, this means that we have to reflect a vector $\mathbf{x}$ along the kth coordinate of the space to which $\mathbf{x}$ belongs. We can easily show, with the help of Figure 4.21, that, if $\mathbf{b}$ is chosen to satisfy:

$$\mathbf{b} = \mathbf{x} + \|\mathbf{x}\|\,\mathbf{E}_K \tag{4.111}$$

we obtain a Householder transformation that reflects the vector $\mathbf{x}$ into the kth coordinate axis of the space. $\mathbf{E}_K$ is the vector of unit norm that lies on the kth coordinate axis. In matrix notation, the vector $\mathbf{b}$, which compresses all the energy of $\mathbf{x}$ into its kth element, is

$$\mathbf{b} = \mathbf{x} + (\mathbf{x^T x^*})^{1/2}\,\mathbf{E}_K$$

$$\mathbf{E}_K^T = [\underbrace{0,\ldots,1}_{K},0,\ldots,0] \tag{4.112}$$

which, when substituted into the mathematical expression of the Householder matrix, gives

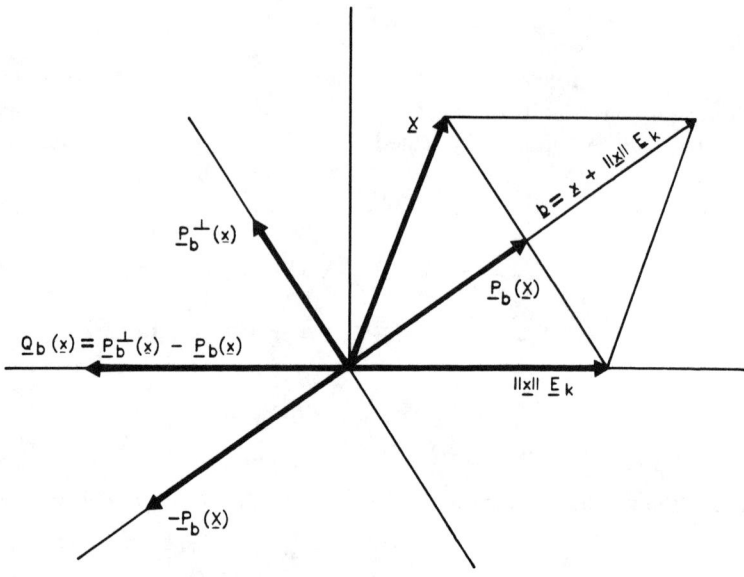

Figure 4.21 Example of the Householder transform that yields sparse reflection for $N = 2$ and real-valued data. (From Steinhardt, IEEE ASSP Magazine, Vol. 5, No. 3, July 1988, pp. 4–12, ©1988 IEEE.)

$$\mathbf{Q_b} = \mathbf{I} - 2\frac{(\mathbf{x} + a\mathbf{E}_K)(\mathbf{x}^H + a\mathbf{E}_K^H)}{(\mathbf{x}^H + a\mathbf{E}_K^H)(\mathbf{x} + a\mathbf{E}_K)} \quad (4.110b)$$

where a represents the norm of $\mathbf{x}$. Calculating the Householder transform of $\mathbf{x}$, we obtain

$$\begin{aligned}
\mathbf{Q_b}\mathbf{x} &= \mathbf{x} - 2\frac{(\mathbf{x} + a\mathbf{E}_K)(\mathbf{x}^H + a\mathbf{E}_K^H)}{(\mathbf{x}^H + a\mathbf{E}_K^H)(\mathbf{x} + a\mathbf{E}_K)}\mathbf{x} \\
&= \mathbf{x} - 2(\mathbf{x} + a\mathbf{E}_K)\frac{\mathbf{x}^H(\mathbf{x} + a\mathbf{E}_K)}{(\mathbf{x}^H + a\mathbf{E}_K^H)(\mathbf{x} + a\mathbf{E}_K)} \\
&= \mathbf{x} - 2(\mathbf{x} + a\mathbf{E}_K)\frac{\mathbf{x}^H\mathbf{x} + a\mathbf{x}^H\mathbf{E}_K}{2\mathbf{x}^H\mathbf{x} + a\mathbf{x}^H\mathbf{E}_K + a\mathbf{E}_K^H\mathbf{x}} \\
&= \mathbf{x} - (\mathbf{x} + a\mathbf{E}_K) = -a\mathbf{E}_K = -(\mathbf{x}^T\mathbf{x}^*)^{1/2}\mathbf{E}_K \quad (4.113)
\end{aligned}$$

We have demonstrated that indeed the aforementioned choice of $\mathbf{b}$ (4.112) compresses the vector energy in its Kth component.

The next step is to apply the HT to the triangularization of a matrix. This can be done by considering the transpose **U** of the first row of the matrix as a column vector to which we apply the HT. All the energy of the row of data is compressed in the first element by selecting $\mathbf{E}_1$ as follows:

$$\mathbf{E}_1^T = [1, 0, \ldots, 0] \qquad (4.114)$$

The vector **b** is defined as:

$$\mathbf{b} = \mathbf{U} + a\mathbf{E}_1 \qquad (4.115)$$

where a is the norm of the vector **U** (i.e., the norm of the first row of the data matrix). Because the original data to transform are along a row, instead of premultiplying a column vector by the Householder matrix, we can, equivalently, postmultiply a row of data by the same Householder matrix. In fact, this operation again compresses the row data energy in the first element:

$$\mathbf{U}^H\mathbf{Q} = -a\mathbf{E}_1^H = (-a, 0, \ldots, 0) \qquad (4.116)$$

Now we have the mathematical tools to solve the batch least-squares problem by means of the HT. As described in Subsection 4.2.3.3, we need to transform the matrix **X** of voltage data received by the antennas of the SLC into one which has a lower triangular form, but the same covariance matrix as the original data matrix. This is shown pictorially in Figure 4.22 for a data matrix of $N = 3$ channels (two auxiliaries and one main) and $n = 4$ data snapshots. The procedure starts by postmultiplying the original data matrix **X** by the Householder matrix $\mathbf{Q}_1$, which operates on the first row of **X**, thus compressing the energy of the first row in the first data element. Subsequently, the transformed data matrix **X'** is postmultiplied by another Householder matrix, $\mathbf{Q}_2$, which operates to compress the energy of the last $n - 1$ elements of the second row of **X'** in the second data element. Proceeding in this way, the original ($N \times n$) data matrix is transformed into an $N \times N$ lower triangular matrix together with the last $(n - N)$ columns that are exclusively zero. These columns of zeros do not contribute to the correlation matrix, and they may be dropped. Finally, the optimum set of SLC weights is found by the ($N \times N$) lower triangular matrix **C** as follows:

$$\dot{\mathbf{W}} = \mathbf{C}^{-H}\mathbf{C}^{-1}\mathbf{E}_1 \qquad (4.117)$$

where the vector $\mathbf{E}_1$ (from (4.114)) is due to the limitation of having a unit weight in the main radar channel. A schematic block diagram of the implementation of the SLC system by means of the HT is shown in Figure 4.23. After recording, the data

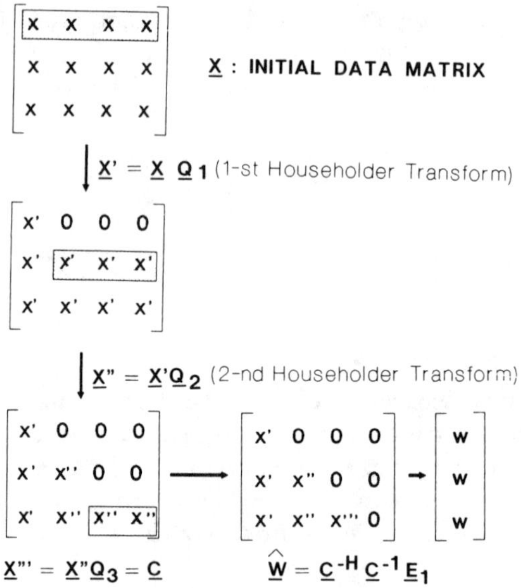

$$\begin{bmatrix} x & x & x & x \\ x & x & x & x \\ x & x & x & x \end{bmatrix}$$ $\underline{X}$: INITIAL DATA MATRIX

$\downarrow \underline{X}' = \underline{X}\,\underline{Q}_1$ (1-st Householder Transform)

$$\begin{bmatrix} x' & 0 & 0 & 0 \\ x' & x' & x' & x' \\ x' & x' & x' & x' \end{bmatrix}$$

$\downarrow \underline{X}'' = \underline{X}'\underline{Q}_2$ (2-nd Householder Transform)

$$\begin{bmatrix} x' & 0 & 0 & 0 \\ x' & x'' & 0 & 0 \\ x' & x'' & x'' & x'' \end{bmatrix} \longrightarrow \begin{bmatrix} x' & 0 & 0 & 0 \\ x' & x'' & 0 & 0 \\ x' & x'' & x''' & 0 \end{bmatrix} \to \begin{bmatrix} w \\ w \\ w \end{bmatrix}$$

$\underline{X}''' = \underline{X}''\underline{Q}_3 = \underline{C} \qquad \hat{\underline{W}} = \underline{C}^{-H}\underline{C}^{-1}\underline{E}_1$

Figure 4.22 Householder least-squares processing for $N = 3$ and $n = 4$. (Adapted from Steinhardt, IEEE ASSP Magazine, Vol. 5, No. 3, July 1988, pp. 4–12, ©1988 IEEE.)

matrix $\mathbf{X}(N,n)$ is transformed into a new matrix $\mathbf{C}(N, N)$ by means of the Householder matrix $\mathbf{Q}$. At this point, the SLC equation:

$$\mathbf{X}\mathbf{X}^H\hat{\mathbf{W}} = \mathbf{E}_1 \tag{4.118}$$

is rewritten as

$$\mathbf{C}\mathbf{C}^H\hat{\mathbf{W}} = \mathbf{E}_1 \tag{4.119}$$

where $\mathbf{C}$ is the lower triangular Cholesky factor of the previous equation. The weights $\hat{\mathbf{W}}$ can be obtained by two backsubstitution steps as follows. Indicating by a new matrix $\mathbf{T}$ the product $\mathbf{C}^H\hat{\mathbf{W}}$, (4.119) becomes

$$\mathbf{CT} = \mathbf{E}_1 \tag{4.120}$$

which can be solved in $\mathbf{T}$ by backsubstitution. Subsequently, the additional equation:

$$\mathbf{C}^H\hat{\mathbf{W}} = \mathbf{T} \tag{4.121}$$

is solved by backsubstitution in $\hat{\mathbf{W}}$.

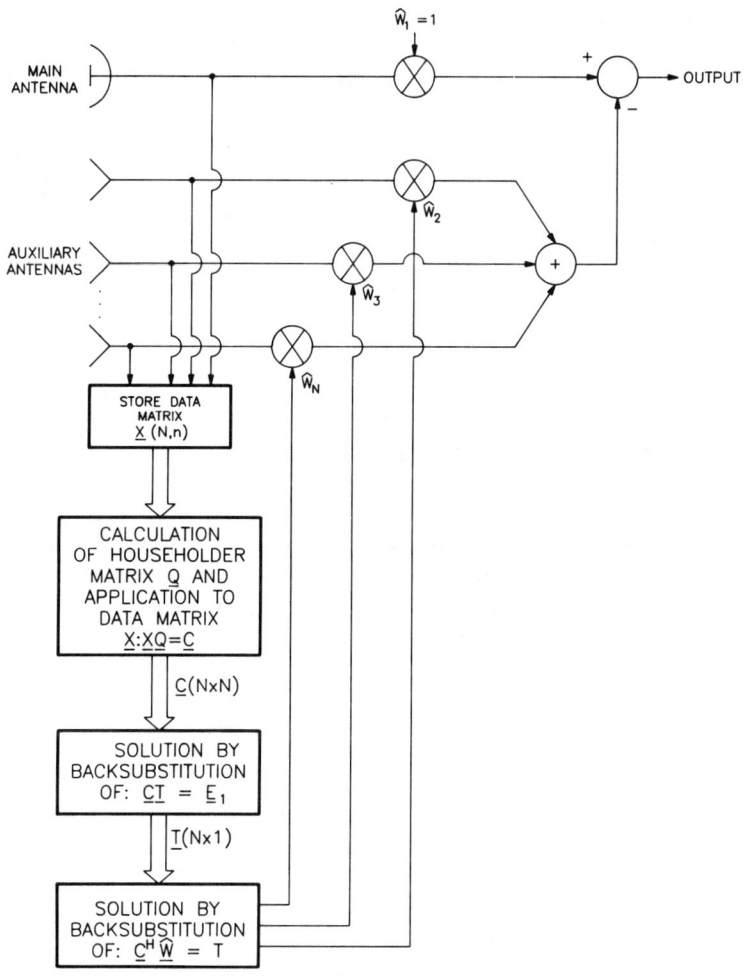

Figure 4.23 Implementation of SLC by Householder transformation.

In summary, the weight vector $\hat{\mathbf{W}}$ has been obtained without forming and inverting any covariance matrix. Finally, we briefly mention the recursive least-squares problem. Recursive processing is needed because the interfering environment is changing with time and the optimum set of weights needs to be updated. As new data are appended to the data matrix and old data are deleted, an interesting approach would be to reuse old computations without starting again. Roughly speaking, if $\mathbf{C}$ is the lower triangular matrix, $\mathbf{Y}$ is the matrix with new data to append, and $\mathbf{Z}$ is the matrix with old data to delete, we wish to compute the new lower triangular matrix, avoiding the explicit calculation of the covariance matrix $CC^H + YY^H - ZZ^H$. We cannot use the Householder matrices here because they are orthonormal and hence preserve only positive sums, while, in the above expression, there is a difference term. When data deletion arises, orthonormal transformations such as the HT do not preserve the covariance matrix. The hyperbolic HT may be the solution to this problem. Details on this quite advanced method can be found in Steinhardt (1988) and Rader and Steinhardt (1986) with a rigorous analysis of the numerical issues and the number of operations to perform.

4.3 PERFORMANCE EVALUATION

This section illustrates how to evaluate the performance of an SLC system under several interesting conditions. First, attention is devoted to the practical causes that limit the SLC's nulling capabilities, which should be taken into account and, when possible, accurately compensated. Section 4.3.1 lists the phenomena that keep the correlation coefficients between the main and auxiliary signals lower than unity, thus causing a corresponding reduction of the achievable CR values. Afterwards, analytical insight into some of the more relevant limitation sources is given in Section 4.3.2. CR values are provided for several cases of practical importance. Methods to compensate for the above-mentioned limitations are also illustrated.

4.3.1 Practical Limitations of SLC Nulling Capabilities

In discussing cancellation performance, an important point to remember is that the SLC cancels an interfering signal by applying a complex weight to it, which is incident on each auxiliary antenna. The auxiliary array then sums the weighted signals and subtracts the result from the radar signal. Thus, if the signals on the various antennas differ from one another in some manner other than a complex multiplicative constant, the SLC will not be able to null the interfering signals perfectly. In more detail, the cancellation performance of the SLC depends on the correlation coefficients of the signal received through the main antenna and those received through the auxiliary antennas. High values of correlation coefficient correspond to remarkable SLC performance. Several factors contribute to keeping the correlation coefficients lower than unity (Carlson *et al.*, 1990; Wardrop, 1983; Fielding *et al.*, 1977; Monzingo and Miller, 1980; Ries and Krucker, 1976; Farina, 1977; Farina, 1987b). Many of these occur in different sections of the SLC system: radio frequency, intermediate frequency, analog-to-digital conversion, and baseband. If each section of this chain is not identical among their main and auxiliary channels, mismatching errors cause the canceler performance to degrade. The purpose of this section is to give an overview of these limitations, and an analysis of their effects on the achievable CR is provided in Section 4.3.2.

4.3.1.1 Path Matching

Cancellation occurs at a summing point somewhere in the processor. To be effective, all signals emanating from the same interference source must arrive at the summing point after undergoing the same delay and not following paths that treat the signals differently. In particular, the frequency responses of the various paths over which the interference has travelled must be similar. A mismatching of frequency responses of the receivers on each channel can cause signal decorrelation and limit the cancellation performance. When the frequency responses do not match exactly, choosing a set of complex weights that will perfectly cancel any interference becomes impossible.

Various factors contribute to poor path matching, the main are listed below.

Antenna Separation

The auxiliary antennas are deliberately separated to have different degrees of freedom to deal with the multiple-jammer case. However, the separation produces different phases of the signals induced by the same jammer in the main and auxiliary channels. The relative phases, which are frequency-dependent, are due to the time difference between the jamming signals, spatially sampled and captured by the main and auxiliary aerials. This phase mismatch results in a degradation of the CR value.

Sidelobe Frequency Sensitivity

The gain in the sidelobes of a directional antenna is frequency-dependent and will generally vary significantly within the frequency band of the radar and from sidelobe to sidelobe. This is usually a limiting factor in the performance of a cancellation system. Another limitation strictly related to this is that the sidelobe structure of the main antenna is different with respect to the patterns of the auxiliaries, thus resulting in different spatial filtering of the same jamming source.

Propagation Paths

The jamming signals impinging on a receiving system can be categorized as coming from either primary or secondary sources. A primary source is a generator of energy, and a secondary source is due to the scattering of energy from a primary source by either a point object, such as a chimney, or from a distributed object, such as a grove of trees. The signals from a secondary source will be correlated with those from its associated primary source, the correlation being dependent on the difference in direct and indirect path lengths from the primary source to the receiver (Wardrop, 1983). The path-length difference is equivalent to a differential delay, and so the correlation between primary and secondary sources can be obtained from the autocorrelation function of the primary source, after filtering by the radar receiver, evaluated at the differential delay (Farina, 1977).

In the initial design of an SLC system, a useful rule of thumb is to estimate the number of jammers (i.e., primary and secondary sources) and to provide as many auxiliaries as the number of jammers. This approach, however, assumes that the jammers are uncorrelated. A better estimate of the required number of auxiliaries will be given by the sum of the primary sources and those secondary sources which are effectively uncorrelated, the allowable decorrelation being a function of the desired jammer suppression and the levels of the indirect signals. In practice, the situation will be further complicated by polarization changes that occur in the scattering process, which will enhance decorrelation. With the current trend toward mobile systems, the designer can expect to have little knowledge about the actual multipath of an operational site. Consequently, the choice of the number of auxiliaries will probably be influenced more by economic factors than by the scenario (Wardrop, 1983).

Paths Internal to the Radar System

These paths include all elements from the antennas at the cancellation point. These paths, of course, may be quite long with stretches of waveguide along the antenna mast. An unavoidable consequence is that amplitude and phase mismatch exists among the global transfer functions of the receiving channels.

Variations in the insertion phase and gain of the main and auxiliary channels can be compensated, provided that suitable phase and amplitude adjusters are available with a complementary measurement system. Unfortunately, the phases and gains are normally equalized at a center frequency; the variations in phase and gain with frequency remain. These variations can arise owing to filter mistuning in the receivers, variations in cable length, and variations of frequency matching. The effects of amplitude and phase mismatch are analyzed in Section 4.3.2. The effect of such mismatch, of course, can be minimized with additional hardware, as shown in Subsection 4.3.2.4.

Crosstalk between Channels

Crosstalk between channels is due to inadequate screening between them or coupling via power rails and local oscillator distribution networks. Undesired signal reflections from cable and device interconnections are reflected back from one channel and enter another channel by mutual coupling. Therefore, there may be many very low-level versions of an interfering signal present at the antenna inputs and these can have a wide variety of time delays (Carlson *et al.*, 1990, p. 302). If the crosstalk path lengths are small compared with the reciprocal of the signal bandwidth, the error can be considered equivalent to additional amplitude and phase errors in the channel, albeit errors which vary with the angular position of the source due to the phasing of the crosstalk components. As this error will vary pseudorandomly, a reasonable assumption is that the error voltage will be N times that of the mean crosstalk between two channels, where N is the number of the auxiliary antennas. Consequently, the acceptable crosstalk can be determined by the acceptable gain and phase errors in a channel. This source of performance degradation can be minimized by care at the design stage.

Polarization Effects

The polarization of the jammer will not likely be matched to that of the SLC antennas. For example, the jammer may be radiating uncorrelated vertically and horizontally polarized components, whereas the radar antenna is designed to receive horizontally polarized signals. Nevertheless, the radar antenna will probably give an attenuated output for vertically polarized signals as well. If the main and auxiliary channels do not have the same polarization behavior, a single auxiliary cannot simultaneously null both polarization components of interference. The application of SLC to a conventional reflector antenna suffers in this respect, as matching the polarization performance of the main and auxiliary antennas is difficult if they are of different construction. In an adaptive phased array, however, the main beam is formed from the same type of elements as used for the auxiliaries. In this case, the polarization performance is more likely to be matched. One technique to cope with randomly polarized jammers is to provide more degrees of freedom (i.e., more auxiliaries). The suggestion that the number of auxiliaries be doubled to cope with the possibility that all jammers will radiate both polarizations is not normally welcomed by the designer (Wardrop, 1983).

A detailed analysis of the effects of element cross-polarization in adaptive antennas is reported in Brown (1981). As already mentioned, this effect is particularly important when an interfering signal is dual-polarized with the two polarizations orthogonal and temporally uncorrelated. In this case, the jammer will appear as two sources and a reduction of CR is experienced unless more degrees of freedom are available.

4.3.1.2 Nonzero Bandwith of the Receiving Channels

An important mechanism by which the received interference signals on the various antennas of the SLC can become uncorrelated is the so-called *signal dispersion*. This occurs when a jammer with nonzero bandwidth arrives at the array from some angle other than the broadside direction. The antenna-to-antenna propagation time delay between each pair of antennas produces a phase shift proportional to the signal frequency. Therefore, the complex weights required to null a jammer at one frequency will be slightly different from those required to null a jammer at another, nearby frequency. If the interfering signal has a significant bandwidth and the jammer arrives from some angle other than broadside, the array requires several, closely spaced nulls (i.e., the use of several degrees of freedom) to null all frequency components simultaneously (Carlson et al., 1990). In fact, a wideband jammer can be perceived as a spatial cluster of narrowband jammers. The equivalent bandwidth Δf of a cluster of jammers expressed as a fraction of the center frequency f_c is (Haykin, 1980, pp. 136–149):

$$\frac{\Delta f}{f_c} = \frac{\Delta(\sin\theta)}{\sin\theta_c} \qquad (4.122)$$

where $\Delta(\sin\theta)$ is the spread of the cluster in the $\sin\theta$ space and $\sin\theta_c$ is the center of cluster. As a consequence, wideband jamming received through a wideband radar receiver requires more auxiliaries; otherwise, the radar suffers severe reductions of cancellation.

In a practical situation involving power minimization in the presence of thermal noise, this would lead to an excessive amount of thermal noise at the array output. Therefore, an appropriate compromise is necessary between nulling the jammers and minimizing the thermal-noise contributions due to the auxiliary array. An alternative approach to achieve broadband nulls is to limit the design to one spatial degree of freedom per jammer, and to provide the required extra degrees of freedom in the form of variable frequency-domain filters associated with the element's output. Such filters can, in principle, be constructed to have varying amplitude and phase characteristics over the bandwidth of interest (Hargrave, 1990). These filters can be implemented by using tapped delay-line techniques, as discussed in Subsection 4.3.2.4.

4.3.1.3 Effect of Target Signal

Normally, a very small SNR is assumed in the auxiliary elements compared to the small SNR in the high-gain main beam. This results in a negligible steering of the auxiliary array toward the target direction and a slight degradation of the jammer cancellation and of the main-beam pattern. However, radars with a large number of auxiliary channels face a reduction of CR. In fact, the target signal reaching the secondary antennas causes a degradation of the main-lobe pattern. In some instances, a detrimental amplitude modulation of the main-lobe signals is experienced. A numerical analysis of this effect is reported in Ries and Krucker (1976). The main-lobe signals, which are the desired signals for the main channel, must be removed from the secondary antenna signals before the null-steering apparatus operates on them. Constrained adaptive processors have been developed for this case, placing a "null" on the auxiliary antennas in the main-beam direction (Kennedy, 1978).

4.3.1.4 Scenario Dependence

As we have learned, the adaptive weights can be calculated by iterative closed-loop techniques. Difficulties arise if the jammer scenario is such that the interference matrix becomes ill conditioned, as then the closed-loop approach requires many iterations before reaching the steady-state.

One scenario, which seems to produce ill conditioning in the matrix, occurs when the auxiliary elements "see" similar phase fronts from several jammers. This arises if two jammers are separated in azimuth by a fraction of a beamwidth, or the jammers are in different grating lobes of the pattern formed by the auxiliaries. The latter case suggests that the selection of the auxiliaries from the elements of a phased-array antenna is of particular importance (Wardrop, 1983; White, 1983).

An excellent analysis of the convergence properties of an SLC subject to two jammers sufficiently close in their angular positions and with different power levels has been done in Mesiwala and Widrow (1979). When two or more jammers with widely different power levels are involved, the time required to null all jammers becomes very long in comparison with the time required to null a single jammer. In particular, a high-power jammer is nulled quickly, but a lower power jammer is essentially unaffected. Briefly, the paper shows the behavior of $\lambda_{MAX}/\lambda_{MIN}$ (4.26c), which is a measure of the speed of convergence of the SLC, as a function of the configuration of the auxiliary array with respect to the main antenna. A number of alternative configurations are examined and compared with the minimal SLC configuration made of two auxiliaries aligned with the main antenna. The alternative configurations comprise more than two auxiliary antennas set up in linear, triangular, hexagonal, and two-concentric-triangles configurations. The addition and judicious placement of a few extra auxiliaries speeds the adaptive nulling process, even in a difficult operational condition such as that constituted by two closely spaced jammers of disparate power levels. The triangular and two-concentric-triangles configurations give better performance than the linear configuration. The two-concentric-triangles configuration has elements spaced at a longer distance from the main antenna, experiencing a larger time difference in the received signals. Intuitively, a large separation between the auxiliary and main antennas (i.e., an overall large aperture) may tend to magnify the apparent angular separation between closely spaced jammers. This magnification tends to reduce the disparity between the eigenvalues of the adaptive null steerer. However, too large a separation between the antenna phase centers is not recommended due to the decorrelation effects on the received jamming signals. A trade-off should be sought among the convergence ratio and the steady-state cancellation value. This is an ever present issue, which is considered in Subsection 4.3.2.5.

Another paper that addresses the topics of auxiliary antenna position is Morooka and Kawabata (1980). The investigations in the above-mentioned papers provide powerful tools to configure the auxiliaries around the main antenna.

4.3.1.5 Clutter

When heavy clutter is present, the SLC will attempt to minimize the power in the adapted output without differentiating between clutter and other forms of interference. In other words, the adapted pattern will contain nulls steered in the direction of the main beam of the antenna. Additionally, the large clutter return, which is contained in the radar channel, is detrimental to the convergence of the estimation procedure of the auxiliary weights of the SLC. As has been shown (Monzingo and Miller, 1980, p. 304):

$$\frac{K_{3dB}}{N_{aux}} = 2 + \left(\frac{P_c}{P_N}\right) \qquad (4.123)$$

where

1. K is the necessary number of independent samples per channel so that the output residue is within 3 dB of the minimum;
2. N_{aux} is the number of the auxiliary channels; and
3. P_c/P_N is the input clutter-to-noise power ratio in the main channel after beamforming.

As an application example of this equation, consider the case of $P_c/P_N = 30$ dB. This means that the SLC has a convergence that requires at least 1000 independent samples (i.e., range cells) per channel. This number is excessive for the majority of practical applications.

A number of techniques may be used to avoid the problems raised by the presence of clutter. A first technique is particularly suitable for low-PRF radar. The influence of the close-in clutter returns on the adaptive weights can be simply avoided by adaptation in only the clutter-free ranges (e.g., at the end of each PRI) and then subsequent freezing of the weights in the remaining part of the range coverage. This technique does not apply to radars operating in high-PRF range-ambiguous modes with significant clutter in all the range cells. In fact, there is a problem in obtaining clutter-free samples for weight calculation. A second technique particularly recommended in this case has been described in Old (1984a; 1984b). The clutter is removed by means of an MTI filter upstream of the SLC. A third technique is mentioned in Fielding *et al.* (1977). Where the interference spectrum is broader than the radar signal bandwidth, the interference can be received outside the clutter-free band and used to control the canceler inside the band. Other techniques to solve clutter problems are mentioned in the following two patents: Goggins (1976) and Soule and Jureller (1976).

In the remaining part of this section, we analyze in some detail the combination of SLC and MTI processing to contrast the simultaneous presence of clutter and jammer (A. Huizing, TNO Defense Research, The Netherlands, Private Communication). We assume that a narrowband noise jammer is present in the radar sidelobes, while the clutter is received by the antenna main beam. Clutter and jammer are both received by the auxiliary antenna. A sidelobe canceler with one auxiliary antenna is used to cancel the jammer, while a two-pulse MTI is adopted to cancel the clutter. The two processing configurations of Figure 4.24(a) and (b) are compared in terms of cancellation ratio. The purpose of the foregoing discussion is to demonstrate the superior performance of the first processing scheme.

To proceed with the calculations, we need to write the mathematical model of the main and auxiliary signals. For the configuration of Figure 4.24(a), we indicate by $V_M(1)$ and $V_M(2)$ the two samples, separated by a pulse repetition time, T, received by the main antenna. Similarly, $V_A(1)$ and $V_A(2)$ are the two samples received by the auxiliary antenna. The following expressions are easily found:

$$V_M(1) = G_M c(1) + G_{SL} j(1) \exp(i\alpha) + n_M(1)$$
$$V_M(2) = G_M c(2) + G_{SL} j(2) \exp(i\alpha) + n_M(2)$$
$$V_A(1) = G_A c(1) + G_A j(1) + n_A(1)$$
$$V_A(2) = G_A c(2) + G_A j(2) + n_A(2) \qquad (4.124)$$

where G_M and G_{SL} are the main-lobe and sidelobe voltage gain of the radar antenna, while G_A is the voltage gain of the auxiliary antenna. Additionally, $c(1)$ and $c(2)$ are the two samples, T seconds apart, of the clutter echoes, while $j(1)$ and $j(2)$ are the two samples of the jammer signal; $n_M(1)$, $n_M(2)$, $n_A(1)$, $n_A(2)$ are the samples of thermal noise in the main and auxiliary channels. The phase value $\exp(i\alpha)$ accounts for the time delay between the jammer signals received by the main and auxiliary antennas. The clutter samples received at the same time instants by the radar and the auxiliary antennas are assumed to have correlation equal to one. The clutter, jammer, and thermal-noise signals are modeled as Gaussian distributed, independent, random variables with zero mean. Other relevant parameters are as follows:

$$E\{j(1)j^*(2)\} = 0$$
$$E\{n_M(1)n_M^*(2)\} = E\{n_A(1)n_A^*(2)\} = 0$$
$$E\{|n_A|^2\} = E\{|n_M|^2\} = P_N$$
$$E\{|j(1)|^2\} = E\{|j(2)|^2\} = P_J$$
$$E\{|c(1)|^2\} = E\{|c(2)|^2\} = P_C$$
$$E\{c(1)c^*(2)\} = \rho P_C \qquad (4.125)$$

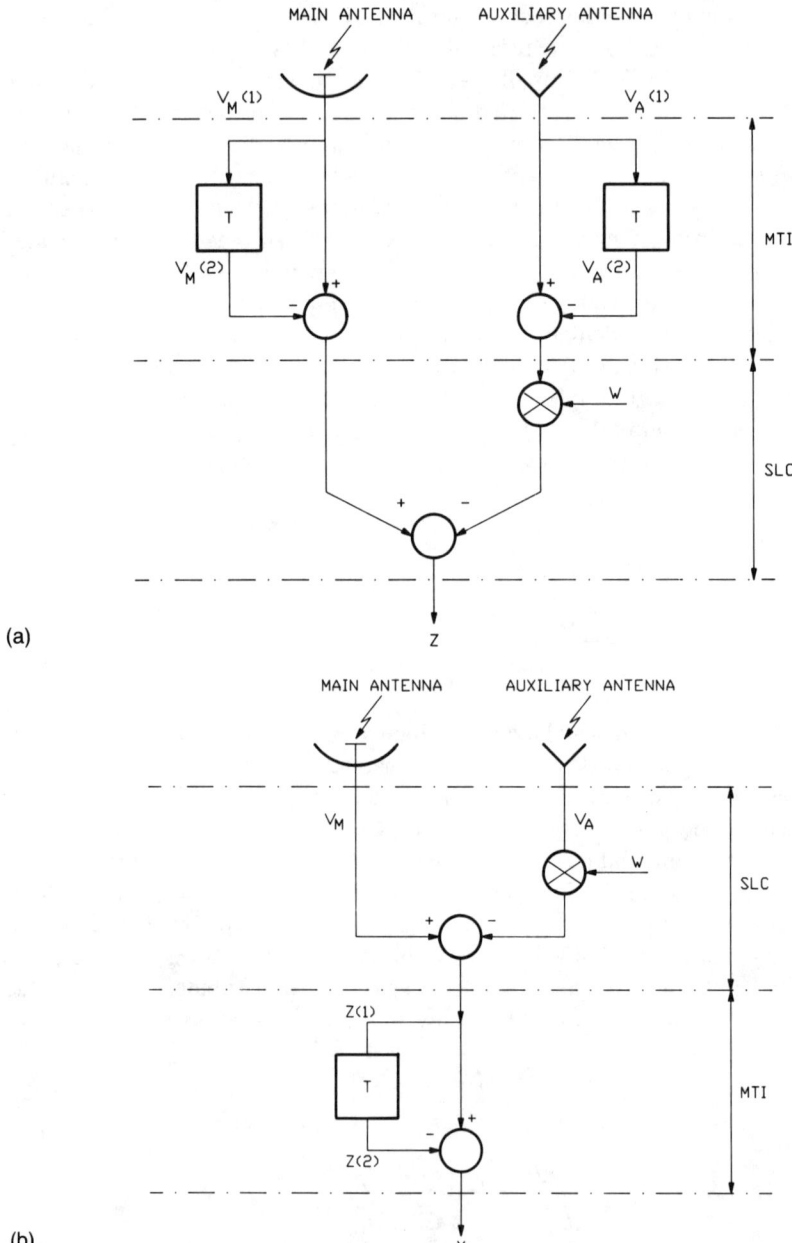

Figure 4.24 (a) Processing configuration with MTI upstream of SLC; (b) processing configuration with SLC upstream of MTI.

where P_J, P_C, and P_N are the power values of the jammer, clutter, and thermal noise, respectively. The parameter ρ is the one-lag temporal correlation coefficient of the clutter.

The following mathematical relations describe the behavior of the scheme of Figure 4.24(a):

$$Z = [V_M(1) - V_M(2)] - W[V_A(1) - V_A(2)]$$
$$W = \frac{E\{[V_M(1) - V_M(2)][V_A(1) - V_A(2)]^*\}}{E\{|V_A(1) - V_A(2)|^2\}} \quad (4.126)$$

where Z is the residue signal of the processor and W is the weight of the SLC system. We note that the SLC weight is calculated by using the clutter-free signals $[V_M(1) - V_M(2)]$ and $[V_A(1) - V_A(2)]$. The system performance is described by the residue power value $E\{|Z|^2\}$, which is a function of the system parameters: the clutter-to-noise ratio, CNR $= P_C/P_N$; the jammer-to-noise ratio, JNR $= P_J/P_N$; the one-lag clutter correlation coefficient, ρ, the gain values of the two antennas, G_M, G_{SL}, G_A; and the parameter α.

The $E\{|Z|^2\}$ values have been plotted *versus* the CNR values, having as parameter the clutter correlation coefficient, as depicted in Figure 4.25. Note that when CNR is very low, the residue power is also very low. However, when the clutter is not negligible, the MTI is unable to suppress the clutter completely and the residue power increases because the SLC weight has a suboptimum value. The reduction of performance is limited if the correlation coefficient ρ of clutter is close to unity.

The processing configuration of Figure 4.24(b) is described by the following equations:

$$Y = Z(1) - Z(2)$$
$$Z(1) = V_M(1) - WV_A(1)$$
$$Z(2) = V_M(2) - WV_A(2)$$
$$W = \frac{E\{V_M(1)V_A^*(1)\}}{E\{|V_A(1)|^2\}} \equiv \frac{E\{V_M(2)V_A^*(2)\}}{E\{|V_A(2)|^2\}} \quad (4.127)$$

Now, we note that the SLC weight, W, is estimated by using jammer and clutter samples mixed together. Also, in this case, we wish to calculate the residue power, $E\{|Z|^2\}$, which is a function of the system parameters already listed for the previous processing configuration. A set of curves for $E\{|Z|^2\}$ is depicted in Figure 4.25, showing worse performance with respect to the set of curves for the processing configuration of Figure 4.24(a).

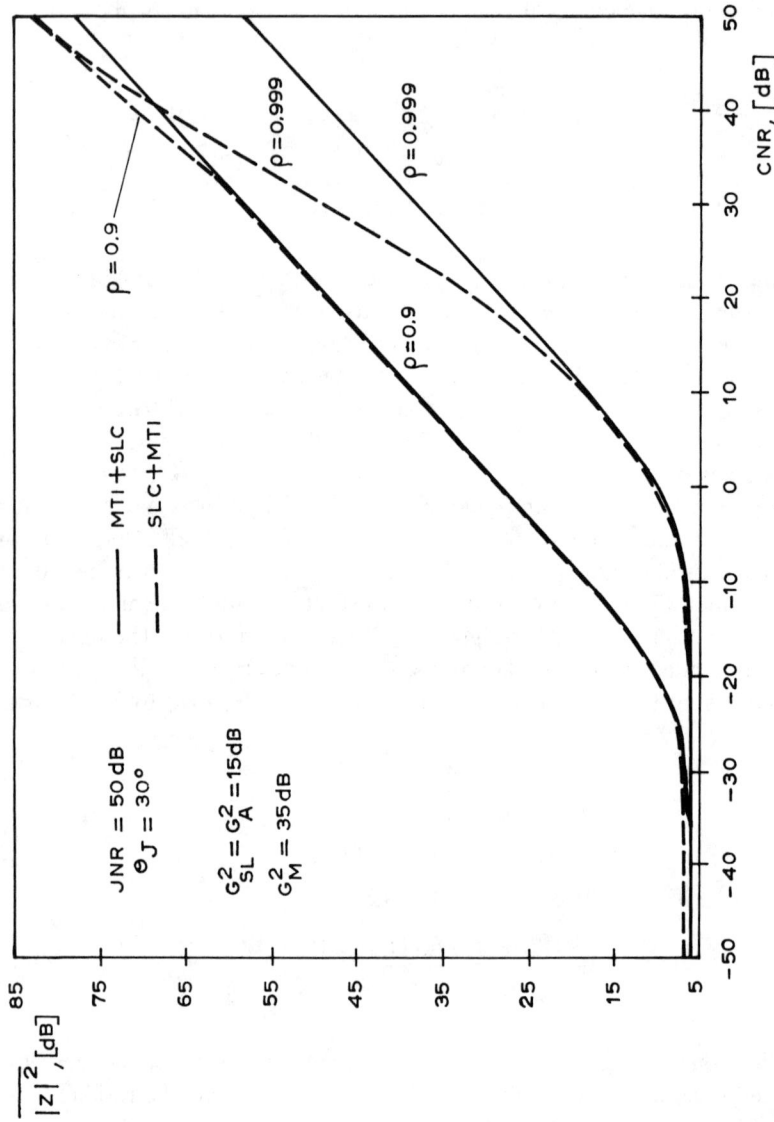

Figure 4.25 Residue power *versus* clutter-to-noise power ratio having as parameter, ρ, the one-lag clutter correlation coefficient.

As a general conclusion, we should remove the clutter from the main and auxiliary channels before sidelobe cancellation to enable an optimum jamming suppression. If the clutter is not removed by an MTI, it perturbs the weight values of the SLC and the jamming is not completely cancelled. This jammer residue at the output of the SLC cannot be removed by the MTI (in the processing configuration of Figure 4.24(b)) because the jamming is uncorrelated on a pulse-to-pulse basis.

The processing scheme of Figure 4.24(a) deserves more comment. In fact, the presence of the MTI in the main radar channel impairs the detection of targets having low radial speed. A better detection performance would be obtained by using, in lieu of a simple MTI, a more sophisticated doppler processor, such as a *moving target detector* (MTD). This choice, however, is not advisable because of the duplication of precious hardware and the need for precisely matching the two doppler processors. A more convenient architecture for the whole system is that shown in Figure 4.26. A conventional pair of MTIs is used to provide clutter-free data for the estimation of the SLC weight, W. Subsequently, the SLC processing is applied to the original main and auxiliary data. Finally, a sophisticated clutter cancellation procedure, performed on the jammer-free residue, allows detecting targets with low radial speed values.

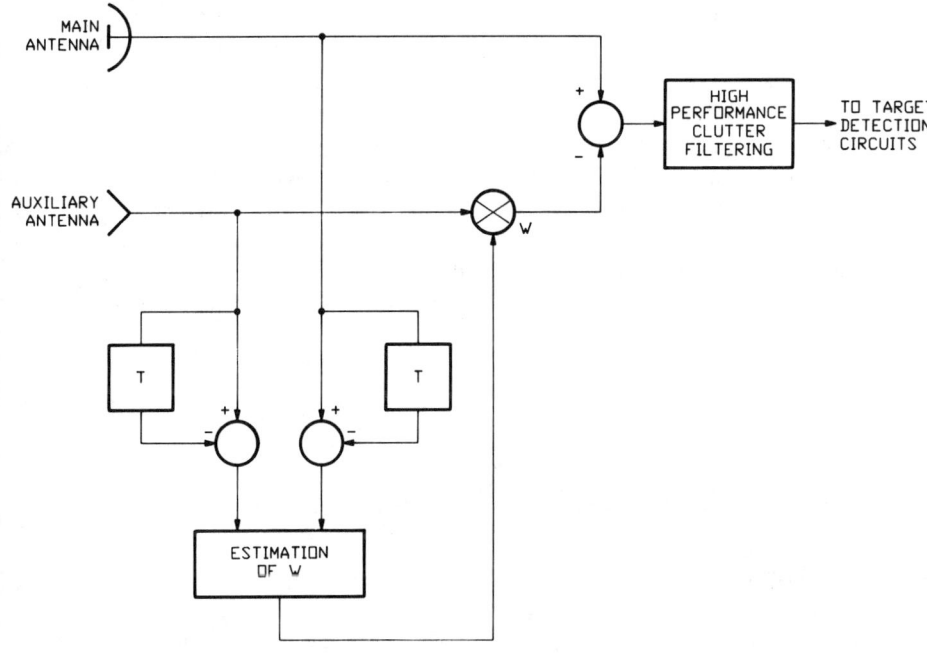

Figure 4.26 Use of MTI and high-performance clutter filtering with the SLC.

4.3.1.6 Loop Noise and Estimation Noise

In addition to the cancellation ratio, another figure of merit for the SLC is the time required to reach the steady-state to which the CR value refers.

There are three possible factors determining the required speed of response. First, the minimum requirement is that the system should act fast enough to follow the antenna movement. Second, if the radar is agile in frequency, the time for cancellation should be small compared with the number of PRIs in which the frequency is constant. Third, if the canceler is required to cancel pulsed interference (Omuro et al., 1984) it should operate within a small fraction of a PRI. This last requirement is not necessarily possible to meet in many circumstances (e.g., a canceler with that rate of response may cancel target echoes). Therefore, SLB (see Chapter 3) is employed against pulsed jammers.

In the closed-loop implementation, the *transient time* is that required for adaptation of all the loops. In the open-loop configuration, the transient time is measured by the number of time samples needed to estimate the jammer covariance matrix. This matrix is not known *a priori* because it would be equivalent to assuming that the jamming locations, effective radiated powers, and main and auxiliary voltage gains at the jammer angles were known *a priori*. The estimation procedures, either related to the closed-loop or open-loop schemes, suffer an estimation noise that is a function of the number of samples used to estimate the required parameters.

Unfortunately, the two above-mentioned figures of merit contradict each other. In the closed-loop approach, the greater the loop bandwidth, the faster is its response to a nonstationary jamming situation, but the poorer is the steady-state CR value (loop-noise phenomenon). In the open-loop approach, fewer time samples to average the estimation of jammer covariance matrix correspond to short reaction to nonstationary phenomena. This, however, is accompanied by poor estimation accuracy of the matrix, with a correspondent loss of the CR value at steady-state (estimation-noise phenomenon). These two aspects will be thoroughly considered in Subsection 4.3.2.5.

4.3.1.7 Miscellany

Signal Quantization and Processing Accuracy in Digital Implementation

The number of bits in the analog-to-digital converter determines the maximum cancellation ratio that an SLC can achieve. In addition, the ADC must have the large dynamic range necessary to handle high levels of jamming as well as clutter. An n-bit ADC divides the output of a phase detector in $2^n - 1$ discrete intervals. The number of bits, n, necessary for a given level of cancellation is approximately given by

$$CR_{dB} = 20 \log_{10} 2^n = 6n \qquad (4.128)$$

Thus, an ADC with 10 bits guarantees a maximum cancellation ratio of 60 dB (Andrews and Gerlach, 1989, p. 472).

The number of bits necessary in the digital signal processor for a required cancellation ratio is a function of the interference environment, thermal-noise level, and type of SLC algorithm adopted.

A simulation exercise has been described in Subsection 4.2.1.3 for the special case of digital implementation of the Howells-Applebaum loop. We showed that the number of bits to represent the signals, circuit parameters, and arithmetic operations affect the SLC cancellation ratio. Results of simulations and practical experiments were also cited for other implementation schemes (e.g., the Kalman filter, Subsection 4.2.2.4).

An analysis, summarized here, has evaluated the precision required in computing the SLC weights (Nitzberg, 1976). In Section 4.1, the SLC residue power $E\{|z|^2\}$ was shown to have the minimum value (e.g., P_{min}) given by (4.6), when the SLC weight vector has the optimum value (4.5b). Due to computational errors, the weight vector $\mathbf{W}$ will differ from the optimum value as follows:

$$\mathbf{W} = \hat{\mathbf{W}} + \mathbf{\Delta} \qquad (4.129)$$

Consequently, the residue power increases with respect to P_{min} as follows:

$$P_r = P_{min} + P_{add} \qquad (4.130)$$

where P_{add} takes into account the effect of the error $\mathbf{\Delta}$ of the SLC weights. Assume that each optimum weight is quantized by using $n - 1$ magnitude bits. The cancellation degradation, expressed as the ratio of P_{add} and P_{min}, is

$$\eta \triangleq \frac{P_{add}}{P_{min}} = \frac{N^2}{2^{2n-1}} R_0 \qquad (4.131)$$

where N is the number of auxiliary antennas, and R_0 is the ratio of the preadapted and postadapted interference-to-noise ratios when the optimum weights are used. The improvement factor, R, when the weights of (4.129) are used, is related to R_0 and η as follows:

$$R = \frac{R_0}{(1 + \eta)} \qquad (4.132)$$

When η is equal to one, the residue power P_r is twice the minimum value P_{min} so that the improvement factor degrades by 3 dB. As a numerical example, consider the case of $N = 8$ auxiliary antennas producing an R_0 value of 40 dB. To obtain the value $R = 37$ dB corresponding to $\eta = 1$, the weights should be quantized to 10 bits.

Finally, we wish to recall additional relevant results from the technical literature (Rader and Steinhardt, 1986; Monzingo and Miller, 1980; Ward et al., 1986) that have been quoted and discussed at length in Subsections 4.2.3.6 and 4.2.2.3. We have indicated the connection between the covariance matrix eigenvalue spread and the number of bits required to invert the matrix. This relationship (4.102a) and the parent equation (4.102b), valid for the data matrix, suggest and motivate the use of data-domain adaptive algorithms in Section 4.2.3.

Offset-Error Problem

Consider the analog implementation of the Howells-Applebaum loop of Figure 4.2. The correlator providing the multiplication between the output signal and auxiliary signal may introduce a drift signal, which passes through the low-pass filter that implements the integrator. A quantitative analysis of this effect has been done (Ries and Krucker, 1976), which shows how critical, even for relatively modest cancellation ratio, are the offset requirements. To avoid the correlator offset problem, the authors suggest use of the IF passband correlator loops with a single-sideband (SSB) weighting filter.

Image Rejection

Desirable from an ECCM viewpoint is to force a jammer to spread its energy over as wide a band as possible (Wardrop, 1983). This, however, causes some difficulties owing to image signals entering the first mixer of the radar receiver. Elementary trigonometry shows that the phase relationships of the IF signals (in a sidelobe canceler) are not the same at the main and image frequencies. In fact, one is the conjugate of the other. This implies that jamming at both the main and image frequencies cannot be simultaneously cancelled, unless additional degrees of freedom are available in the canceler.

The image signals can be suppressed by a number of means. One technique is to have a high first IF, and use an RF filter prior to the mixer, which allows only the main frequencies to pass, rejecting the image frequencies. The agile frequency band is then constrained to be less than twice the IF value. This approach has the disadvantage that the RF filters are an additional source of phase and gain tracking errors. A second approach is to use image cancelling mixers. These are relatively complex and costly, but avoid the use of RF filters. The image cancellation required should be greater than or equal to the required jammer cancellation. At present, 20–25 dB of image cancellation over a 10% bandwidth at S-band is typical for a good mixer. Image frequency jamming is clearly a serious problem.

Nonideal Quadrature Phase Detector

This is the last source of nulling limitation that we shall consider. The ensuing analysis is devoted to the calculation of the CR degradation due to the nonideal behavior of the quadrature phase detector in the radar channel (A. Huizing, TNO Defense Research, The Netherlands, Private Communication). The analysis is limited to the case of one auxiliary channel. The IF signal $V_{M,IF}(t)$ is down-converted by two mixers to the in-phase signal $V_{M,I}(t)$ and the quadrature signal $V_{M,Q}(t)$, respectively. The intermediate signal is represented by

$$V_{M,IF}(t) = A(t) \sin(\omega_{IF} t + \phi) \qquad (4.133)$$

and the LO signal by

$$V_{LO}(t) = \sin(\omega_{IF} t) \qquad (4.134)$$

The expressions of the in-phase and quadrature signal components are

$$V_{M,I}(t) = V_{M,IF}(t)\sin(\omega_{IF} t)$$
$$V_{M,Q}(t) = V_{M,IF}(t)\sin\left(\omega_{IF} t + \frac{\pi}{2}\right) \qquad (4.135)$$

Assume that the nonideal phase detector is characterized by a difference in gain for the two mixers and a phase shift not equal to $\pi/2$. We can attribute the nonideal behavior to the quadrature channel. After low-pass filtering, the in-phase and quadrature signals can be written as

$$V_{M,I}(t) = \frac{A(t)}{2} \cos\phi$$
$$V'_{M,Q}(t) = g \cos(\delta) V_{M,Q}(t) - g \sin(\delta) V_{M,I}(t) \qquad (4.136)$$

where g is the mixer gain different from one, δ is the phase shift from $\pi/2$, and $V_{M,Q}$ is the quadrature signal for ideal conditions (i.e., $g = 1$, $\delta = 0$):

$$V_{M,Q}(t) = \frac{A(t)}{2} \sin\phi \qquad (4.137)$$

The complex-valued baseband signal in the radar channel has the following expression:

$$V_M(t) = V_{M,I}(t) + j[g\cos(\delta)V_{M,Q}(t) - g\sin(\delta)V_{M,I}(t)] \quad (4.138)$$

while the baseband auxiliary channel is

$$V_A(t) = V_{A,I}(t) + jV_{A,Q}(t) \quad (4.139)$$

Having stated the mathematical model of the main and auxiliary signals, the analysis proceeds with the calculation of the SLC's weight and output. This is done by evaluating the power and cross-power values of the main and auxiliary signals:

$$\begin{aligned}
\overline{|V_M|^2} &= \overline{|V_{M,I}|^2}\,[1 + g^2\sin^2(\delta)] + \overline{|V_{M,Q}|^2}\,[g^2\cos^2(\delta)] \\
\overline{|V_A|^2} &= \overline{|V_{A,I}|^2} + \overline{|V_{A,Q}|^2} \\
\overline{V_M V_A^*} &= \overline{V_{M,I}V_{A,I}} + \overline{V_{M,Q}V_{A,Q}}\,g\cos(\delta) - j\,\overline{V_{M,I}V_{A,Q}}\,g\sin(\delta)
\end{aligned} \quad (4.140)$$

In the presence of jammer and noise, these expressions become

$$\begin{aligned}
\overline{|V_M|^2} &= \frac{1}{2}(1 + g^2)(G_{SL}^2 P_J + P_N) \\
\overline{|V_A|^2} &= G_A^2 P_J + P_N \\
\overline{V_M V_A^*} &= \frac{1}{2}G_{SL}G_A P_J\, e^{j\alpha}(1 + g\cos\delta - jg\sin\delta)
\end{aligned} \quad (4.141)$$

where G_{SL} and G_A are the voltage gain values of the radar antenna sidelobe and auxiliary antenna, respectively. P_J and P_N are the free-space jammer power and thermal-noise power in the SLC channels, respectively. Finally, $e^{j\alpha}$ accounts for the time delay of the jammer signals in the main and auxiliary channels. As usual, the output signal from the SLC is

$$Z = V_M - \frac{\overline{V_M V_A^*}}{\overline{|V_A|^2}} V_A \quad (4.142)$$

For large values of jammer-to-noise power ratio, the residual power normalized to the thermal-noise power is

$$\frac{\overline{|Z|^2}}{P_N} = \frac{1}{2}(1 + g^2)\left(1 + \frac{G_{SL}^2}{G_A^2}\right) + \frac{1}{4}[1 - 2g\cos\delta + g^2]G_{SL}^2 \frac{P_J}{P_N} \quad (4.143)$$

The detrimental effect of the nonideal quadrature phase detector leads to a residue that exceeds the ideal value $(1 + G_{SL}^2/G_A^2)$ due to thermal-noise only. The second term on the right-hand side of the previous equation gives the measure of the SLC performance degradation. As a numerical example, consider the case of $G_{SL}^2 \, P_J/P_N$ = 30 dB, phase error $\delta = 5°$, and gain mismatch $g^2 = 0.5$ dB. The second term on the right-hand side amounts to 4.6 dB, compared with the ideal value (e.g., 3 dB) of the first term on the right-hand side when $G_{SL} = G_A$ and $g = 1$.

4.3.2 Calculation of Cancellation Ratio (CR)

Cancellation ratio is expressed by (4.7). In this section, the performance index is evaluated under a number of meaningful conditions. This problem has been widely discussed in a number of publications, such as Compton (1988), and Monzingo and Miller (1980), among the others.

4.3.2.1 Effects of Antenna Phase Centers Separation and Receiving Channel Bandwidth

In this case, the decorrelation effects between the main and auxiliary signals are due to the time difference experienced by the jammer signals received by the main and auxiliary antennas. The jammer spectrum is assumed to be wide enough to occupy the whole bandwidth of the main and auxiliary receiving channels. Mismatching effects due to the main and auxiliary antenna patterns, and phase and gain differences between the main and auxiliary channels are considered afterward. Also, multipath effects and other detrimental phenomena are disregarded for the moment.

To apply (4.7), the matrix **M** and the vector **R** must be evaluated. One way to do this is to divide the jamming spectrum into a number of spectral lines spaced in frequency in such a way that they do not crosscorrelate (Gabriel, 1976). The covariance matrix of a monochromatic (i.e., when the jamming bandwidth is much less than c/L, where L is the spatial extension of the main and auxiliary array) jamming signal is

$$\mathbf{M} = \{\mathbf{V}*\mathbf{V}^T\} = P_N \mathbf{1} + P_J \mathbf{S}*\mathbf{S}^T \qquad (4.144a)$$

where P_N is the noise power (the noise is uncorrelated between the channels), and **S** is

$$\mathbf{S}^T = \left\{ \exp\left[-j\, 2\pi\left(\frac{d}{\lambda}\right) \sin\theta_J\right], \right.$$
$$\left. \exp\left[-j\, 2\pi\left(\frac{2d}{\lambda}\right) \sin\theta_J\right], \ldots, \exp\left[-j\, 2\pi\left(\frac{Nd}{\lambda}\right) \sin\theta_J\right] \right\} \qquad (4.144b)$$

where θ_J is the jammer incident angle with respect to the boresight of the main and auxiliary array. We assume that the antennas are lined up with an interspace, d, between their phase centers. The main antenna is the first or last element of the array. The jammer power is P_J. The covariance matrix for k uncorrelated jamming sources is

$$\mathbf{M} = P_N \mathbf{I} + \sum_{i=1}^{k} P_i \mathbf{S}_i^* \mathbf{S}_i^T \qquad (4.144c)$$

where P_i and $\mathbf{S}_i$ are the respective power and incidence vector for the ith jamming source. Similiar calculations can evaluate the covariance vector $\mathbf{R}$ (4.2b). The deterioration in output noise power, as the constant power of the interference source is spread over an increasingly wider bandwidth, is shown by Gabriel (1976) for the adaptive array case. Here, we prefer to do the calculations as follows.

Due to the wideband condition, the jamming signals in the different channels are partially correlated. As a consequence, the generic (h, k) element of the covariance matrix $\mathbf{M}$ of (4.144a) is modified as follows:

$$M_{hk} = P_N \delta(h - k) + E\{V_h V_k\}(\mathbf{S}^* \mathbf{S}^T)_{hk} \qquad (4.144d)$$

where $\delta(\cdot)$ is the Kronecker operator and the term $E\{V_h V_k\}$, representing the correlation between the jamming signal amplitudes at the hth and kth auxiliary antennas, differs from the jamming power P_J for a multiplicative correlation coefficient $\rho_{hk} \leq 1$. Under the assumptions that the channel transfer functions (auxiliary and main) are identical, the crosscorrelation function $E\{V_h V_k\}$ at a given time instant, t, between the two signals' amplitudes (either two auxiliaries, or the main and an auxiliary) equals the autocorrelation function of the signal amplitude on a generic channel between the time instants t and $t + |h - k|T_a$, where T_a is the time delay between the signals received by the two channels under consideration (4.16). Given these hypotheses, the matrix $\mathbf{M}$ and vector $\mathbf{R}$ may be expressed by means of a correlation coefficient ρ.

Different values of ρ are obtained in relation to the low-pass equivalent IF filter shape. Three types of IF filter are considered (Farina, 1977):

1. Rectangular, to which the following correlation coefficient corresponds:

$$\rho(X) = \frac{\sin \pi |X|}{\pi |X|} \qquad (4.145a)$$

with $X = BT_a$, where B is the filter bandwidth and T_a is given by (4.16). X is known to be the signal wavefront dispersion (time-bandwidth product); see Subsection 4.3.1.2.

2. One-pole filter, to which the following correlation coefficient corresponds:

$$\rho(X) = \exp(-\pi|X|) \qquad (4.145b)$$

3. Gaussian-shaped filter (approximating the transfer function of a cascade of several one-pole filters), to which the following correlation coefficient corresponds:

$$\rho(X) = \exp\left[-0.5\left(\frac{\pi X}{1.17}\right)^2\right] \qquad (4.145c)$$

In calculating the different expressions for ρ, the 3 dB bandwidth of the filters is assumed to be coincident. Figure 4.27(a) illustrates the correlation coefficient ρ versus the time-bandwidth product $|X|$ for the three filters considered. We immediately recognize that higher ρ values, which correspond to higher CR performance, are obtained by the rectangular-shaped filter. The Gaussian-shaped filter also provides good performance. The one-pole filter does not generally provide good results. Independently of the filter shape, high ρ values are obtained by reducing the phase center's distance, d, by reducing the bandwidth of the receiving channels, and for jammer incident angles close to boresight.

For the case of an SLC with one auxiliary antenna, several values of CR are shown in Figure 4.27(b) versus the jammer-to-noise power ratio value, JNR $\triangleq P_J/P_N$ having the correlation coefficient, ρ, as a parameter. The maximum achievable CR is limited by the JNR value (as already indicated by (4.8h)).

A simple mathematical expression for the CR can be obtained under the following hypotheses: (1) one auxiliary antenna, (2) JNR $\gg 1$, and (3) rectangular spectra of the receiving channels. The following formula is found by (4.15) and (4.145a). (See also Andrews and Gerlach, 1989, p. 469):

$$\text{CR} = \frac{1}{1 - [\text{sinc}(\pi X)]^2} \approx \frac{3}{(\pi BT_a)^2}, \quad X = BT_a \ll 1 \qquad (4.146)$$

For the case of more auxiliaries, the CR is obtained by substituting the proper wideband expressions of **M** and **R** in (4.7). The curves of Figure 4.28 refer to the cancellation of only one jammer versus the number of auxiliaries (assumed to be equispaced) and for some values of the radar bandwidth B/f_0 (normalized to the carrier frequency f_0) and the JNR. The curves refer to the case of a Gaussian-shaped filter for the main and auxiliary receiving channels. In this case, the $(i - j)$ element of the matrix **M** has the following expression:

$$(\mathbf{M})_{ij} = P_N \delta(i - j) + P_J (\mathbf{S}^*\mathbf{S}^T)_{ij}\, \rho^{(i-j)^2} \qquad (4.147)$$

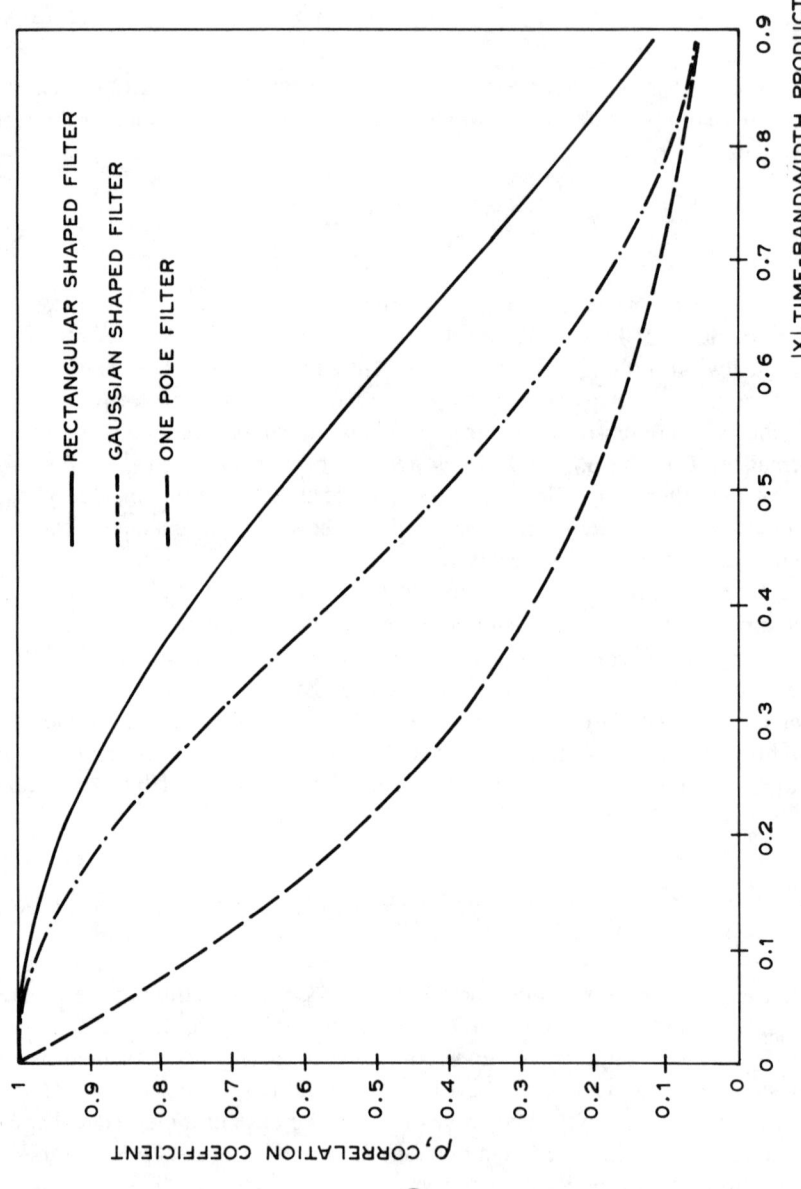

Figure 4.27 (a) Correlation coefficient *versus* time-bandwidth product; (b) cancellation ratio of an SLC with one auxiliary antenna.

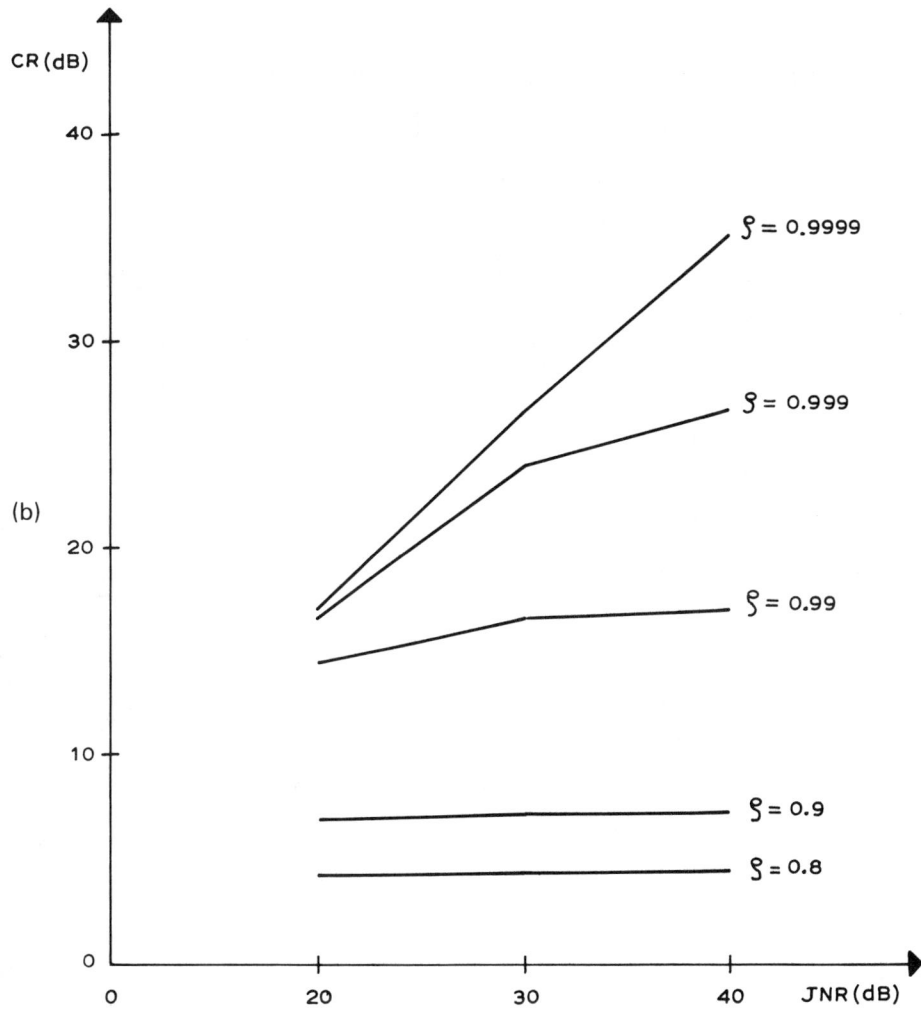

Figure 4.27 Continued

where ρ is the correlation coefficient between two contiguous auxiliary antennas, calculated from (4.145c), taking for $X = B \sin\theta/(2f_0)$ as the proper value corresponding to the numerical parameters of the application example. We assume that $d = 0.5\lambda_0$. A similar expression can be written for the elements of the crosscorrelation vector **R**. Different mathematical expressions apply for the rectangular-shaped and one-pole filters, which are not considered here. From Figure 4.28, we see that the maximum value of CR is limited to the JNR value; when CR approaches JNR,

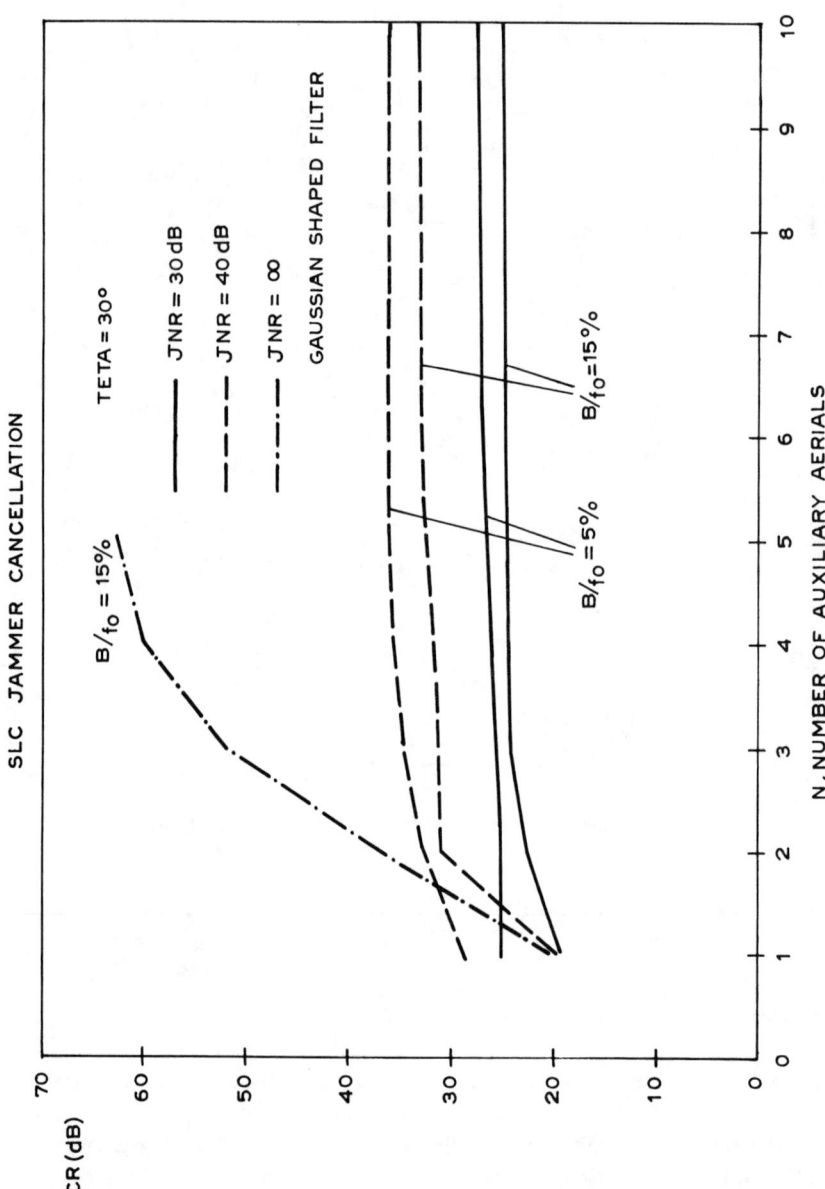

Figure 4.28 Cancellation ratio of the jammer *versus* the number of auxiliary antennas (θ is the jammer incidence angle).

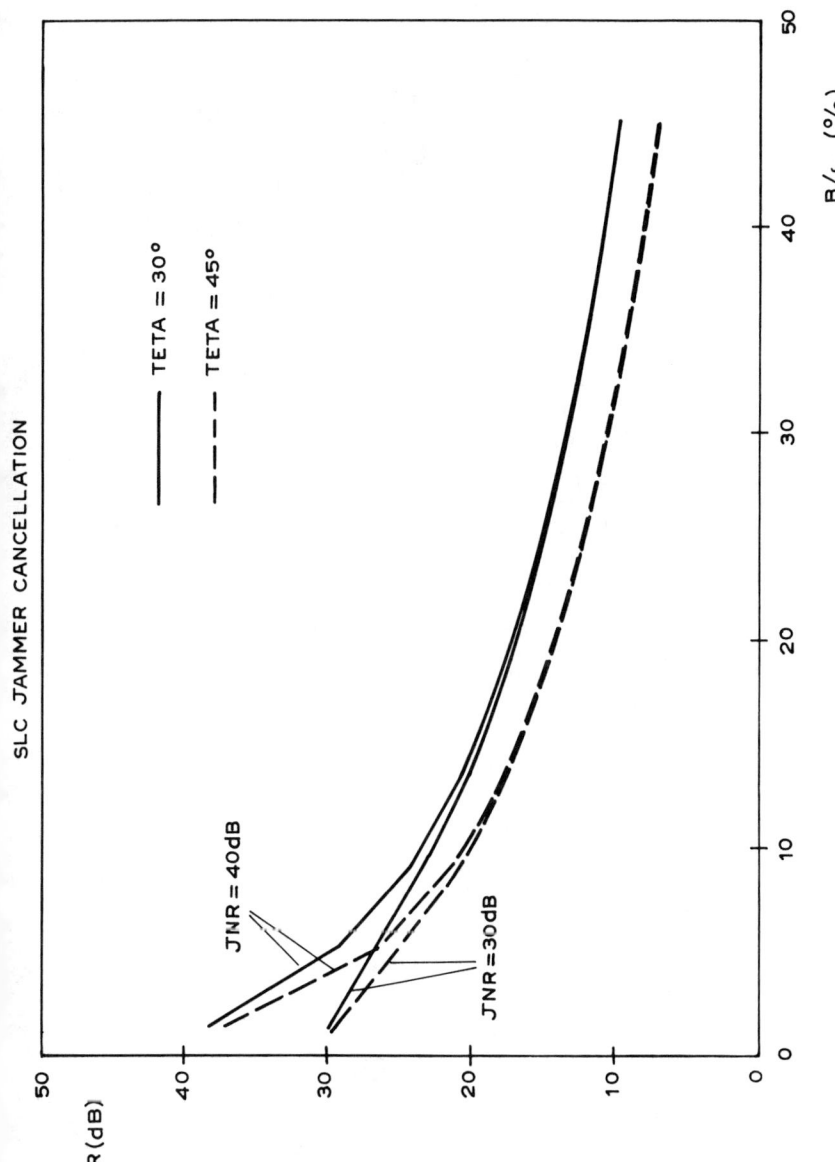

Figure 4.29 Jamming bandwidth effect on the CR for an SLC with one auxiliary aerial antenna (θ is the jammer incidence angle).

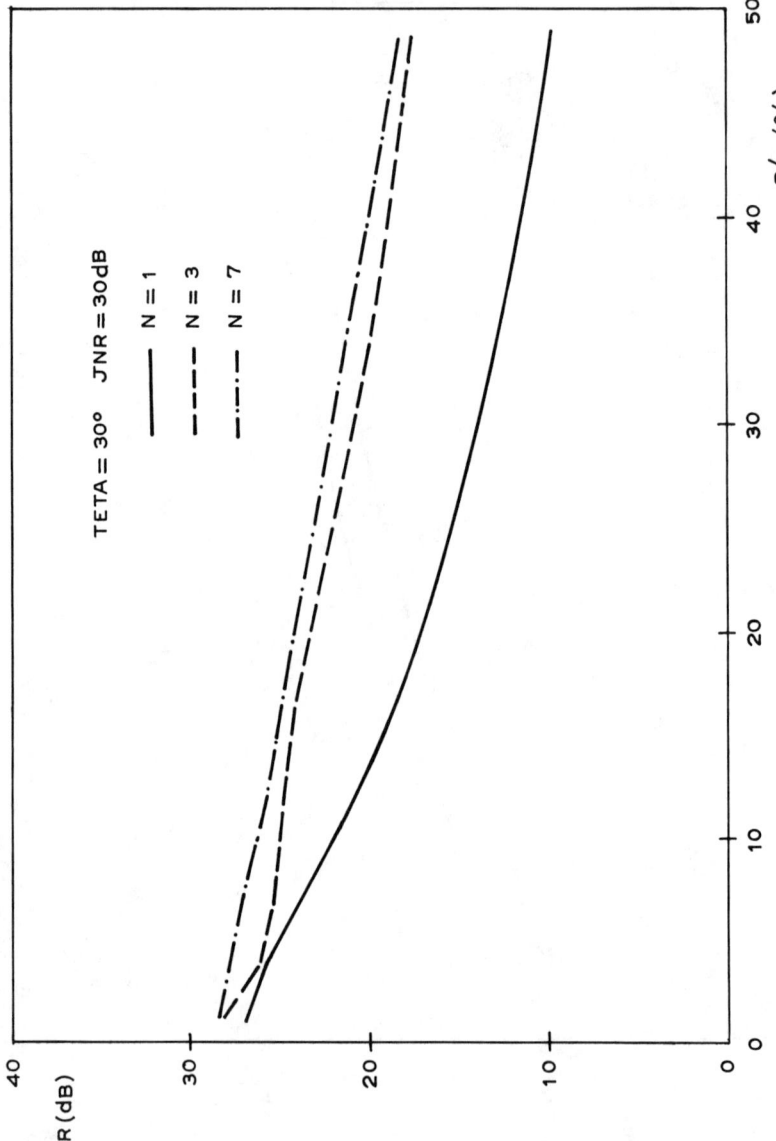

Figure 4.30 Jamming bandwidth effect on the CR for an SLC with more auxiliaries (θ is the jammer incidence angle).

the jammer power has been reduced under the system noise level. Also note that, to achieve good performance, more than one auxiliary antenna is required, expecially for high values of JNR. The channel bandwidth plays a key role as a limiting factor to the achievable CR. Therefore, additional spatial degrees of freedom are useful, resulting in the assignment of additional zeros distributed over the RF bandwidth.

For better understanding of the way in which we can compensate for the bandwidth effect, let us consider the case of an SLC system equipped with one auxiliary. Figure 4.29 illustrates the degradation of CR as a function of the bandwidth of the receiving channels (a Gaussian-shaped transfer function has been assumed). The curves refer to two different values of the jamming incident angle and two different values of JNR. Figure 4.30 illustrates the improvement of CR obtained by using more than one auxiliary. This is motivated by the capability to form more nulls, which are close to each other, thus producing a wide nulling action.

Another way to compensate for the bandwidth effects is to use a tapped delay line downstream of each auxiliary. This approach has been considered in Monzingo and Miller (1980), in White (1983, 1987), and will be discussed in Subsection 4.3.2.4.

4.3.2.2 Effects of Channels Mismatching

Let us now consider the effects on CR of the amplitude and phase mismatching between the transfer functions of the main and auxiliary channels. This problem has been considered in: Monzingo and Miller, 1980; Compton, 1988; Nitzberg, 1989; Carlson et al., 1990; Farina and Giusto, 1981; Teitelbaum, 1991. Here, for example, we consider the scheme of Figure 4.31(a), which refers to an SLC with one auxiliary. The main and auxiliary antennas are assumed to be omnidirectional; this assumption will be removed afterward. The transfer functions of the main and auxiliary channels are illustrated in Figure 4.31(b), where the amplitude and phase mismatching are shown. Also note that other mismatching models may be adopted, for example, the sinusoidal ripple function.

Let us evaluate the CR for the case of Figure 4.31, assuming, for simplicity, JNR = ∞. The CR value is

$$\text{CR} = (1 - |\rho|^2)^{-1} \quad (4.148\text{a})$$

$$\rho = \frac{\overline{n_0 n_1^*}}{\sqrt{\overline{|n_0|^2}\ \overline{|n_1|^2}}} \quad (4.148\text{b})$$

where n_0 and n_1 are the main and auxiliary signals, respectively.

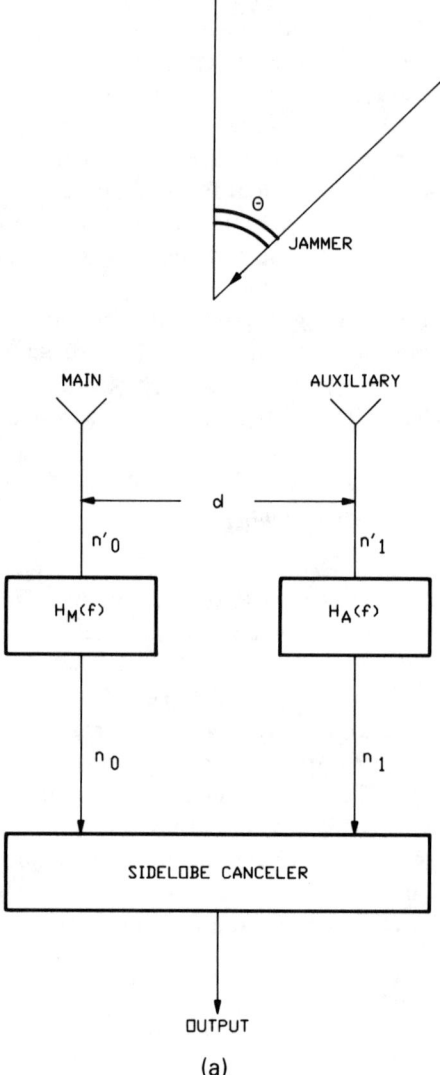

Figure 4.31 (a) Scheme for the calculation of the channel mismatching effects; (b) amplitude and phase mismatching models.

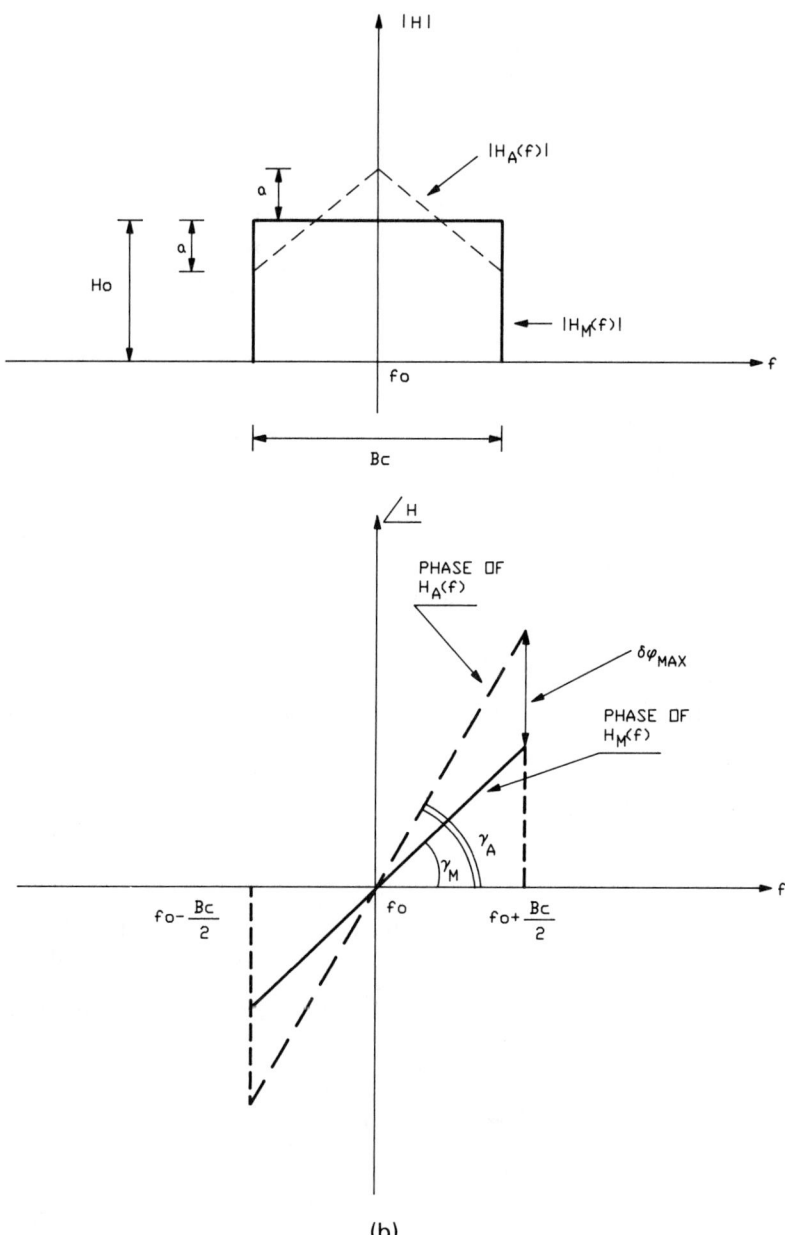

Figure 4.31 Continued

The following expressions are found for $\overline{|n_0|^2}$, $\overline{|n_1|^2}$ and $\overline{n_0 n_1^*}$, respectively:

$$\overline{|n_0|^2} = \int_{f_0-B_c/2}^{f_0+B_c/2} |H_M(f)|^2 \, df = H_0^2 B_c \tag{4.149a}$$

$$\overline{|n_1|^2} = \int_{f_0-B_c/2}^{f_0+B_c/2} |H_A(f)|^2 \, df = H_0^2[1 + (a')^2/3]B_c \tag{4.149b}$$

where

$$a' = a/H_0$$

$$\overline{n_0 n_1^*} = \int_{f_0-B_c/2}^{f_0+B_c/2} S_{n'_0 n'_1}(f) H_M(f) H_A^*(f) \, df \tag{4.149c}$$

where $S_{n'_0, n'_1}$, is the cross-spectral density of the jamming signal at the antennas, i.e., upstream of the channel filters. Assuming a white process for the incoming jammer with unitary power spectral density, the cross-spectrum is

$$S_{n'_0 n'_1}(f) = e^{-j2\pi f T_a} e^{-j2\pi f_0 T_a} \tag{4.149d}$$

because n'_0 differs from n'_1 for the delay $T_a = d \sin\theta/c$. The following expression is found for $\overline{n_0 n_1^*}$.

$$\overline{n_0 n_1^*} = H_0 \, e^{-j4\pi f_0 T_a} \left\{ \frac{2(H_0 + a) - 4a}{2\pi T_a - m_M + m_A} \sin[(2\pi T_a - m_M + m_A)B_c/2] \right.$$

$$\left. + \frac{8a/B_c}{[2\pi T_a - m_M + m_A]^2} [\cos[(2\pi T_a - m_M + m_A)B_c/2] - 1] \right\} \tag{4.149e}$$

where $m_M = \tan\gamma_M$ and $m_A = \tan\gamma_A$ (see Figure 4.31(b)). By combining (4.149a), (4.149b), and (4.149e), the coefficient ρ is

$$\rho = \frac{(1 - a')\dfrac{\sin(\pi B_c T_a + \delta\phi_{MAX})}{\pi B_c T_a + \delta\phi_{MAX}} + a'\left[\dfrac{\sin(\pi B_c T_a + \delta\phi_{MAX})/2}{(\pi B_c T_a + \delta\phi_{MAX})/2}\right]^2}{\sqrt{1 + 1/3(a')^2}} \tag{4.150}$$

where $\delta\phi_{MAX}$ is the maximum phase mismatching between the channels, as indicated in Figure 4.31(b). Note that ρ, and hence CR, depends on the following key parameters: the time-bandwidth product $B_c T_a$, amplitude mismatching a', and phase mismatching $\delta\phi_{\text{MAX}}$. By assuming $B_cT_a = 0$ (as is the case for a broadside jammer, i.e., $\theta = 0$) and $\delta\phi_{\text{MAX}} = 0$, the corresponding CR value is

$$\text{CR}(B_cT_a = 0, \delta\phi_{\text{MAX}} = 0) = 1 + 3/(a')^2 \qquad (4.151)$$

From this equation, we immediately see how critical is the parameter a'. For instance, a mismatching of $a' = 10\%$ corresponds to 24.7 dB of CR. Figure 4.32 illustrates contours of constant CR values as a function of a' and $\delta\phi_{\text{MAX}}$ for different values of the parameter B_cT_a. Note that the requirements of the amplitude and phase mismatching become more critical as B_cT_a becomes greater than zero, as is usually the case.

4.3.2.3 Effects of the Differences between Main Antenna Sidelobes and Auxiliary Patterns

This subsection illustrates the degradation of CR due to the difference between the main antenna sidelobes with respect to the pattern of the omnidirectional aerial antennas (Farina and Giusto, 1981). This difference produces a different spatial filtering of the same jammer. In addition, the amplitude mismatch between the main and auxiliary channels is also taken into account according to the approach followed in the previous subsection. Several examples are considered, which refer to different numbers of auxiliaries and jamming sources.

The system considered is depicted in Figure 4.33. The main antenna is planar, having a circular shape and uniform illumination; three auxiliary omnidirectional antennas are placed around the main antenna. The positions of the auxiliaries are given by the angles $\phi_i (i = 1, 2, 3)$. The direction of incidence of a jammer is determined by the two angles (θ_J, ϕ) (elevation and azimuth). The signals received through the main and auxiliary antennas are down-converted at an appropriate frequency f_0, and then filtered through the receiving channels having transfer functions $H_M(f)$ and $H_A(f)$. Assume that all the auxiliary channels have the same transfer function $H_A(f)$, which is different from the transfer function $H_M(f)$ of the main channel. As an example, the mismatching in the transfer function amplitudes is again shown in Figure 4.33, and no phase mismatch is assumed.

Calculate the jammer CR of the SLC system when only one auxiliary antenna is activated, the system being subject to just one jammer. The cancellation ratio is given by (4.148a) and (4.148b). Assuming a unity jammer density power over the bandwidth B_c, the mean power $|n_0|^2$ has the following expression:

$$\overline{|n_0|^2} = H_0^2 \int_{f_0 - B_c/2}^{f_0 + B_c/2} F_\phi^2(f) \, df \qquad (4.152a)$$

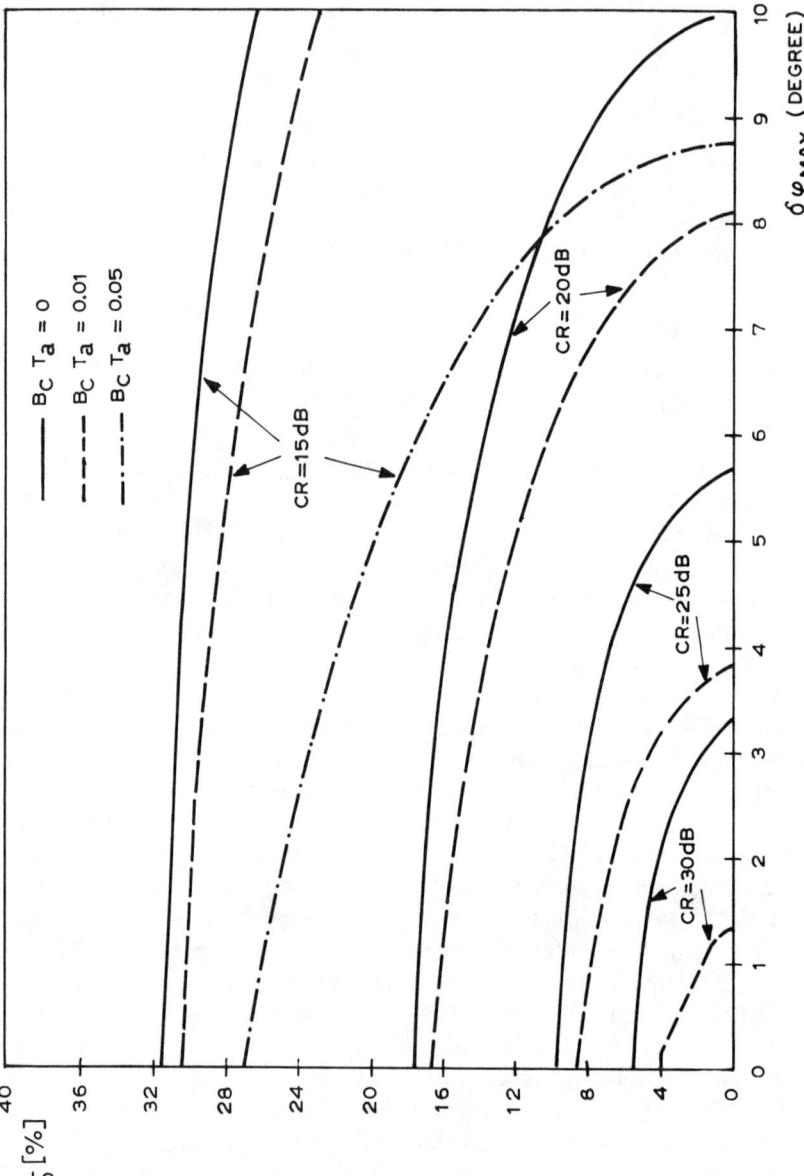

Figure 4.32 CR value as a function of amplitude and phase mismatching of receiving channels.

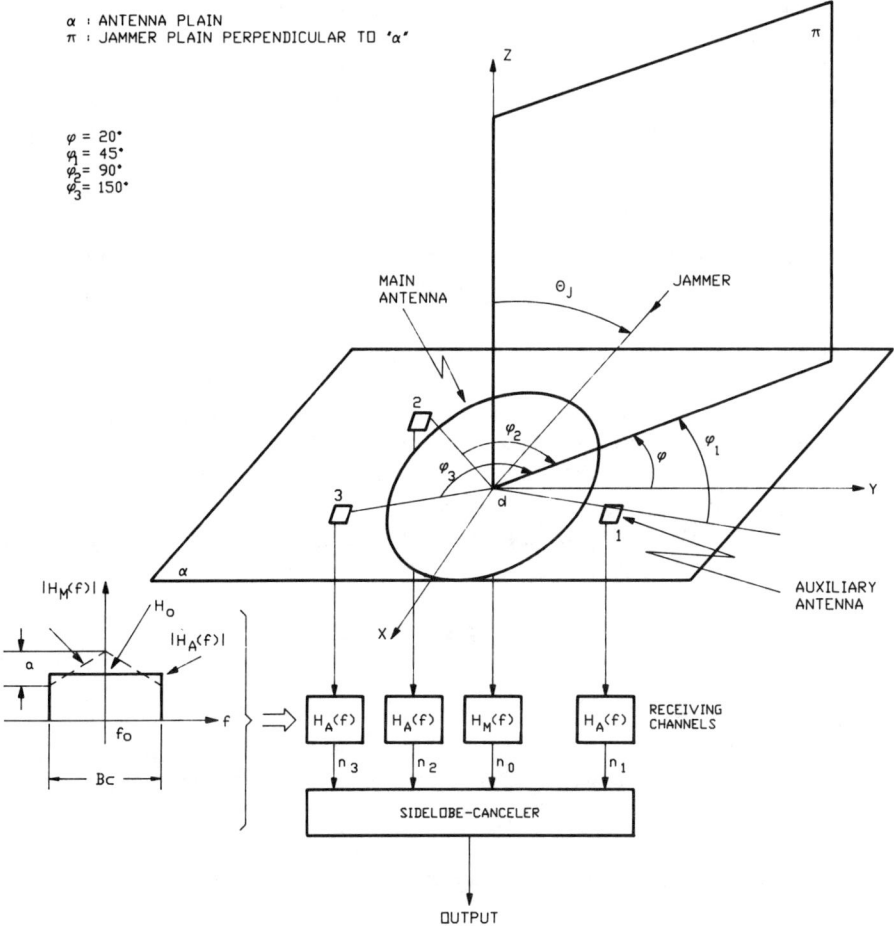

Figure 4.33 The SLC scheme considered for a calculation concerning a mismatch between the patterns of the antennas and the receiving channels.

where F_ϕ is the pattern of the main antenna in the jammer plane π, defined by the angle ϕ. The power $\overline{|n_1|^2}$ is given by (4.149b).

The cross-power $\overline{n_1^* n_0}$ is

$$\overline{n_1^* n_0} = H_0^2 \int_{-B_c/2}^{B_c/2} F_\phi(f)(1 + a' - 4a'/B_c|f|)e^{-j\Delta_{10}}df \quad (4.152b)$$

where $\Delta_{10} = d(\sin\theta_J \cos\phi_1)(f_0 + f)2\pi/c$; d is the distance between the antenna phase centers; c is the velocity of light; and Δ_{10} is the phase difference between the signals received by the main and auxiliary antennas.

CR is drawn in Figure 4.34(a) with respect to different values of the jammer elevation angle θ_J and having the mismatching quantity $a' = a/H_0$ as parameter. We assume $B_c = 1$ MHz and $f_0 = 5$ GHz. We can see that the maximum CR value (obtained when $\theta_J \to 0$) is limited by the mismatching parameter a'. In addition, CR is reduced, corresponding to the deep nulls of the main antenna pattern. However, in this case, the jamming input power also is reduced.

Let us now consider the cancellation of two uncorrelated jammers in the plane π, one placed in the direction of the first sidelobe peak of the main antenna, and the second with a variable elevation angle θ_J. Figure 4.34(b) shows the cancellation versus θ_J and with $a' = 0.05$ (curve CR_2). The cancellation is equal to that of Figure 4.34(a) (curve CR_1), when both jammers have almost the same direction; otherwise, the cancellation strongly decreases. Finally, consider three jammers, the first two placed in the direction of the two first adjacent sidelobes and the third with variable direction. We can show that the cancellation is nearly zero throughout (the cancellation curve has not been drawn).

Consider an SLC with two auxiliaries. In this case, CR is given by the following equation:

$$CR = \frac{\overline{|n_0|^2}}{\overline{|n_0 - \mathbf{W}^T \mathbf{n}|^2}} \quad (4.153)$$

where $\mathbf{n}^T = [n_1, n_2]$ and $\hat{\mathbf{W}} = \left[\overline{\mathbf{n}^* \mathbf{n}^T}\right]^{-1} \cdot \overline{\mathbf{n}^* n_0}$. The power $\overline{|n_0|^2}$ is given by (4.152a), while $\overline{|n_1|^2}$ and $\overline{|n_2|^2}$ have the same expressions as (4.149b). The cross-power $\overline{n_1^* n_0}$ is equal to that of (4.152b), while $\overline{n_2^* n_0}$ is obtained by substitution of Δ_{10} with Δ_{20}:

$$\Delta_{20} = (f_0 + f) \, d(\sin\theta_J \cos\phi_2) 2\pi/c \quad (4.154)$$

where Δ_{20} is the relevant phase difference between the main and second auxiliary. The cross-power $\overline{n_1 n_2^*}$ is

$$\overline{n_1 n_2^*} = H_0^2 \int_{-B_c/2}^{B_c/2} (1 + a' - 4a'/B_c|f|)^2 e^{-j\Delta_{12}} df \quad (4.155)$$

where $\Delta_{12} = (f_0 + f) d \sin\theta_J (\cos\phi_1 - \cos\phi_2) 2\pi/c$.

The cancellation CR is drawn in Figure 4.34(c) for different jamming environments with $a' = 0.05$. We can see that the curve CR_1 is relevant to only one jammer and is almost the same as the corresponding case of Figure 4.34(b). The curve CR_2 shows the improvement obtained against two jammers with an additional auxiliary antenna with respect to the corresponding case of Figure 4.34(b). Finally, when three jammers are considered, the curve CR_3 shows a strongly reduced performance, except when the direction of the mobile jammer coincides with that of one of the two fixed jammers.

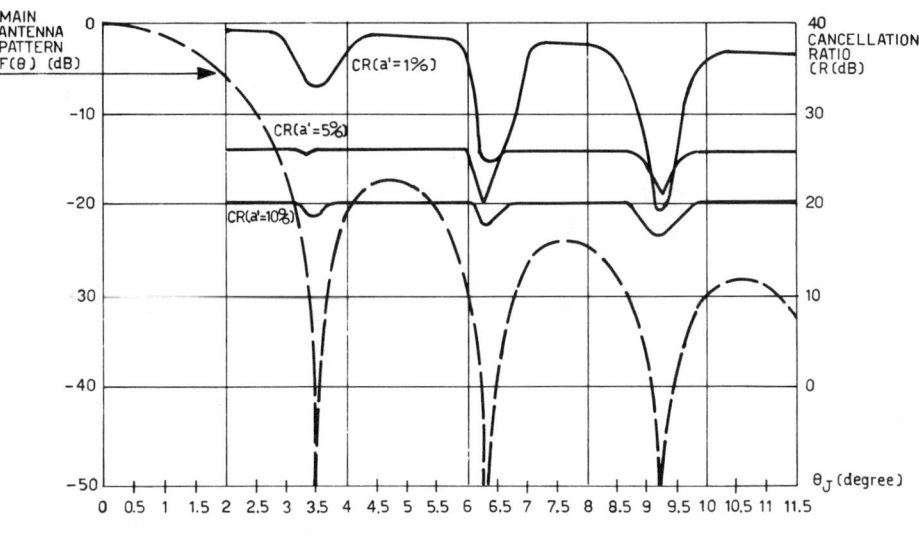

(a)

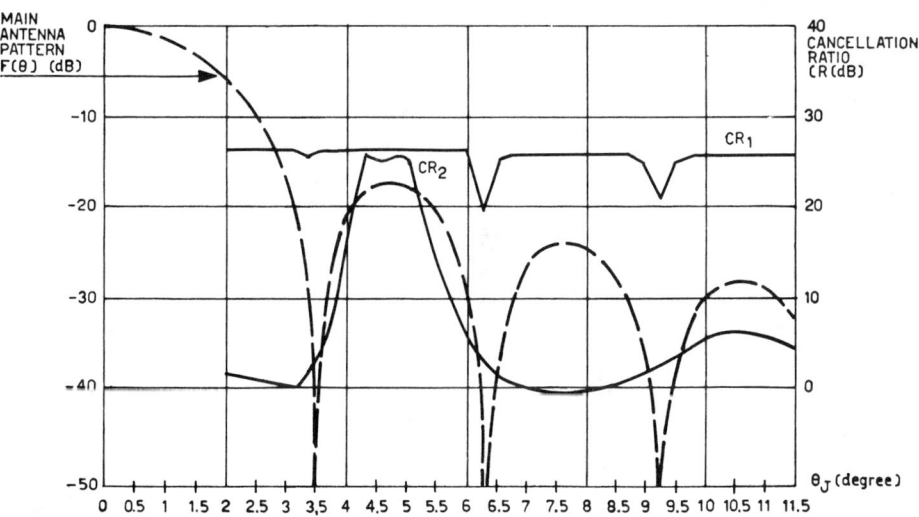

(b)

Figure 4.34 (a) The CR and the main-beam antenna pattern: one auxiliary antenna and one jammer; (b) the CR and the main-beam antenna pattern: one auxiliary antenna and two jammers;

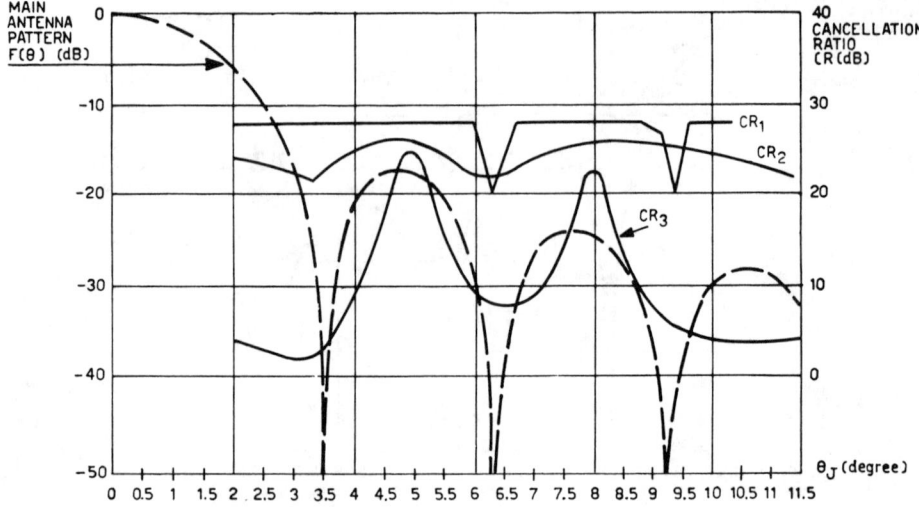

(c)

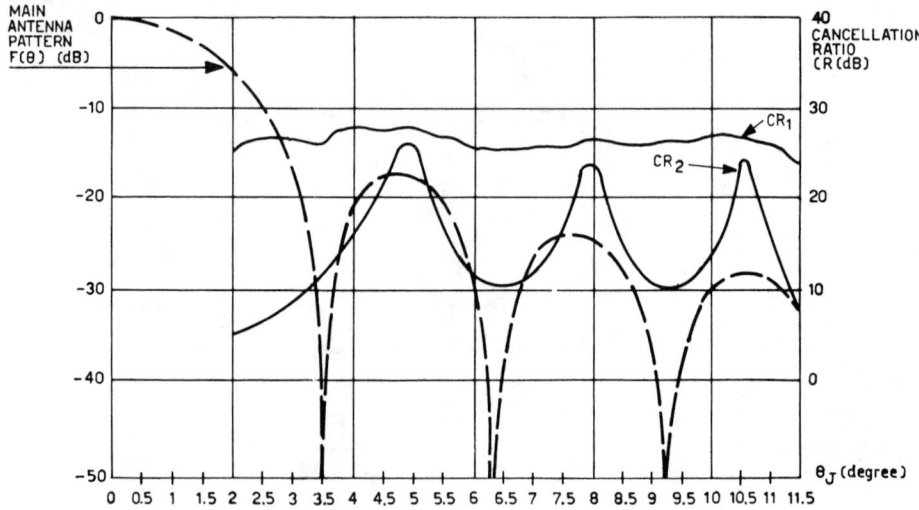

(d)

Figure 4.34 (c) The CR and the main-beam antenna pattern: two auxiliary antennas; (d) the CR and the main-beam antenna pattern: three auxiliary antennas.

Finally, consider the canceler with three auxiliary antennas. The CR is obtained by following the same mathematical procedure as used for the two-auxiliary case. The computation results are shown in Figure 4.34(d). The cancellation of one or two jammers is quite similar to CR_1, which refers to the cancellation of three jammers. When there are four jammers, the cancellation is strongly reduced, except when the mobile jammer comes from a direction coincident with that of one fixed jammer. This behavior is shown by the curve CR_2.

4.3.2.4 Compensation of Bandwidth, Channel Mismatching, and Multipath Effects

The purpose of this subsection is to illustrate a common approach based on the use of a tapped delay line in the SLC to compensate for degradations due to bandwidth, channel mismatching, and multipath phenomena. This problem has been widely discussed in Monzingo and Miller, 1980; Cantoni and Godara, 1982; Compton, 1988; Ko, 1989b; 1990; Carlson *et al.*, 1990; Teitelbaum, 1991.

Consider the problem of compensating for bandwidth effects. In the event that the interfering signal cannot be adequately characterized by a single frequency and its frequency content encompasses a significant spectrum, a complex-valued weight **W**, appropriate for one frequency, f_1 (narrowband case), will not be appropriate for a different frequency, f_2, because the array pattern nulls shift as the value of the frequency changes. This leads to the conclusion that different complex-valued weights are required at different frequencies if an array null is to be maintained in the same direction for all frequencies of interest. A simple and effective way of obtaining different amplitude and phase weightings at a number of frequencies over the band of interest is to replace the quadrature hybrid circuitry of Figure 4.35(a), adopted for narrowband processing, by a transversal filter having the transfer function $h(f)$. The scheme of Figure 4.35(a) realizes one complex weight $(W_1 + jW_2)$, which operates on the in-phase and quadrature components of the signal received by the array element. The transversal filter can be realized by a tapped delay line having L complex weights, as shown in Figure 4.35(b). A tapped delay line has a periodic transfer function, with a period $1/\Delta$ along the frequency axis. If the tap spacing, Δ, is sufficiently close and the number, L, of taps is large (the frequency resolution of the filter is $1/(L\Delta)$), this network controls gain and phase at each frequency within the band of interest. An upper limit on the tap spacing is given by the desired array cancellation bandwidth, B, as $B \leq 1/\Delta$. Fast Fourier Transform (FFT) techniques also have been exploited to replace the time-domain processor of Figure 4.35(b) by an equivalent frequency-domain processor using the frequency-domain equivalent of the time domain. The advantage of such approach lies in alleviating the hardware problem.

This preamble brings us to the modified scheme of Figure 4.36(a) considered for compensation purposes (Monzingo and Miller, 1980, pp. 451–453). The SLC

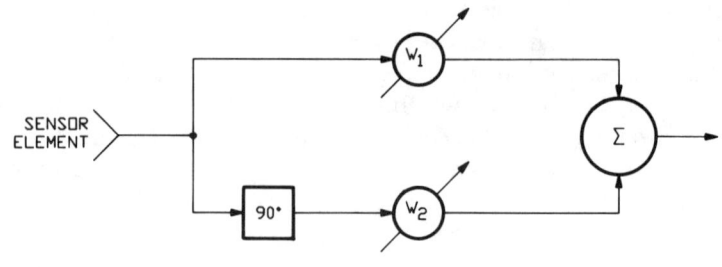

(a)

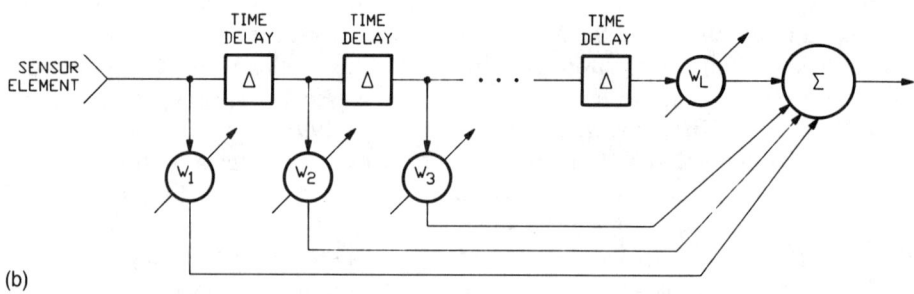

(b)

Figure 4.35 (a) Realization of a complex weight by means of a quadrature hybrid circuit. (From Monzingo and Miller, *Introduction to Adaptive Arrays*, p. 60, ©1980, reprinted by permission of John Wiley & Sons, Inc.) (b) Transversal filter realized with a tapped delay line having L complex weights. (From Monzingo and Miller, *Introduction to Adaptive Arrays*, p. 61, ©1980, reprinted by permission of John Wiley & Sons, Inc.)

with one auxiliary is considered for the sake of simplicity. In addition, the two-array element model exhibits all the salient characteristics of a complex system with more auxiliaries. The jamming signals $V_M(t)$ and $V_A(t)$ pass through the corresponding channel filters, $H_M(f)$ and $H_A(f)$, which may be different in general. To compensate for the channel mismatch and the nonzero bandwidth, a tapped delay line is inserted into the auxiliary channels, and the weights $\mathbf{W}^T = [W_1, W_2, \ldots, W_L]$ on these taps are adjusted so that the power of the output $Z(t)$ is minimized. The main channel is delayed so that the center tap of the auxiliary channel corresponds to the output of the delay D in the main channel, thereby permitting compensation for both positive and negative values of the off-broadside angle, θ. Let us define the complex signal vector:

$$\mathbf{X}^T(t) = [X_1(t), X_2(t), \ldots, X_L(t)] \tag{4.156}$$

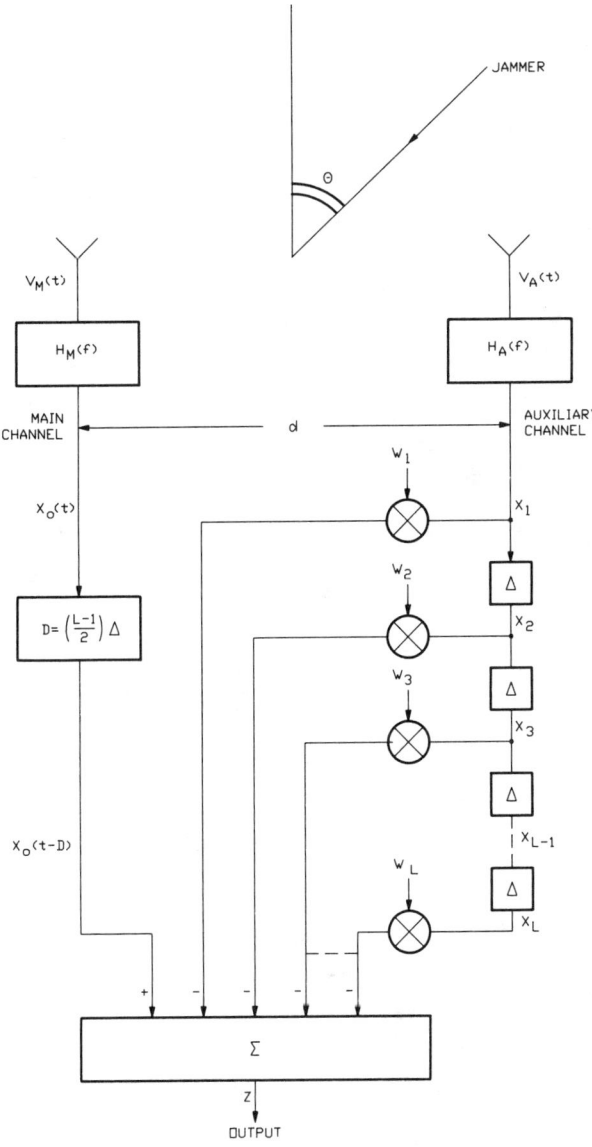

(a)

Figure 4.36 (a) The SLC scheme with tapped delay line (figure adapted from Monzingo and Miller, *Introduction to Adaptive Arrays*, p. 452, ©1980, reprinted by permission of John Wiley & Sons, Inc.,); (b) loci of constant CR values: triangular amplitude and linear phase mismatching, one auxiliary antenna, two taps, $B_c T_a = 0.01$, $B_c \Delta$ as parameter; (c) loci of constant CR values: triangular amplitude and linear phase mismatching, one auxiliary antenna, four taps, $B_c T_a = 0.01$, $B_c \Delta$ as parameter.

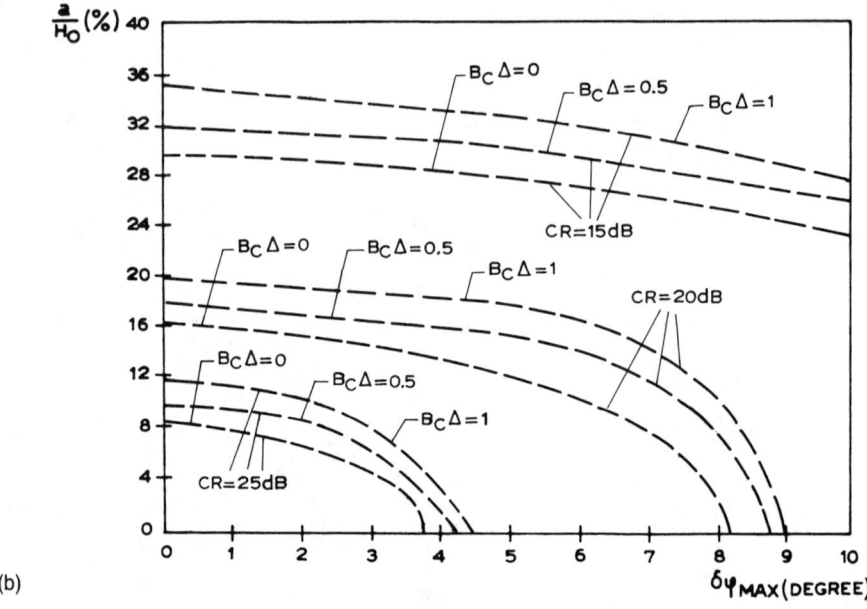

(b)

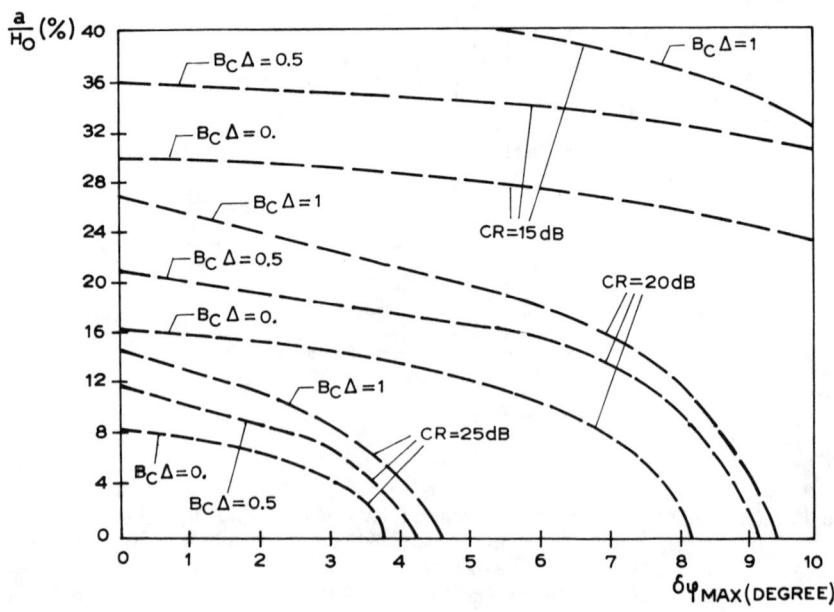

(c)

Figure 4.36 Continued

with

$$X_2 = X_1(t - \Delta)$$

$$X_2 = X_1(t - 2\Delta)$$

$$X_L = X_1[t - (L - 1)\Delta]$$

where $X_1(t)$ is the signal shaped by the transfer function $H_A(f)$ of the auxiliary channel. The output signal $Z(t)$ is given by

$$Z(t) = X_0(t - D) - \mathbf{W}^T \mathbf{X}(t) \quad (4.157)$$

We understand that the optimum set $\hat{\mathbf{W}}$ of weights (i.e., that which minimizes the residue power $E\{|Z|^2\}$), is found to be

$$\hat{\mathbf{W}} = \mathbf{A}^{-1}\mathbf{b} \quad (4.158)$$

where $\mathbf{A} = E\{\mathbf{X}^*(t)\mathbf{X}^T(t)\} = \mathbf{R}_{XX}(0)$ and $\mathbf{b} = E\{\mathbf{X}^*(t)X_0(t - D)\} = \mathbf{r}_{XX_0}(-D)$; $\mathbf{R}_{XX}(0)$ is the autocorrelation matrix of $\mathbf{X}$, evaluated at time instant $t = 0$ and $\mathbf{r}_{XX_0}$ is the crosscorrelation vector between $\mathbf{X}$ and X_0. The residue power is

$$\begin{aligned} E\{|Z(t)|^2\} &= E\{|X_0|^2\} - \mathbf{r}_{XX_0}^H(-D)\mathbf{R}_{XX}^{-1}(0)\mathbf{r}_{XX_0}(-D) \\ &= E\{|X_0|^2\} - \mathbf{b}\mathbf{A}^{-1}\mathbf{b} \end{aligned} \quad (4.159)$$

Concerning the compensation of the bandwidth effects, detailed investigations have been made (Bowers and Perry, 1980; Ahmed and Evans, 1983; White, 1983; Monzingo and Miller, 1980, pp. 429–451; Compton, 1988, pp. 120–137). In particular, Bowers and Perry illustrate simulation results of an SLC with one auxiliary antenna and four taps. The analysis refers to the influence of the delay, Δ, on the CR. A detailed investigation of the effectiveness of several processing schemes is due to Rodgers and Compton (1979); see also (Monzingo and Miller, 1980, pp. 429–451); (Compton, 1988, pp. 120–137), which illustrate the relevance of the number of taps in the processors.

The transversal filter not only is useful for providing the desired adjustment of gain and phase over the frequency band of interest for wideband signals, but is also well suited to provide array compensation for the effects of multipath, and interchannel mismatch effects. The compensation of interchannel mismatch effects has been investigated (Lewis et al., 1986, pp. 163–183; Monzingo and Miller, 1980, pp. 461–475). The simulation studies show that the CR is a function of the product $B_c\Delta$ between the channel bandwidth and the tap delay, the number of taps, and the type of mismatch between the channels. Monzingo and Miller (1980, p. 469) found that the number of sufficient taps, assuming for the amplitude mismatch model a sinusoidal ripple, is

$$L \approx \left(\frac{N_r - 1}{2}\right)[7 - 4(B_c\Delta)] + 1 \qquad (4.160)$$

where N_r is the number of half-cycles of the ripple appearing in the mismatch model. An optimum procedure to select the above-mentioned parameters is also described in Lewis et al. (1986, pp. 163–193). Intuitively desirable is to select the product $B_c\Delta$ to match the amplitude mismatch model, i.e., $B_c\Delta$ equal to number cycles of ripple mismatch (Monzingo and Miller, 1980, p. 471). However, when the number of cycles of mismatch ripple exceeds unity, it is best to set $B_c\Delta = 1$ (Monzingo and Miller, 1980, p. 472).

In the following, additional results of a simulation exercise are reported, showing the benefit of using the processor of Figure 4.36(a) to compensate for the mismatching between the main and auxiliary channels, as depicted in Figure 4.31(b) (i.e., triangular mismatch for the amplitude; linear mismatch for the phase). The results are shown as contours of constant CR values for the case of $B_cT_a = 0.01$ (dashed curves of Figure 4.32), which means off-boresight jammer direction of arrival. Specifically, Figure 4.36(b) shows the benefit of using two taps in the processor. Nine contour lines are indicated, three contours for three values of CR. For each value of CR, three curves differ for the value of the tap delay, Δ. The value $B_c\Delta = 0$ means that we do not use any taps. The corresponding contour lines are the dashed lines depicted in Figure 4.32. We note an increase in performance for $B_c\Delta = 1$. Figure 4.36(c) differs from the previous one (b) because four taps are used. A comparison between the two figures shows the advantage of using more taps.

Normally channel mismatch requires higher degrees of freedom to match the complexities of the (usually) high-order IF filters within the receiver channels than is required to correct for time-delay distortion produced by off-boresight wavefront arrivals. Furthermore, the update rates needed for channel equalization are usually very low. Therefore, a reasonable approach is to update the larger-order, channel-equalization calculations separately, but much less often.

A very interesting experimental analysis, described (Carlson et al. (1990, pp. 302–305) and (Teitelbaum, 1991), reports on the equalization of four-channels of an adaptive array operating at UHF. The center frequency of the IF section is 1 MHz and the bandwidth is 200 kHz. The difference between two typical analog channels in the IF band is on the order of -30 dB, with typical maximum-to-minimum spacing of 50 kHz of the amplitude ripple. Digital channel equalization has been used to correct this type of match between analog channels. A 32-tap transversal digital equalizer has been inserted at each receiver output. The initial injection of a calibration signal into all channels adjusts the equalizers to match each channel's frequency response. For this number of taps, a rough empirical estimate of the CR limit due to imperfect equalization is 76 dB. However, the ultimate limit of performance is established by the accuracy of the digital equalizer and, in practice, is set by the number of bits of the ADC. For the experimental system, the CR limit has been reported to be 55 dB. (See also Johnson et al., 1991.)

In multipath situations, the cancellation features of the SLC are limited because one jammer may produce more than one correlated jamming source. A tapped delay-line processor combines delayed and weighted replicas of the input signal to form the filtered output signal, and thereby has the potential for providing compensation for multipath effect because multipath rays also consist of delayed and weighted replicas of the direct path ray. An important point to note is that the canceler is ineffective unless $B_c\, t_{\rm spec} \ll 1$, where $t_{\rm spec}$ is the delay of the specular jammer multipath. In fact, direct and multipath jammer signals need a high degree of correlation for effective cancellation. For jammers at long distances and radars with large bandwidths, it may be impossible to achieve $B_c\, t_{\rm spec} \ll 1$. It is therefore desirable to partition the total bandwidth B_c into N sub-bands of width ΔB such that $\Delta B t_{\rm spec} \ll 1$ even though $B_c\, t_{\rm spec} \gg 1$. This separation can be achieved either by using an FFT processor in the SLC channels or by resorting to the tapped delay line of Figure 4.36a (Fante, 1991). A detailed analysis of the compensation of the multipath by use of Figure 4.36(a) has been done in Monzingo and Miller (1980, pp. 451–461), where indications are given for selecting the number of taps and their time duration. To maintain the value of intertap delay close to the delay time associated with the multipath ray was found to be desirable (Monzingo and Miller, 1980, p. 476). The number of taps required for one indirect ray is limited to 5 for 30 dB of cancellation and to 7 for 40 dB of cancellation (Monzingo and Miller, 1980, p. 459). The problem of direct and reflected paths may involve rapid updates of the adaptive weights to track the variations of the interfering environment when the platform carrying the jammer moves with respect to the radar.

4.3.2.5 Trade-off of Steady-State Cancellation and Convergence Time

This trade is a fundamental issue for all adaptive systems and, in particular, adaptive arrays, including SLC. As already stated, two figures of merit define the SLC performance: the power cancellation ratio at the steady-state and the time required to compute the estimated weights through direct or indirect estimation of the jammer and noise covariance matrix. Unfortunately, these figures compete. Consequently, the following problems arise: (1) to find that adaptation algorithm in which this interdependence is weak and (2) to select the algorithmic parameters to trade-off the canceler performance in steady-state and transient conditions.

In the past, great emphasis was given to this trade-off problem, especially for the analog closed-loop implementation (Section 4.2.1). Because these methods are not appropriate for practical problems involving large degrees of freedom, attention is given to the steady-state and transient trade-off for the direct methods (Sections 4.2.2 and 4.2.3). Nevertheless, we still find it instructive to review the control-loop noise issue, i.e., the phenomenon by which the variance of the array element weights produces an additional noise in the array output. This phenomenon was discovered for the closed-loop algorithms by Widrow (1966) and applied to the adaptive antennas in (Brennan et al., 1971). From this analysis, the output power of the adaptive array becomes

$$\begin{pmatrix} \text{output power} \\ \text{of} \\ \text{adaptive array} \end{pmatrix} = \begin{pmatrix} \text{output power} \\ \text{for} \\ a\ priori\ \text{known weights} \end{pmatrix} \left[1 + \left(\frac{GT_s}{\tau}\right) \sum_{n=1}^{N} \lambda_n \right] \quad (4.161)$$

where G is the gain of the amplifier in the adaptive loop (see Figure 4.7); τ/T_s is the time constant of the low-pass filter as determined by the sampling rate of the system; $\lambda_n (n = 1, 2, \ldots, N)$ is the nth eigenvalue of the covariance matrix of the interference. As is also well known, the array convergence rate is determined by the same eigenvalues of the disturbance covariance matrix. The eigenvalues link the steady-state and transient behaviors. Also shown is that, when the ratio of the sum of the eigenvalues to the smallest eigenvalue is large, there is no set of loop parameters that provides rapid low-noise convergence of the system.

Equation (4.161) has been obtained by assuming that the control-loop bandwidth is much smaller than that of the input jamming signals. The assumption implies that the instantaneous weighting vector is independent of the instantaneous jamming signal. Hence, the expected value of their product (which occurs in the implementation of the algorithm) can be separated into a product of expected values ((4.9f) and (4.31)). This simplifies the analysis of the steady-state and transient behaviors. Another approach suggested (Gerlach and Lang, 1986) removes this assumption to include the input noise, which can vary at any rate. The technique models the jamming sources as continuous-state "jump" Markov processes, which modulate a carrier frequency. By resorting to the theory of stochastic differential equations, the first moment of $\mathbf{W}(t)$ is found. This result allows us to find the null depth and the loop time constant as a function of the adaptive loop parameters and, in particular, that of the time constant of the low-pass filter. These relationships let us select the key system parameters, trading off steady-state and transient performance.

The above-mentioned research applies to the adaptive array case, but is not easily transferred to the SLC system. A mathematical approach to the SLC problem was undertaken by Nitzberg (1981; 1985; 1986) for the digital implementation of the correlation feedback loops of Figure 4.4. The author found that the weight vector $\mathbf{W}$ converges to the optimum value when the gain value of the amplifier in the correlation loop is less than 2 divided by the trace of $\mathbf{M}$ (1986). The steady-state residue power is greater than the optimum residue power (i.e., that corresponding to the case of *a priori* known **W**) because of the effect of the estimated noise. The result is that

$$\begin{pmatrix} \text{output power} \\ \text{of} \\ \text{adaptive SLC} \end{pmatrix} = \begin{pmatrix} \text{output power} \\ \text{for} \\ a\ priori\ \text{known weights} \end{pmatrix}$$

$$\cdot \left\{ 1 - \sum_{n=1}^{N} [G\lambda_n/(2 - G\lambda_n)] \right\}^{-1} \quad (4.162)$$

which is a bit different from the equivalent equation (4.161) for the adaptive array.

Another analysis of the SLC trade-off was done in Farina and Studer (1984) for the continuous-time correlation feedback loops. This analyzed the case of an SLC with either one or two auxiliary antennas. The calculations were made by evaluating the second-order statistics of the weights. In fact, the fluctuation of the weights around their mean values generates the loop-noise effect. For the SLC with one auxiliary, the following expression was found for the steady-state cancellation ratio:

$$\text{CR} = \{(1 - |\rho|^2) + (1 + |\rho|^2) F(\alpha)\}^{-1} \quad (4.163)$$

where $F(\alpha)$ is a function that depends on the key parameter:

$$\alpha = B_L/2B_c \quad (4.164)$$

which is the ratio of the closed-loop bandwidth B_L to twice the jammer bandwidth. The jammer is assumed to occupy the whole radar channel bandwidth. The ideal CR is $(1 - |\rho|^2)^{-1}$ (4.15). The CR, taking into account the loop-noise effect, is shown in Figure 4.37. The figure illustrates the importance of keeping α as low as possible to obtain high cancellation performance, even at the expense of slower circuit response. When $\alpha \to 0$, the CR approaches the ideal value because an infinite time of observation is used to obtain an accurate estimate of the correlation coefficient, ρ. Also note that the higher the correlation coefficient, ρ, between $V_M(t)$ and $V_A(t)$ (i.e., the greater is the ideal CR value), the steeper is the reduction of the SLC cancellation for a given increase in loop bandwidth.

The analysis has been extended to the case of an SLC with two auxiliary antennas. A mathematical expression of the CR has been found, which appears similar to the one previously obtained for the single auxiliary case. The difference is in the ideal cancellation and loop-noise terms. In particular, the loop-noise terms incorporate the fluctuations of both adaptive weights. An illustration similar to Figure 4.37 has been found to show the dependence of CR on the bandwidth ratio, α, and the correlation coefficient, ρ. SLC systems equipped with one

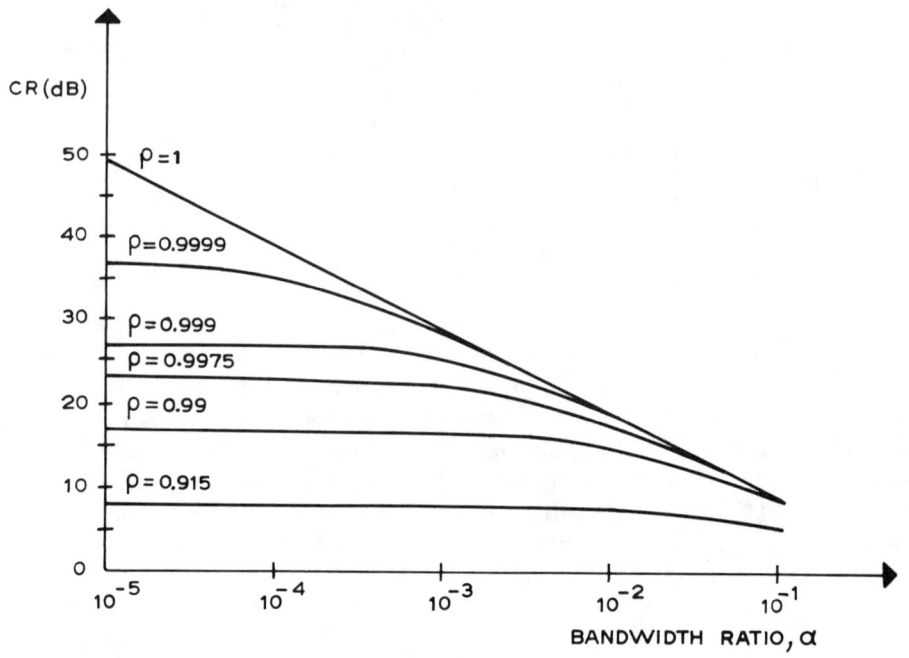

Figure 4.37 The CR for an SLC with one auxiliary, taking into account the loop-noise effect. (From Farina and Studer, IEE Proc., Vol. 129, Pt.F, No. 1, February 1982, pp. 52–58.)

and two auxiliaries have also been compared. For $\alpha \to 0$, the CR of the SLC with two auxiliaries is greater than the value obtained with one auxiliary. When α increases, an SLC with two auxiliaries is more sensitive to the loop-noise effect than is an SLC with one (because two loops contribute to the overall loop noise); therefore, a worst-case CR might result. Consequently, two regions, A_1 and A_2, can be defined on the "$\alpha-\rho$" plane. In region A_1, the SLC with one auxiliary performs better than the SLC with two auxiliaries, while, in region A_2, the situation is reversed. The two regions are illustrated in Figure 4.38. The figure helps selection of the number of auxiliaries and loop bandwidth of the adaptive circuits when one jamming source is to be suppressed. The comparison applies when one jamming source is considered. Results may be very different in the case of more jammers.

A further analysis of the trade-offs between steady-state and transient behaviors for an SLC uses the digital implementation of the Howells-Applebaum system (see Figure 4.7). A comparison has been also made with an SLC using the same number of auxiliaries, but employing the Gram-Schmidt algorithm for the adaptive weight setting (Bucciarelli et al., 1982). In Figure 4.39, the cancellation of one jammer is

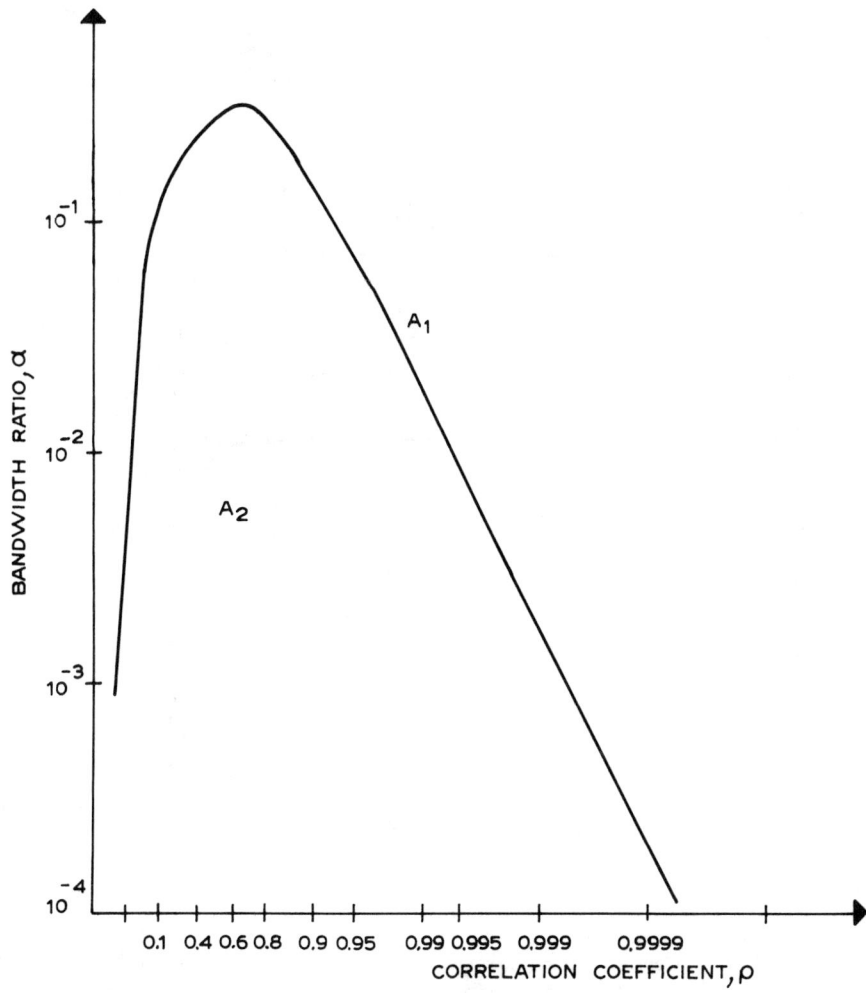

Figure 4.38 Comparison between SLC systems with one and two auxiliary antennas. (From Farina and Studer, IEE Proc., Vol. 129, Pt. F, No. 1, February 1982, pp. 52–58.)

drawn *versus* the bandwidth ratio, α, and for several combinations of N auxiliaries and the correlation coefficient, ρ. When α is small (slow adaptation of the weights), the steady-state jammer cancellation is equal to the ideal value, corresponding to the interference covariance matrix known *a priori*. However, when a faster reaction time is required, the cancellation dramatically decreases due to the strong fluctuation of the weight estimate.

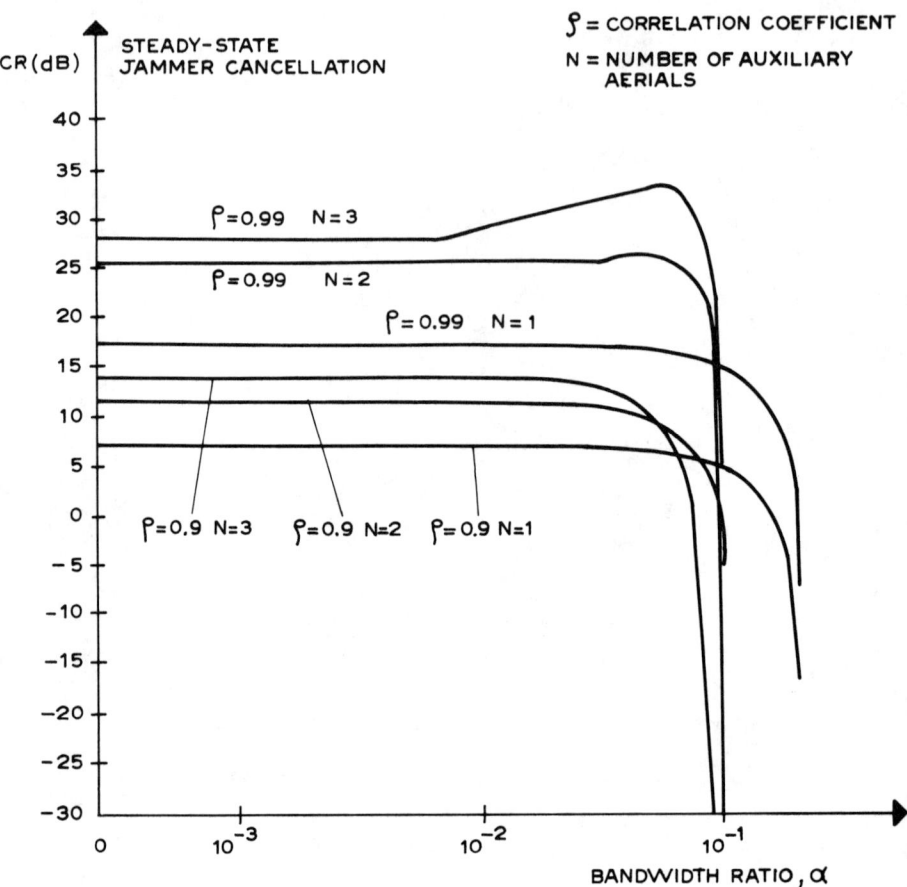

Figure 4.39 Steady-state cancellation of the Howells-Applebaum system. (From Bucciarelli *et al.*, IEE Intl. Radar Conf., Radar 82, London, 18–20 October 1982, pp. 486–490.)

Figure 4.40 instead shows the number of time samples needed to reach the steady-state cancellation value with a 10% error. This number of samples increases with the number N of auxiliaries, the correlation coefficient ρ, and as the bandwidth ratio α decreases. We easily recognize that nonstationary jamming signals cannot be adequately cancelled with the Howells-Applebaum system. Good cancellation requires values of ρ close to unity, a large number of auxiliaries, and small bandwidth ratio, α. The consequence is a very slow adaptation of the weights. An alternative adaptation scheme is based on the Gram-Schmidt algorithm (see Figure 4.11). Recall that the Howells-Applebaum and Gram-Schmidt have the same steady-state CR. In other words, the Gram-Schmidt cancellation *versus* α is that shown in Figure 4.39

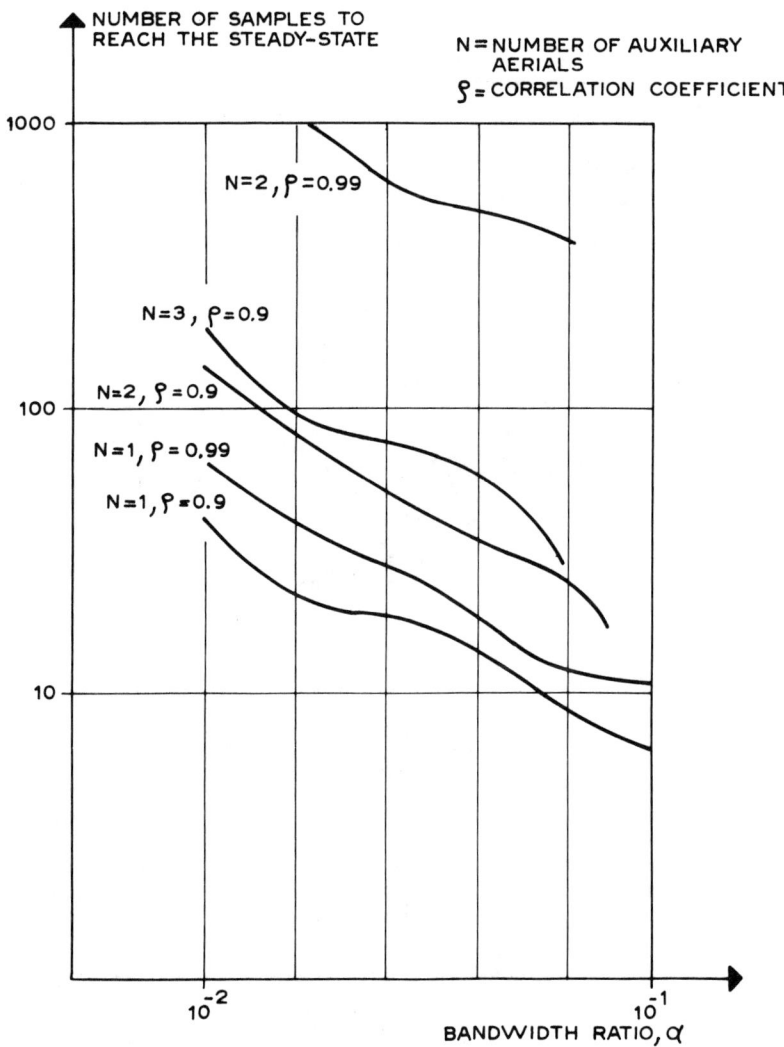

Figure 4.40 Number of samples to reach the steady-state in the Howells-Applebaum canceler. (From Bucciarelli et al., IEE Intl. Radar Conf., Radar 82, London, 18–20 October 1982, pp. 486–490.)

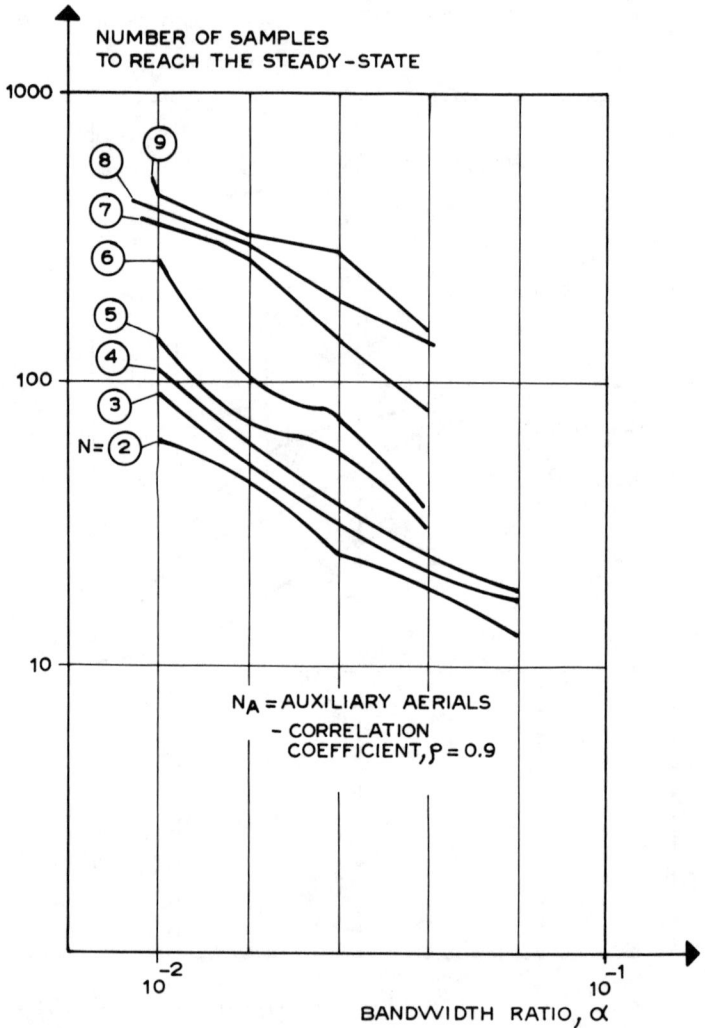

Figure 4.41 Number of samples to reach the steady-state in the Gram-Schmidt canceler. (From Bucciarelli *et al.*, IEE Intl. Radar Conf., Radar 82, London, 18–20 October 1982, pp. 486–490.)

The main difference between the two systems refers to their transient behavior. Figure 4.41 shows the number of time samples needed to reach the steady-state in the Gram-Schmidt case. These results refer to up to $N = 9$ auxiliaries and a correlation coefficient, $\rho = 0.9$. Comparison can be made with the curves of Figure 4.40 for up to $N = 3$ auxiliaries. The comparison cannot be extended to a greater number of auxiliaries due to the slow response of the Howells-Applebaum system, which involves very long computer time for the Monte Carlo simulation. The comparison shows that the Gram-Schmidt system has excellent transient response, thus alleviating the trade-off issue with the steady-state performance. If the same transient time were required of the Howells-Applebaum system, a reduction of the steady-state CR value would result.

4.4 CONCLUDING REMARKS

This chapter has described the SLC system for jammer cancellation. After reviewing the working principle of the SLC, many types of implementation schemes were considered. They can be grouped as closed-loop and open-loop methods, according to the procedures for estimating the covariance matrix of the interference and then evaluating, on a real-time basis, the corresponding weight vector. The benefits and drawbacks of each scheme were indicated. Because of cost and complexity, analog loops, especially in the IF region, were of more interest in the past than digital realizations at baseband. Requirements for many degrees of freedom, accuracy of weights, and fast reaction time recommend digital solutions. A recent breakthrough in SLC technology is represented by the application of linear algebra methods, such as the QR algorithm, which operates directly on the data matrix, thus avoiding the need to calculate the sample covariance matrix. Also relevant is the possibility of mapping these algorithms onto parallel computers such as the systolic arrays. The expected result is the capability of real-time operation in a fast changing interference environment. As a consequence, the preferred technique today should be the open-loop data-domain approach, implemented with digital technology.

Subsequently, we turn our attention to the assessment of the SLC's performance in terms of steady-state CR and transient time response. An extensive list of all error sources that can limit the achievable CR value has been compiled and described. A mathematical analysis has been done for some errors. In addition, methods to compensate for the limitation sources have been discussed. In this case, the trend is toward the use of more auxiliaries with respect to the expected number of jammers. This approach considers the wideband jamming and multipath problems. An additional trend is toward the use of space-time processors with an adaptive tapped delay line downstream from each auxiliary antenna. This configuration is an alternative way to compensate for wideband jammer, channel mismatching, and multipath. The final decision to use more antennas instead of more taps, or a suitable mix of antennas and taps, depends on the specific environment in which the SLC is expected to operate and the costs of the solutions.

Guidelines for the design of a specific SLC system are also found in this chapter. However, the system designer should keep in mind that extensive simulation, followed by verification on the field, is the sole viable approach to design so complex a system, which is affected by several, hardly predictable, limiting phenomena.

Several unclassified experimental results are discussed in the open literature. The reader may refer to the special issues on adaptive arrays (IEE, 1980, 1983; IEEE, 1976a; 1976b; 1982; 1986). More specifically we mention the following papers (Ward *et al.*, 1986; Carlson *et al.*, 1990; Rader and Steinhardt, 1986; Ries and Krucker, 1976; Webb, 1983; Bristow, 1983; Wardrop, 1983; Lacomme, 1984; Old, 1984; Tanaka *et al.*, 1984; Johnson *et al.*, 1991). The success of these experiments justifies the optimism for the SLC, which should be considered as a key ECCM technique. The SLC system is certainly an area of fruitful research and development as testified by a long list of patents (e.g., Brennan *et al.*, 1979; Dollinger, 1977; Downie, 1986a; Farina, 1985, 1986; Hauptmann, 1983; Howard, 1969; Lewis and Olin, 1976; Masak, 1979, 1983).

An additional bonus of the SLC is that it is compatible with other ECCM and processing techniques, such as the frequency agility (to some extent), the sidelobe blanking (Chin *et al.*, 1989), the MTI, and the monopulse processing.

MTI should operate upstream of the SLC. Otherwise, the clutter samples would impair the estimate of the SLC weights. This requires an MTI device for each auxiliary channel in addition to the MTI along the main channel. A more general approach, in which both the spatial (antenna pattern) and temporal (doppler frequency filtering) responses of the system are adaptively controlled, is also possible (Brennan and Reed, 1973). In this case an adaptive space-time system, (of the type discussed in Subsection 4.3.2.4), senses the angular-doppler distribution of the external noise field (both clutter and jammer) and adjusts the adaptive weights to maximize the signal-to-interference ratio and the optimum detection performance. The space-time algorithm can be formulated to exhibit block Toeplitz partitions in the sample covariance matrix as described in Marple (1987, p. 454). Here the covariance matrix is a matrix of block submatrices, which are assumed to contain the time-domain covariances. The periodic samplings of the scattered radar signal in the time-domain produces a time sample covariance matrix which is Toeplitz. Therefore, we can use more efficient algorithms, such as the Levinson lattice and the FFT.

Monopulse processing is still possible, provided that the jammer is cancelled from the sum-and-difference channels (Lin and Kretschmer, 1990). The same set of auxiliary antennas can be used for the SLC systems in the sum-and-difference channels. However, degradation of the estimated accuracy of the angular coordinates of the target should be expected. As a matter of fact, the jammer cancellation in the "difference" channel may produce a shift of the null.

To sum adaptive nulling is an especially attractive option for cases in which the required long-term illumination tolerances for fixed networks of ULSAs (Chapter 2) become impractical, or when the beamwidth increase and loss associated with severe aperture tapers cannot be accommodated. When carefully implemented adaptive nulling can achieve extremely deep nulls at locations of interfering signals (e.g., 68 dB is reported for an experimental system described by Carlson *et al.* (1990)). The nulls produced can be much deeper than the sidelobe level that can be achieved in practice with an amplitude taper alone. The price of the superior performance is the increased complexity of the antenna and signal processing systems. The adaptive antenna system is capable of automatically moving the nulled positions as those of the interfering sources move with respect to the antenna. In addition, adaptive nulling can automatically compensate for component aging, temperature effects, and mutual coupling among individual antenna elements (Carlson *et al.*, 1990).

Many research trends can be envisaged for the SLC system. The following is a sample of interesting topics:

- SLC should operate in connection with extremely wideband radar;
- Very fast adaptation algorithms can cope with rapidly varying jamming and clutter environments as is expected of airborne radars;
- SLC can simultaneously suppress impulse (such as due to deception jammers) and continuous jamming (Downie, 1986b);
- SLC with many auxiliaries can cope with many jammers (White, 1983, 1987; Yuen, 1989);
- SLC provides nulling of main-beam jammers (White, 1987);
- Nonstationary adaptive algorithms (Kalman) use more *a priori* information on the kinematics of moving platforms carrying the radar;
- Under certain conditions, SLC is compatible with pulse compression devices (Gerlach, 1991);
- SLC is compatible with CFAR thresholding (Kelly, 1986; Hendon and Reed, 1990);
- SLC provides cancellation of correlated jammers and jammer multipath;
- SLC provides near-field adaptive nulling (Fenn and Tsandoulas, 1989).

Main-lobe notchers deserve additional consideration (Lewis *et al.*, 1986). They are used to cancel high-duty-cycle interference entering the main lobe of a radar. Main-lobe notchers are implemented in the same way as coherent sidelobe cancelers, except that the auxiliary antenna of the main-lobe notcher must have a much higher gain than that of a sidelobe canceler. This requirement is due to the fact that the radar antenna's main lobe has a much higher gain than its sidelobes. Thus, to keep the notcher weight near unity, the auxiliary antenna must have a gain close to that of the radar antenna's main lobe. If the weight is kept near unity, the canceler is prevented from amplifying the thermal noise in the auxiliary receiver and adding it to the radar signals. If the weight is allowed to become large, the canceler effectively jams the radar with thermal noise. An example of a high-gain auxiliary antenna is the twin feed of a monopulse antenna (Farina, 1985). A further example of a high-gain auxiliary antenna is to use a cross-polarized feed as the auxiliary. With this arrangement, the canceler steers a polarization null rather than a directional null, and targets will be accepted, provided that the polarization of the skin echo return differs significantly from that of the source of interference (Fielding, 1976). A problem associated with the main-lobe cancellation is that a null steered in the direction of the main lobe causes cancellation of target echoes as well as interference. However, if the null width is narrower than the main lobe and, provided that the target and source of interference are not in exactly the same direction, cancellation of the interference and reception of the target echo can be achieved, albeit with some reduction of the received target signal power.

Chapter 5
Adaptive Arrays

5.1 INTRODUCTION

A view of an adaptive array is shown in Figure 5.1(a). A linear array (a planar or conformal array could be also considered) of N antennas captures the electromagnetic energy from a source with direction θ. A far-field plane wave is assumed, while a spherical wavefront should be taken into account for near-field sources. Each antenna is equipped with its own receiver (RCVR) and an analog-to-digital converter (ADC). The set of signals, called *snapshots,* received by the antennas at a certain time, after digitization, is sent to a digital signal processor. As the wavefront continues to impinge on the array, more snapshots are collected and used by the signal processor to extract information on the sources present in the environment. Pertinent information is

1. Direction-of-arrival (DOA);
2. Power;
3. Frequency spectrum;
4. Extent in range;
5. Crosscorrelation between the sources;
6. The polarization properties;
7. Evolution of these quantities with time.

The basic purpose of the adaptive array is twofold, namely:

1. The detection of echoes scattered from targets of interest against unwanted clutter echoes and jammer energy;
2. The estimation of the power distribution in the environment (the so-called *spatial spectrum pattern*) with the purpose of understanding, classifying, and tracking the relevant sources of energy surrounding the array.

These are the two major topics discussed in this chapter.

To grasp the working principle of the adaptive array, consider again the scheme of Figure 5.1(a) in which a wavefront, corresponding to a source direction θ from

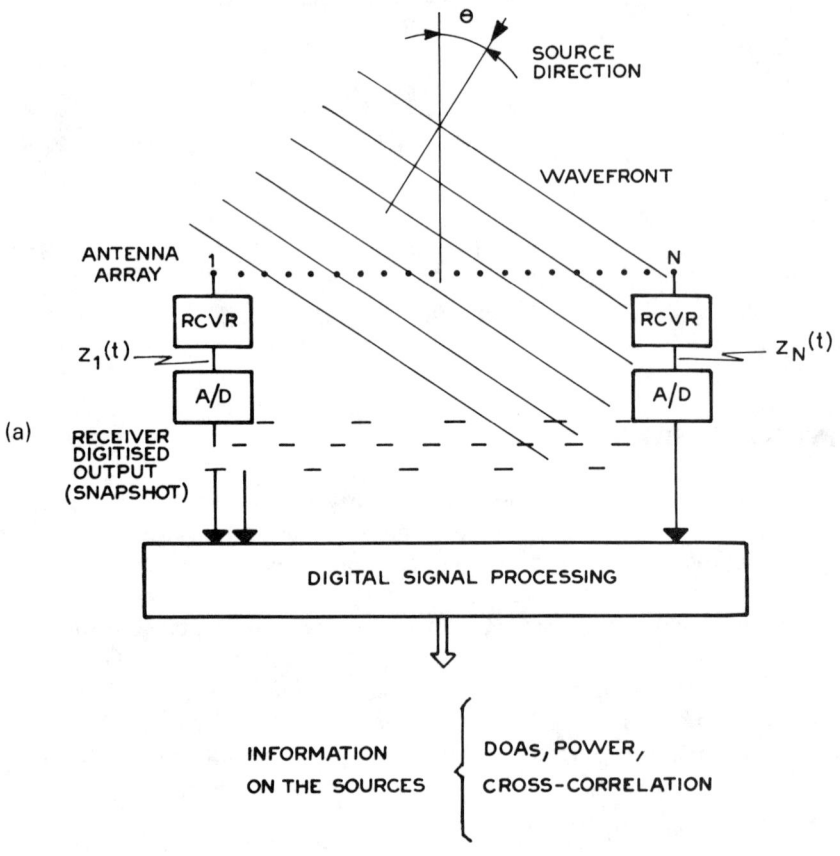

Figure 5.1 (a) The basic concept of an adaptive array of antennas; (b) the adaptive array scheme for jammer cancellation and detection of a useful target.

the boresight, impinges on the array. The phase ϕ_k of the received signal at the kth antenna element is (ωx_k) in radians; x_k is the location of the elemental phase center with respect to the midpoint of the array in wavelengths (λ):

$$x_k = \frac{d}{\lambda}\left(k - \frac{N+1}{2}\right), \quad k = 1, \ldots, N \qquad (5.1)$$

where d is the interelement distance, constant throughout the array, λ is the wavelength of the radiating source, and N is the number of elements in the array; ω is defined as

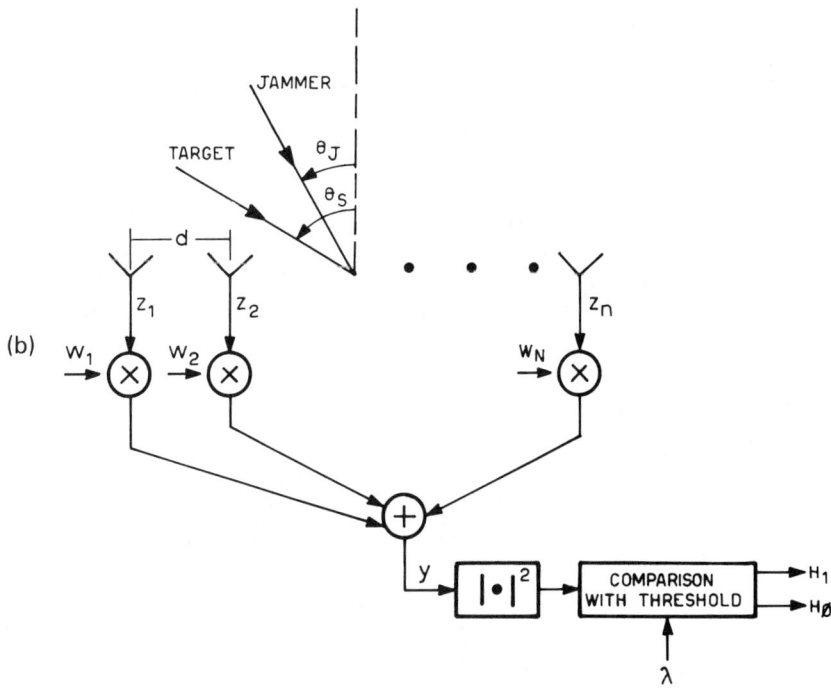

Figure 5.1 Continued

$$\omega = 2\pi \sin\theta \qquad (5.2)$$

We can see that the phase values throughout the array elements change as a function of source direction, θ. In particular, when $\theta = 0$, the phase $\phi_k = 0$ for each $k \in [1,N]$, and the received signal is constant along the array aperture. Instead, when θ is different from zero, the phase ϕ_k varies linearly along the array. Therefore, the received signal changes throughout the array aperture as a sinusoid with a *spatial frequency* equal to $\sin\theta$ (note that sometimes the spatial frequency is also defined as $\sin\theta/\lambda$), i.e., strictly related to the source direction, θ. At this point, the problem of interfering source cancellation or estimation of source parameter (DOA, power, *et cetera*) can be cast in the framework of filter theory. In fact, there is a duality between the spatial domain (i.e., the array aperture) and the more commonly used time domain. For instance, the problem of estimating the source DOA can be handled as the classical problem of estimating, from a limited number of data samples, the frequencies of multiple sinusoids embedded in noise, whereas the problem of jammer cancellation can be seen as an equivalent frequency filtering problem.

The material of the chapter is organized as follows. Section 5.2 introduces the model of the snapshot subsequently used to derive several processing schemes and evaluate their corresponding performance. The jammer cancellation and signal enhancement from a desired target is the theme of Section 5.3. Here the usual array manifold[1] is replaced by the digital signal processing function, which, in its basic form, is a weighting and summing network for the received snapshots. The weights are adjusted adaptively to the environment to desensitize the array along the jammer DOAs and to emphasize the wanted signals at the array output (Applebaum, 1976; Applebaum and Chapman, 1976; Gabriel, 1976; Marr, 1986; IEEE, 1976a, 1976b, 1982, 1986; IEE, 1980, 1983; Brennan and Reed, 1973). The adaptive array looks as a generalization of the SLC system concept described in the previous chapter.

However, arrays with a large number of receiving elements need some form of processing reduction. One approach to partial adaptivity is to arrange the array elements in subgroups, which form the inputs of the adaptive processor (Section 5.4). Careful selection of the subgroup elements is necessary to avoid echelon lobes. Other simplifications of the full adaptive array can be made by deterministic spatial filtering (Section 5.5) and phase-only nulling techniques (Section 5.6).

Attention then focuses on the use of an adaptive array to obtain a high-quality estimate of the spatial spectrum pattern. In particular, the basic elements of theory to achieve superresolution in angle (i.e., to distinguish two sources within the main beam of the array) are discussed in Section 5.7. Finally, the chapter concludes with a review of experimental systems and realizations, as they appear in the open technical literature (Section 5.8) with an indication of possible future trends of research and applications of concepts in operational radar systems (Section 5.9).

5.2 MODEL OF THE ENVIRONMENT

The purpose of the section is to introduce a mathematical model of the snapshot signal received by the array (Gabriel, 1986). Consider again the linear array of N elements spaced $d = 0.5\ \lambda$.[2] The signal environment consists of I narrowband plane waves arriving from distinct directions θ_i, $i = 1, 2, \ldots, I$, where $I < N$. The RF phase at the kth antenna element as a result of the ith source is $\omega_i\ x_k$, where x_k is the location of the phase center of the kth element with respect to the array midpoint in wavelengths and ω_i is

---

[1]Concerning the term *manifold*: array antennas are typically constructed of a collection of slotted waveguides (sometimes called "sticks"); these sticks may be coupled in to another waveguide, which is often referred to as a "manifold" and some articles on digital beam-forming refer to the algebraic summation as digital manifold. See, for instance (Li and Vaccaro, 1990, p. 978).

[2]This condition is dictated by the sampling theorem applied to the spatial domain; in fact, the sampling frequency should be at least two times the maximum spatial frequency $1/\lambda$. Here, we have taken the spatial frequency as equal to $(\sin\theta/\lambda)$.

$$\omega_i = 2\pi \sin\theta_i, \quad i = 1, 2, \ldots, I \tag{5.3}$$

The tth time sampled signal at the kth element of the array is

$$Z_k(t) = n_k(t) + \sum_{i=1}^{I} p_i(t) g_k(\theta_i) \exp(j\omega_i x_k), \quad k = 1, \ldots, N \tag{5.4}$$

where p_i is the complex amplitude of the ith source, $g_k(\theta_i)$ is the pattern response of the kth array element in the direction θ_i, $n_k(t)$ is the tth time sample of the Gaussian receiver noise from the kth array element (the receiver noise component is a random variable, independent of both the time index t and the element index k). The previous equation can be put in the following vectorial form:

$$\mathbf{Z}(t) = \mathbf{V}\mathbf{p}(t) + \mathbf{n}(t) \tag{5.5}$$

which, in expanded form, is

$$\begin{bmatrix} Z_1(t) \\ Z_2(t) \\ \vdots \\ Z_N(t) \end{bmatrix} = \begin{bmatrix} v_{11} & v_{12} & \cdots & v_{1I} \\ & & & \\ v_{N1} & v_{N2} & \cdots & v_{NI} \end{bmatrix} \begin{bmatrix} p_1(t) \\ p_2(t) \\ \vdots \\ p_I(t) \end{bmatrix} + \begin{bmatrix} n_1(t) \\ n_2(t) \\ \vdots \\ n_N(t) \end{bmatrix}$$

$$= \begin{bmatrix} \mathbf{v}_1 & \mathbf{v}_2 & \cdots & \mathbf{v}_I \end{bmatrix} \begin{bmatrix} p_1(t) \\ p_2(t) \\ \vdots \\ p_I(t) \end{bmatrix} + \begin{bmatrix} n_1(t) \\ n_2(t) \\ \vdots \\ n_N(t) \end{bmatrix} \tag{5.6}$$

with $v_{ki} = g_k(\theta_i) \exp(j\omega_i x_k)$. The N-dimensional vector $\mathbf{Z}(t)$ is the tth snapshot, i.e., a simultaneous signal sampling across the N elements of the array at the tth time instant. The $N \times I$ matrix $\mathbf{V}$ of source direction is slowly changing, while the I-dimensional vector $\mathbf{p}$ rapidly changes with time, and needs to be described in statistical terms. The N-dimensional vector $\mathbf{n}(t)$ accounts for the noise in the receiving channels. Assuming that the mean values of $\mathbf{p}(t)$ and $\mathbf{n}(t)$ are zero, the covariance matrix of $\mathbf{Z}(t)$ is given by:[3]

---

[3] As in Chapter 4, we take the following definition of the covariance matrix, $\mathbf{M} = E\{\mathbf{Z}^*\mathbf{Z}^T\}$. However, another definition, $\mathbf{M} = E\{\mathbf{Z}\mathbf{Z}^H\}$, where H stands for conjugate transpose, is also common. The two definitions are equivalent, provided that we adhere to the selected definition throughout. In this chapter, we select the first definition for the cancellation problems; in fact, this is the same definition adopted in the basic references on adaptive arrays. We resort to the second definition for superresolution topics (from Section 5.7.4 onward) because it is more common to do so in the literature.

$$\mathbf{M} = \mathbf{V}^*\mathbf{P}\mathbf{V}^T + \sigma^2 \mathbf{I} \tag{5.7}$$

where $\mathbf{P} = E\{\mathbf{p}^*(t)\,\mathbf{p}^T(t)\}$, σ^2 is the noise variance, $\mathbf{I}$ is the N-dimensional identity matrix. The diagonal elements of $\mathbf{P}$ represent the ensemble average power levels of the sources, while the off-diagonal elements give the existing correlation between the sources. Correlated far-field sources are due to specular reflection or diffraction multipath, or may be produced by smart jammers. Before proceeding, let us write explicit expressions of $\mathbf{M}$ for a few explanatory cases.

Example 1: Consider the case of only one source, i.e., $I = 1$, and isotropic receiving elements of the array, i.e., $g_k(\theta) = 1$, $k = 1, 2, \ldots, N$. Equations (5.6) and (5.7) become

$$\begin{bmatrix} Z_1(t) \\ \vdots \\ Z_N(t) \end{bmatrix} = \begin{bmatrix} \mathbf{v}_1 \end{bmatrix} p_1(t) + \begin{bmatrix} n_1(t) \\ \vdots \\ n_N(t) \end{bmatrix} \tag{5.8}$$

with

$$\mathbf{v}_1 = \begin{bmatrix} \exp(j\omega_1 x_1) \\ \vdots \\ \exp(j\omega_1 x_N) \end{bmatrix} = \begin{bmatrix} \exp\left[j2\pi \frac{d}{\lambda} \sin\theta_1 \left(\frac{1}{2} - \frac{N}{2}\right)\right] \\ \vdots \\ \exp\left[j2\pi \frac{d}{\lambda} \sin\theta_1 \left(\frac{N}{2} - \frac{1}{2}\right)\right] \end{bmatrix}$$

$$\mathbf{M} = \overline{|p_1|^2} \begin{bmatrix} 1 & & \exp\left[+j2\pi \frac{d}{\lambda} \sin\theta_1(N-1)\right] \\ \vdots & 1 & \vdots \\ \exp\left[-j2\pi \frac{d}{\lambda} \sin\theta_1(N-1)\right] & & 1 \end{bmatrix} + \sigma^2 \mathbf{I}$$

$$= \overline{|p_1|^2}\, \mathbf{v}_1^*\mathbf{v}_1^T + \sigma^2 \mathbf{I} \tag{5.9}$$

where $\overline{|p_1|^2}$ is the average power of the source.

Example 2: Consider the case of I uncorrelated sources and isotropic receiving elements. The covariance matrix becomes

$$\mathbf{M} = \sum_{i=1}^{I} \overline{|p_i|^2}\, \mathbf{v}_i^*\mathbf{v}_i^T + \sigma^2 \mathbf{I} \tag{5.10}$$

where $\overline{|p_i|^2}$ is the average power of the ith source and $\mathbf{v}_i$ is

$$\mathbf{v}_i = \begin{bmatrix} \exp(j\omega_i x_1) \\ \vdots \\ \exp(j\omega_i x_N) \end{bmatrix} \quad (5.11)$$

Example 3: The explicit expression of the covariance matrix **M** is now obtained for $I = 2$ correlated sources and isotropic receiving elements. See also (Kwok and Brandon, 1979a,b).

$$\mathbf{M} = \overline{|p_1|^2}\, \mathbf{v}_1^* \mathbf{v}_1^T + \overline{|p_2|^2}\, \mathbf{v}_2^* \mathbf{v}_2^T + \sqrt{\overline{|p_1|^2}\,\overline{|p_2|^2}}\, |\rho_{12}|[e^{-j\phi_{12}} \mathbf{v}_1^* \mathbf{v}_2^T + e^{+j\phi_{12}} \mathbf{v}_2^* \mathbf{v}_1^T] + \sigma^2 \mathbf{I} \quad (5.12)$$

where $\rho_{12} = |\rho_{12}|\, e^{j\phi_{12}}$ is the complex-valued correlation coefficient between the two sources.

The $(i - j)$th element of **M** can be explicitly written in a simple form for the case of two sources impinging on the array from two directions, symmetric with respect to the boresight (i.e., $\theta_1 = -\theta_2$):

$$\mathbf{M}_{ij} = \sigma^2 \delta(i,j) + \overline{|p_1|^2} \exp[-j2\pi \sin\theta_1(x_i - x_j)] + \overline{|p_2|^2} \exp[-j2\pi \sin\theta_2(x_i - x_j)]$$
$$+ 2\sqrt{\overline{|p_1|^2}\,\overline{|p_2|^2}}\, |\rho_{12}| \cos[\phi_{12} + 2\pi(x_i \sin\theta_1 - x_j \sin\theta_2)] \quad (5.13)$$

where $\delta(i,j)$ is the Kronecker operator, which is one if $i = j$ and zero otherwise. The first three terms describe the structure of the array covariance matrix when uncorrelated sources are present, and the last term represents the effect of correlation that induces a modulation on the elements of **M**. The phase of modulation for the (ij)th element is given by

$$\phi \triangleq \phi_{12} + 2\pi(x_i \sin\theta_1 - x_j \sin\theta_2) = \phi_{12} + 2\pi(x_i + x_j)\sin\theta_1 \quad (5.14)$$

This term may produce a degradation in the performance of superresolution algorithms. Section 5.7.6 shows how to reduce this modulation by a spatial averaging process.

Example 4: Now let us evaluate the covariance matrix **M** for a wideband source with a flat spectrum in the bandwidth of the receiving channels of the array. Assume, for convenience, identically shaped filters of the RCVRs downstream from the antennas of the array (see Figure 5.1(a)). The hypothesized shape is Gaussian. To calculate the $(h - l)$ element of matrix **M**, we need to calculate the mean value of the quantity $[Z_h(t)\, Z_l^*(t)]$. This is obtained by noting that $Z_h(t)$ and $Z_l(t)$ are delayed versions of the same process: the source signal filtered by the receiving channel. Consequently, the crosscorrelation, $E\{Z_h(t) Z_l^*(t)\}$, is equal to the autocorrelation of $Z(t)$ evaluated at the time instant $(h - l) d(\sin\theta)/\lambda$, where θ is the DOA source. Finally, the $(h - l)$th element of the matrix $\mathbf{M}$ is found to be

$$\mathbf{M}_{hl} = \overline{|p_1|^2} \rho^{(h-l)^2} \exp\left[+j2\pi \frac{d}{\lambda}(l - h) \sin\theta\right] + \sigma^2\delta(h, l) \quad (5.15a)$$

where ρ is the one-lag autocorrelation coefficient (4.145c):

$$\rho = \exp\left[-\frac{1}{2}\left(\frac{\pi}{1.17}\frac{B}{f_0}\right)^2 \left(\frac{d}{\lambda}\sin\theta\right)^2\right] \quad (5.16)$$

where B is the 3 dB bandwidth of the receiving channels and f_0 is the center frequency. In the monochromatic jammer case, $\rho = 1$ and $\mathbf{M}$ reduces to (5.9).

Another way to calculate the crosspower $\mathbf{M}_{hl}$ of (5.15a) is by integration of the interference power spectral density over the receiver bandwidth. Assuming a flat spectrum, P_J, of the interference in front of the array, taking into account the phase due to the differential path $d(\sin\theta)(l - h)$, and indicating with $H(f)$ the receiver transfer function, we have

$$\mathbf{M}_{hl} = \int_{f_0-B/2}^{f_0+B/2} P_J e^{+j2\pi(d/c)(l-h)f \sin\theta} |H(f)|^2 \, df \quad (5.15b)$$

Assuming that $|H(f)|^2$ is Gaussian shaped, the equation becomes:

$$\mathbf{M}_{hl} = \int_{f_0-B/2}^{f_0+B/2} P_J e^{+j2\pi(d/c)(l-h)f \sin\theta} e^{-\frac{1}{2}\left(\frac{f-f_0}{\sigma_f}\right)^2} \, df \quad (5.15c)$$

where σ_f and the 3 dB bandwidth, B, of the filter are related by $2\sigma_f = B/1.17$. Solving the integral (5.15c), we obtain the original (5.15a), where $\overline{|p_1|^2} = P_J B$.

5.3 JAMMER CANCELLATION AND ENHANCEMENT OF TARGET SIGNAL

This section covers the first problem area of an adaptive array: the detection of useful echoes from desired targets against receiver noise and directional disturbance produced by jammers. The section is divided into four subsections. The basic processing scheme is introduced first. An explanatory set of numerical examples illustrates the performance of the adaptive array in terms of improvement factor and receiver operating characteristics. Then, mathematical tools for the eigenanalysis of the covariance matrix of the disturbance are developed. The analysis allows us to understand the intimate nature of the adaptive working principle. The main limitations of the adaptive array performance are considered next. Numerical examples are presented for the mismatching between the receiving channels downstream of each radiating element of the array. Finally, the effects that are produced by the correlation between the useful signal and jammer or among the jammers themselves are investigated at the end of the section.

5.3.1 Basic Principles

This subsection reviews the principles of the adaptive array, whereby the angular pattern of the antenna is adaptively controlled. According to the scheme of Figure 5.1(b), the snapshots are linearly combined by means of the set of weights $\mathbf{W}^T = [W_1, W_2, \ldots, W_N]$. The adaptive array senses the angular distribution of the external disturbance field and adjusts the array weights for maximum signal-to-interference ratio and optimum detection performance. The problem of finding the optimum set $\hat{\mathbf{W}}$ of the weights has been considered by Brennan and Reed (1973) and Applebaum (1976). The ensuing material also follows the theory developed in Monzingo and Miller (1980).

Suppose we have a target at an angle θ_S from the array boresight. Denote the expected echoes from the target by the signal vector:

$$\mathbf{S}^T = [S_1, S_2, \ldots, S_N] \tag{5.17}$$

where S_i is a complex number of the following expression:[4]

$$S_i = A \exp\left[-j2\pi \frac{d}{\lambda}(i-1)\sin\theta_S - j\phi\right] \tag{5.18}$$

with A denoting the *a priori* known amplitude and ϕ is the phase of the target signal. The phase is random and evenly distributed in $[0, 2\pi]$. Denote the jammer and receiver noise samples on the array elements by the vector:

$$\mathbf{N}^T = [n_1, n_2, \ldots, n_N] \tag{5.19}$$

---

[4]For the sake of simplicity, we have assumed equally spaced elements in the array. The theory is quite general and also applies to unequally spaced elements. Also, the reference element is at the extreme left side of the array.

In general, **N** is assumed to have a Gaussian pdf with **0** mean value and covariance matrix $\mathbf{M} = E\{\mathbf{N}^*\mathbf{N}^T\}$. An estimate of the covariance matrix **M** can be obtained by the following expression:

$$\mathbf{M} = \frac{1}{m} \sum_{l=1}^{m} \mathbf{N}^*(l)\mathbf{N}^T(l) \qquad (5.20)$$

where $\mathbf{N}(l)$ is the lth snapshot of the interference vector and m is the total number of independent snapshots averaged. The target signal plus the interference and the system noise is represented by the vector:

$$\mathbf{Z} = \mathbf{S} + \mathbf{N} \qquad (5.21)$$

To detect the signal **S** in a certain range cell, the received samples are linearly combined with a set of weights **W**, giving rise to the following scalar output:

$$y = \mathbf{W}^T\mathbf{Z} \qquad (5.22)$$

The envelope of y is compared with a suitable threshold λ to ascertain which hypothesis (i.e., H_1: target signal present, or H_0: interference alone) actually occurs. Of course, different weights **W** of the spatial filter give rise to different detection performance. Brennan and Reed (1973) show that the filter which gives the maximum probability of detection P_D for a prescribed probability of false alarm P_{FA} is:[5]

$$\hat{\mathbf{W}} = \mu \mathbf{M}^{-1}\mathbf{S}^* \qquad (5.23)$$

where μ is a nonzero complex number. The detection probability for this optimum filter is

$$P_D = Q\left(\sqrt{\mathbf{S}^T\mathbf{M}^{-1}\mathbf{S}^*}, \sqrt{2 \ln \frac{1}{P_{FA}}}\right) \qquad (5.24)$$

---

[5] An interesting generalization of this equation for a CFAR detection scheme has been proposed by Kelly (1986; 1989), and Hendon and Reed (1990). Zunich and Griffiths (1991) propose yet another generalization.

where $Q(\cdot)$ is the Marcum function. The set of weights (5.23) provides the maximum value of the improvement factor, IF, which is defined as follows:

$$\text{IF} = \frac{\text{(signal-to-interference-plus-noise power ratio at the output } y)}{\text{(signal-to-interference-plus-noise power ratio at the input)}} \quad (5.25)$$
$$= \frac{(\text{SINR})_O}{(\text{SINR})_I}$$

The signal-to-interference-plus-noise power ratio at the input, $(\text{SINR})_I$, is measured at the input of an element of the array. The IF value corresponding to the optimum set (5.23) of weights is:

$$\text{IF} = \frac{\mathbf{S}^T \mathbf{M}^{-1} \mathbf{S}^*}{(\text{SINR})_I} \quad (5.26a)$$

The IF is better suited than the CR, adopted in the SLC, to represent the performance of the adaptive array. In fact, in the adaptive array the useful signal is integrated in space, and simultaneously the interference is cancelled.

It is very instructive to include the derivation of the signal-to-interference-plus-noise ratio SINR_O at the output:

$$(\text{SINR})_O = \frac{|\mathbf{W}^T \mathbf{S}|^2}{|\mathbf{W}^T \mathbf{N}|^2} = \frac{|\mathbf{S}^T \mathbf{W}|^2}{\mathbf{W}^H \mathbf{M} \mathbf{W}} \quad (5.26b)$$

where H stands for complex conjugate and transpose. By substituting (5.23) into (5.26b), we obtain the maximum $(\text{SINR})_O$:

$$(\text{SINR})_O = \frac{|\mu|^2 |\mathbf{S}^T \mathbf{M}^{-1} \mathbf{S}^*|^2}{(\mu^* \mathbf{S}^T \mathbf{M}^{-1}) \mathbf{M} (\mu \mathbf{M}^{-1} \mathbf{S}^*)} = \mathbf{S}^T \mathbf{M}^{-1} \mathbf{S}^* \quad (5.26c)$$

The constant μ of (5.23) has a value that depends on the particular criterion being used. In the important special case where we impose the constraint that the array gain in the look direction be unity, i.e., $\mathbf{S}^t \dot{\mathbf{W}} = 1$, the constant μ is $(\mathbf{S}^T \mathbf{M}^{-1} \mathbf{S}^*)^{-1}$.

Several numerical applications of (5.24) and (5.26a) are now given. We assume one or more uncorrelated wideband jammers impinging on the array. A jammer is characterized by its DOA and the jammer-to-noise power level, JNR, at the input of each antenna of the array. The mathematical expression of the covariance matrix

M is given by (5.15) for only one jammer. The equation also accounts for the monochromatic jammer case, which corresponds to the one-lag autocorrelation coefficient ρ equal to unity.

First, evaluate the optimum IF value of the adaptive array against a monochromatic jammer. Direct application of (5.26a) gives the following result:

$$\text{IF} = (1 + \text{JNR}) \left\{ N - \frac{\text{JNR}}{1 + N \cdot \text{JNR}} \left| \left[\frac{1 - (-1)^N}{2} \right] \right. \right.$$
$$\left. \left. + 2 \sum_{K=1}^{\text{int}(N/2)} \cos \left[\left(K - \frac{N+1}{2} \right) (\phi_S - \phi_J) \right] \right|^2 \right\} \quad (5.27)$$

where int($N/2$) is the closest integer value of $N/2$, and $\phi_S = 2\pi (d/\lambda) \sin\theta_S$, $\phi_J = 2\pi (d/\lambda) \sin\theta_J$. The curves of Figure 5.2 illustrate the improvement factor, IF, against a monochromatic jammer, with a JNR value of 30 dB, *versus* the jammer incidence angle θ_J. The interspace d of the array is 0.5 λ. The curves refer to two different values of the target incidence angle θ_S and two values of the number N of the antennas in the receiving array: $N = 2$ and $N = 10$. The maximum IF value (expressed in dB) is the sum of maximum CR in dB (which is equal to JNR (dB)) plus the spatial integration factor (which is equal to N (dB)). The possibility of a spatial integration of the useful target signal differentiates the adaptive array from the SLC; in the SLC, the target enhancement is provided by the main beam of the radar antenna. The figure shows that the improvement factor greatly decreases as the angular distance between the jammer and the target approaches zero. Figure 5.3 shows the improvement factor against a monochromatic jammer *versus* the number of antennas in the receiving array. The target signal is at broadside. Again, the maximum IF value does not exceed the values of [JNR(dB) + N(dB)]. The improvement increases as the jammer moves away from the target.

Figures 5.4 and 5.5 illustrate the receiver operating characteristics (5.24) of the adaptive array against a monochromatic jammer. The receiver operating characteristics (ROCs) are more relevant than the IF. They describe the processor performance from the input to the output of the detector while the parameter IF describes the performance up to the summer of the array. The probability of false alarm is set to 10^{-6}. The target is received at broadside, while different incidence angles are considered for the jammer. The curves are independent of JNR if JNR $\gg$ 1. The curves have the same shape, resulting from a shift of the same curve by an amount of SNR equal to the IF required under the different jamming conditions. As usual, the detection performance improves by increasing the number N of the array antenna, the SNR, and the angular distance between the DOAs of the jammer and target.

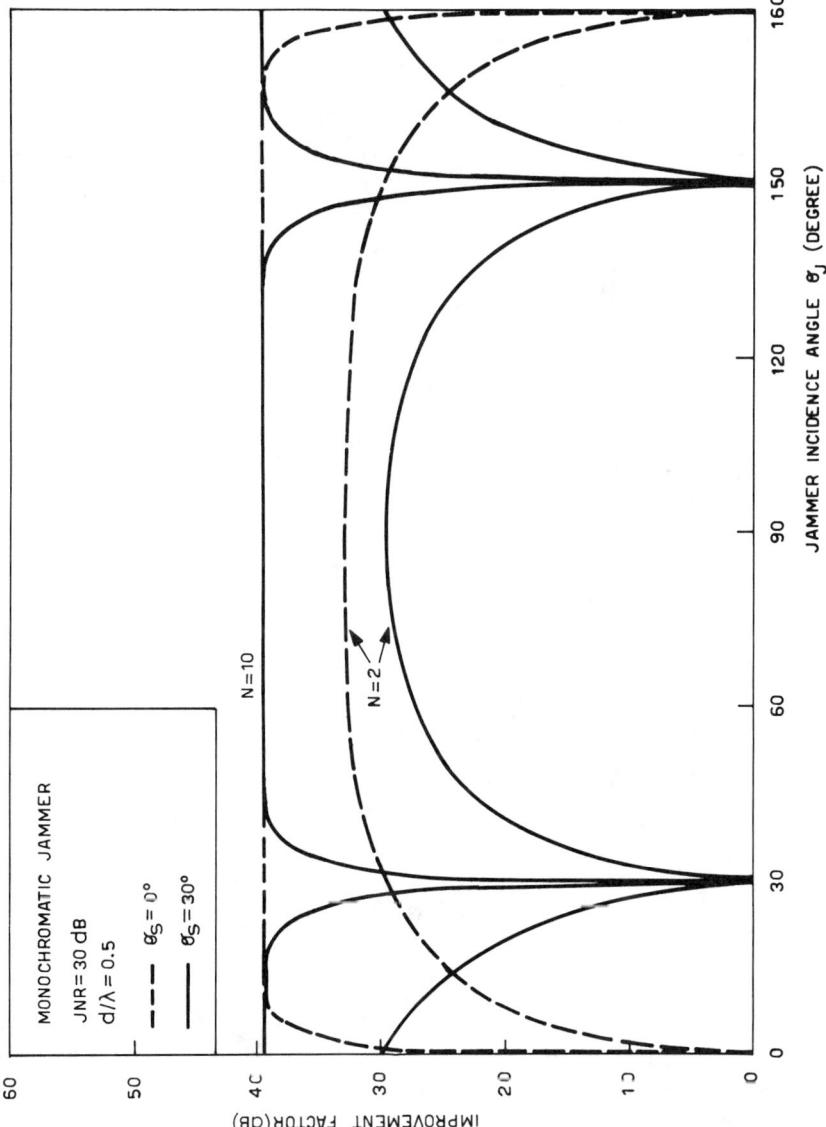

Figure 5.2 Improvement factor of an adaptive array against a monochromatic jammer.

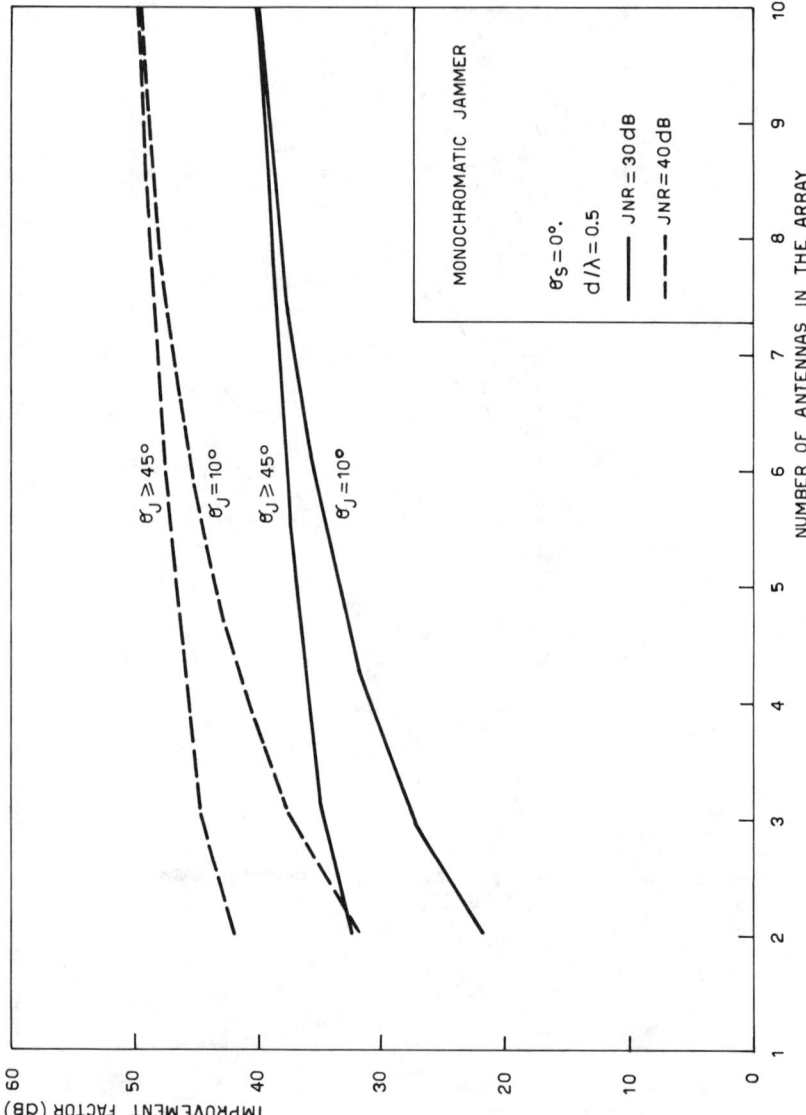

Figure 5.3 Improvement factor against a monochromatic jammer *versus* the number of antennas in the array.

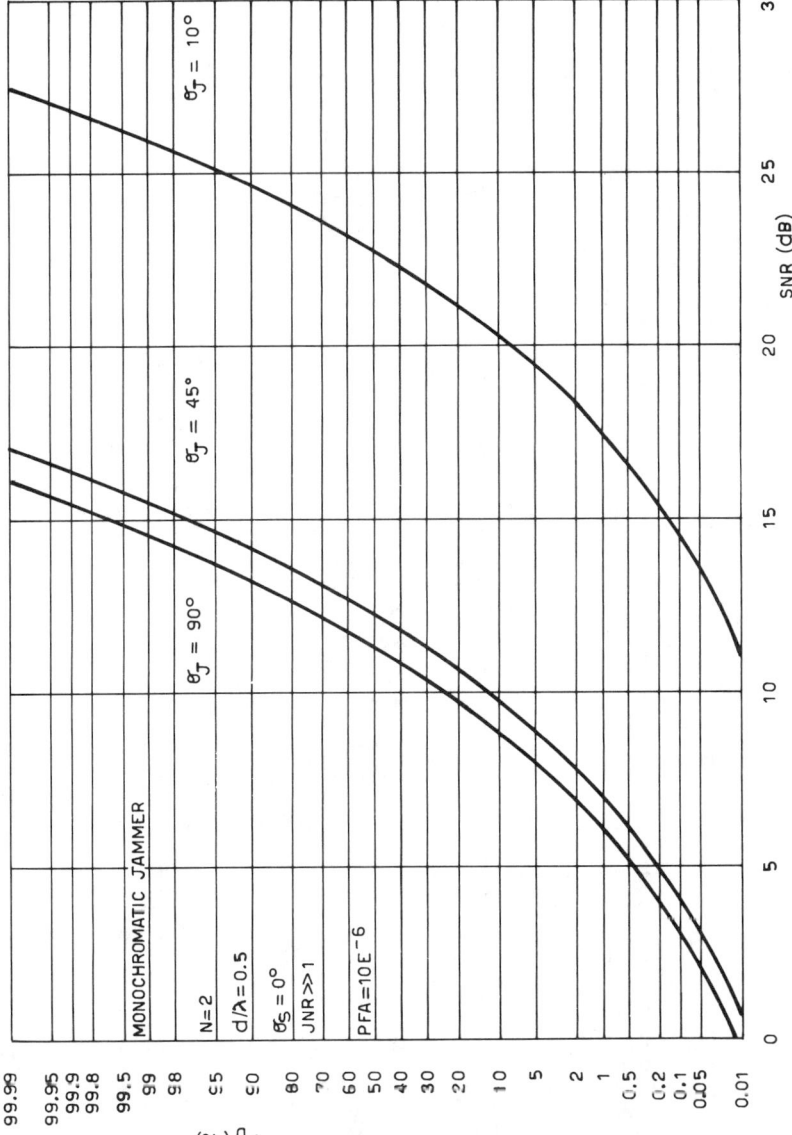

Figure 5.4 Detection probability against a monochromatic jammer, $N = 2$.

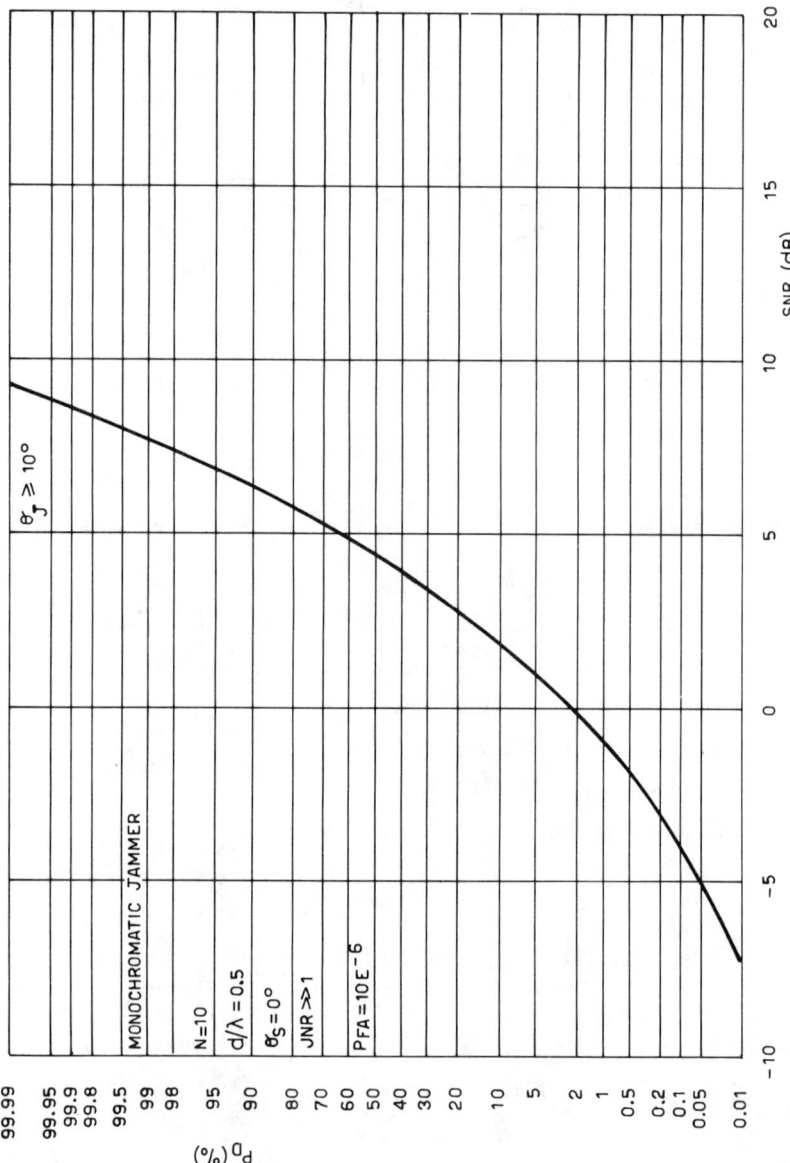

Figure 5.5 Detection probability against a monochromatic jammer, $N = 10$.

Consider now the broadband jamming case. We begin by noting that the bandwidth issue for cancellation is the final bandwidth within which snapshots are taken (usually restricted to the signal bandwidth), whereas the jammer bandwidth is generally many times larger. The IF for the simplest adaptive array with $N = 2$ is shown to be

$$\text{IF} = 2 \, \frac{1 + \text{JNR}\{1 - \rho \cos[(2\pi d/\lambda)(\sin\theta_J - \sin\theta_S)]\}}{1 + 2\text{JNR} + \text{JNR}^2(1 - \rho^2)} (1 + \text{JNR}) \quad (5.28)$$

This IF is drawn in Figure 5.6 *versus* the jammer incidence angle. Two values of the bandwidth (B/f_0), normalized to the carrier frequency f_0, are considered: 1% and 10%; in addition, three different target incident angles are taken into account. A remarkable loss (even 10 db) is suffered in the presence of a wideband jammer $(B/f_0 = 10\%$ with respect to $B/f_0 = 1\%)$. In the broadband jamming case, the IF depends on the additional parameter, ρ (5.16), which, in turn, is related to the jamming incident angle. Thus, the improvement factor does not always increase with the increase of angular distance between the target and jammer (see, for example, the curve (a) of Figure 5.6). In fact, when B/f_0 is very small (e.g., 0.01) the correlation coefficient ρ (5.16) is very close to one, regardless the value of θ_J. The IF is dominated by the angle difference between the target, which is at broadside, and the jammer DOA, as indicated by (5.27). However, when B/f_0 is larger (e.g., 0.1), the correlation coefficient, ρ, decreases with the increase of the jammer incident angle; thus, the adaptive array is less effective as shown by the small cavity of curve (a). Figure 5.7 illustrates the IF *versus* the number of antennas in the array for different values of the JNR and jammer incidence angle; the target is at broadside. Comparing this figure with that for the monochromatic case (see Figure 5.3), a loss is noted due to the nonzero bandwidth of the jammer.

Finally, the detection probability *versus* SNR is calculated for the broadband jamming case (i.e., $B/f_0 = 10\%$). Figure 5.8 illustrates the case of an array with two antennas, for JNR = 30 dB and different incidence angles of the jammer; the target is at broadside. The parameters are the same as curve (a) of Figure 5.6. In the broadband case, the ROC are dependent on JNR for any value of JNR. The prescribed value of PFA is 10^{-6}. Again, from this figure, the detection performance appears to worsen as the jammer moves away from the target incident angle. Figure 5.9 provides the ROC curves for a number $N = 10$ of antennas in the array.

To complete our description of the performance, consider the radiation pattern of the adaptive array. An example is shown in Figure 5.10, which refers to an array with $N = 10$ elements. The signal incident angle is $\theta_S = 0°$, while the jammer incidence angle is $\theta_J = 10°$ and the JNR is set to 30 dB. Two cases are shown: monochromatic jamming and wideband jamming with $B/f_0 = 10\%$. The narrow null corresponds to the monochromatic jammer, while the wider null applies to the wideband jamming. In the latter case, note that the main beam is off-boresight and the sidelobe level is higher with respect to the monochromatic jammer.

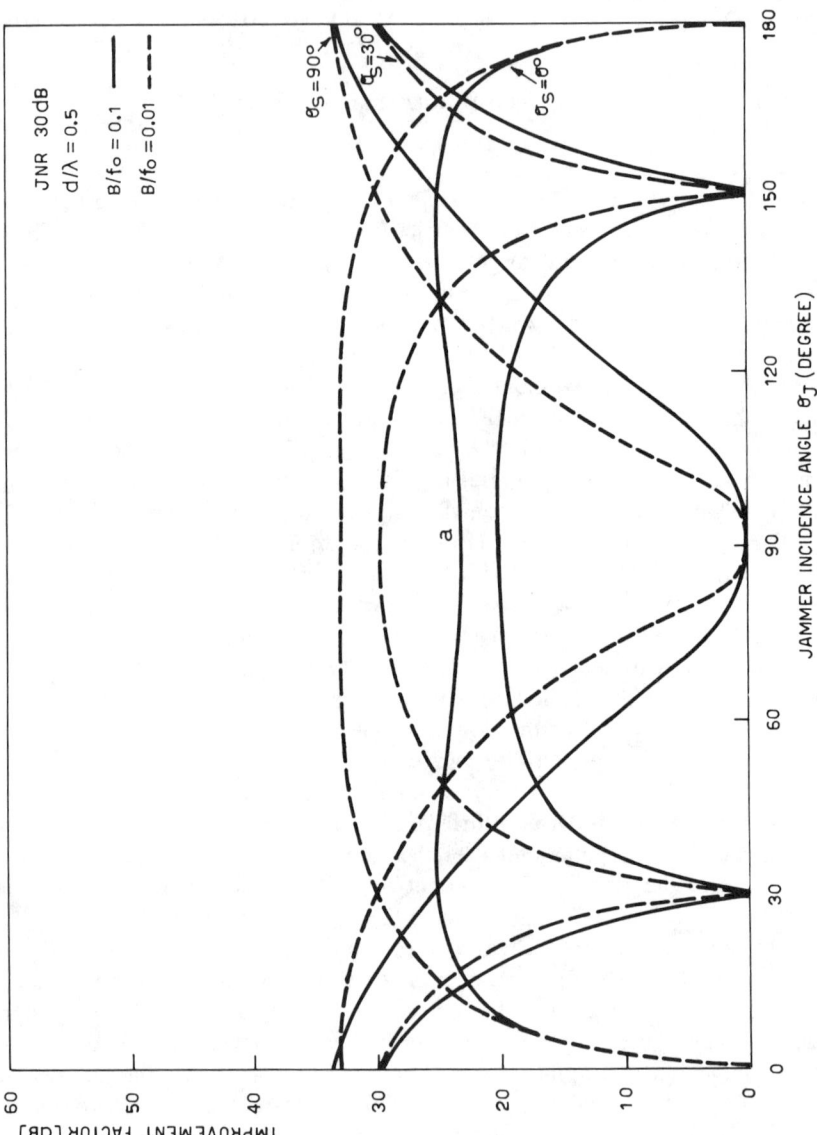

Figure 5.6 Improvement factor in the broadband case for two antennas in the array.

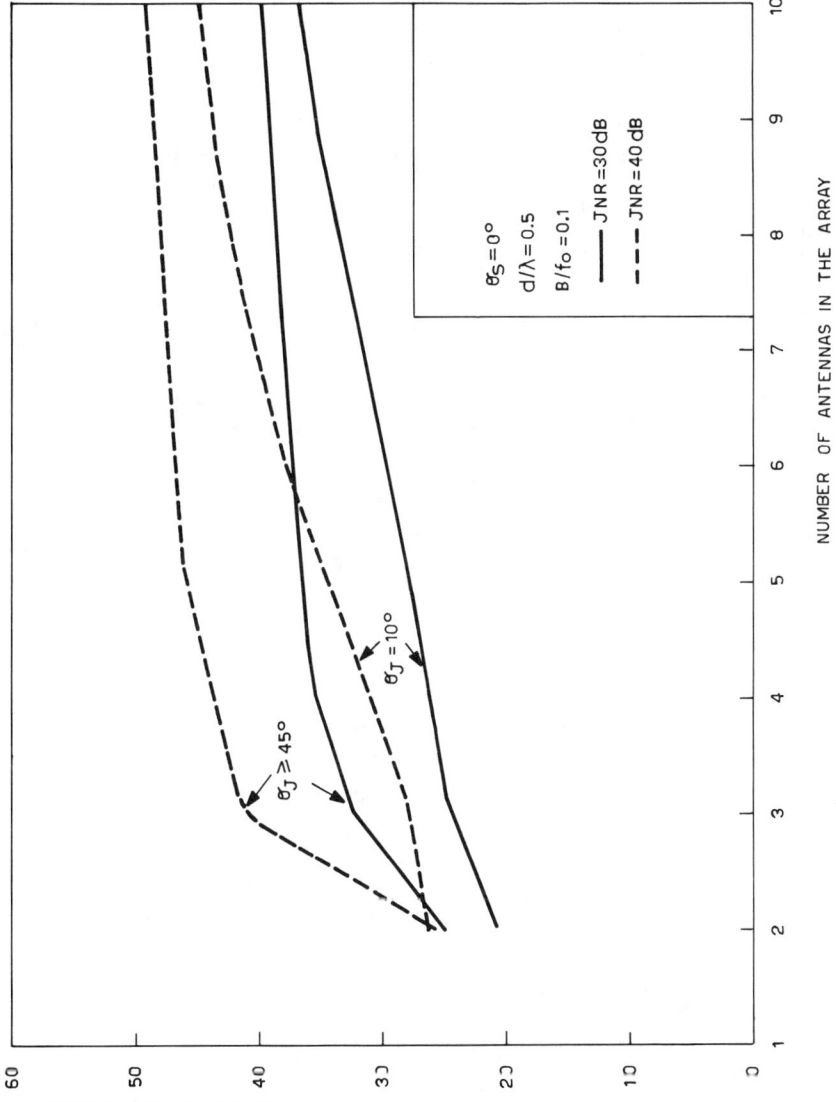

Figure 5.7 Improvement factor in the nonmonochromatic case *versus* the number of antennas in the array.

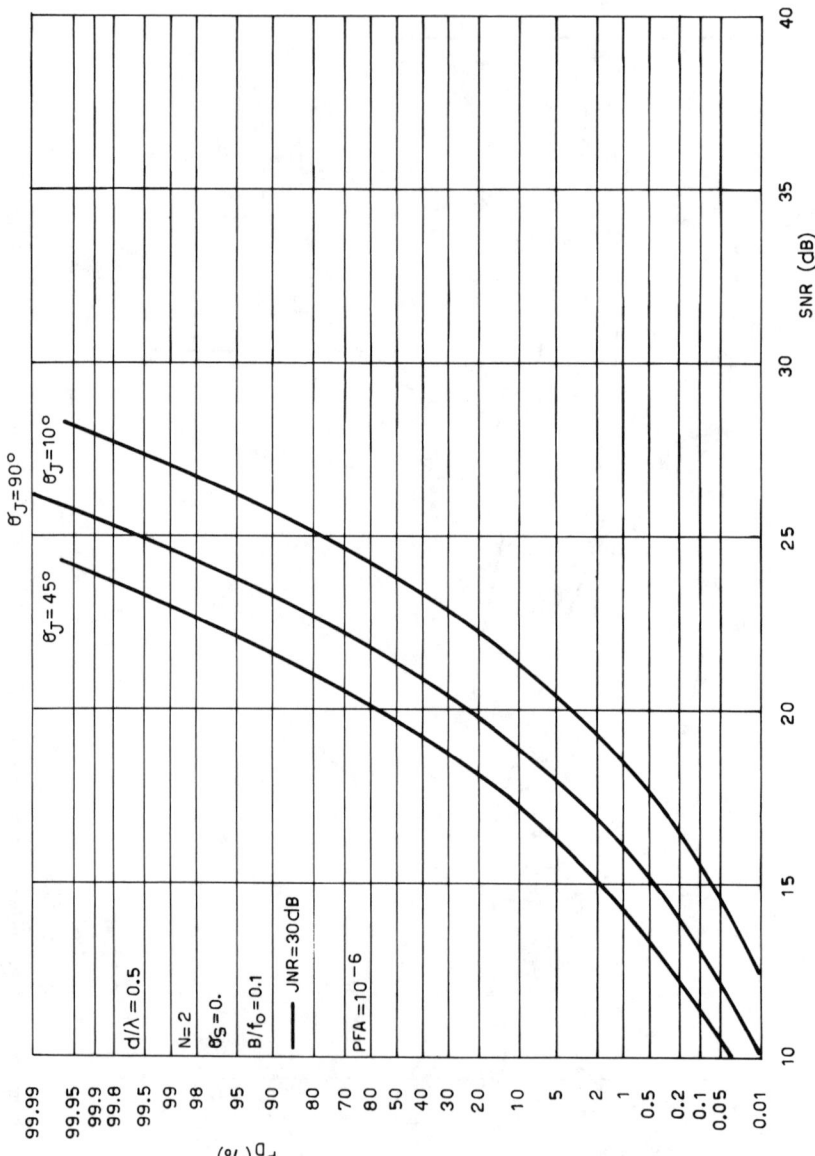

Figure 5.8 The detection probability for the nonmonochromatic jammer case, $N = 2$.

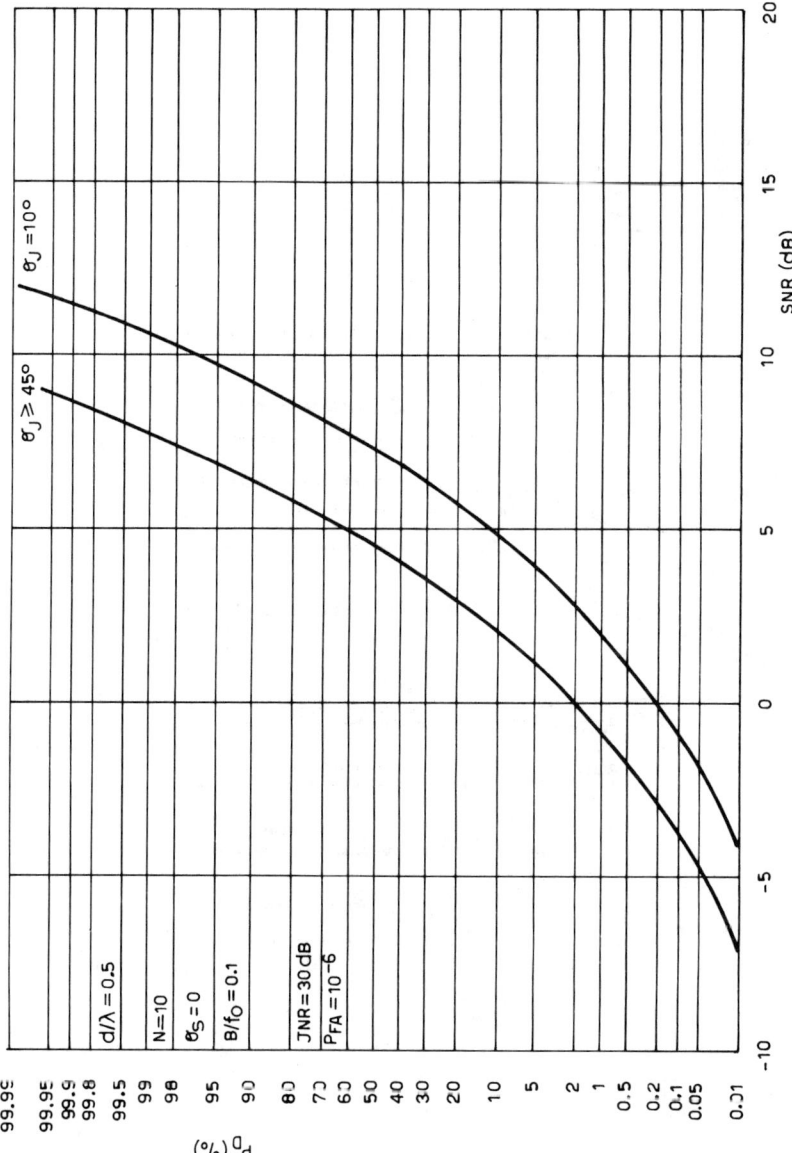

Figure 5.9 The detection probability for the nonmonochromatic jammer case, $N = 10$.

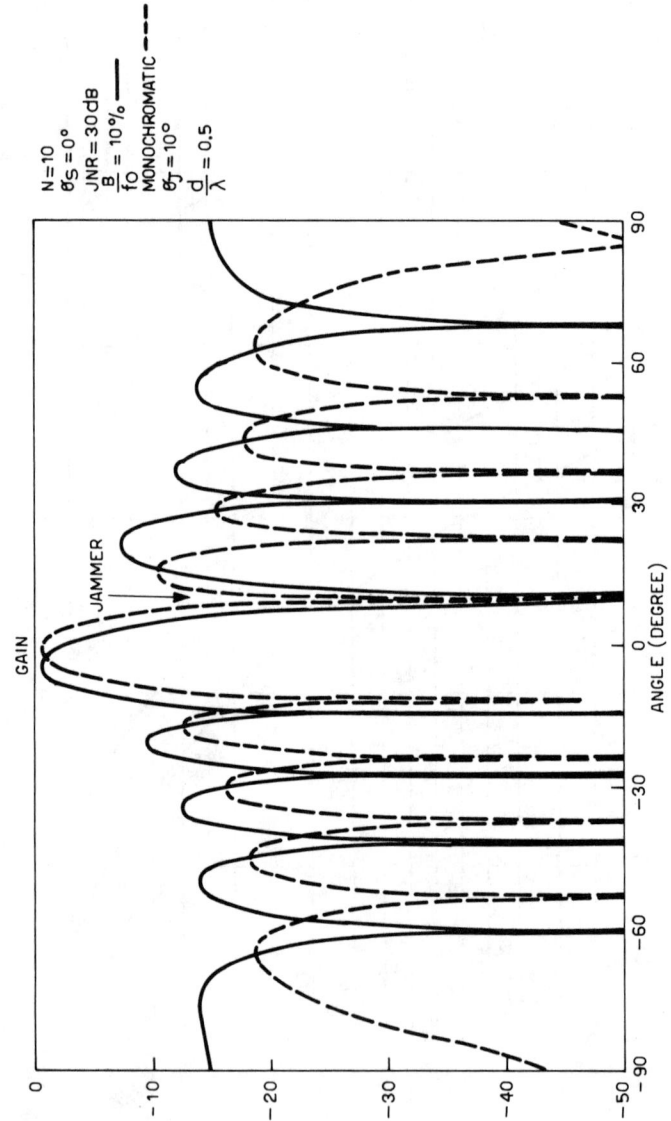

Figure 5.10 Radiation pattern of the adaptive array.

5.3.2 Eigenanalysis Applied to Adaptive Arrays for Jammer Cancellation

The N-dimensional covariance matrix $\mathbf{M}$ is positive definite Hermitian[6] and can be decomposed into eigenvectors and eigenvalues (Marple, 1987, pp. 74–75; Haykin, 1986, pp. 53–66; Gabriel, 1986, Appendix B) as follows:

$$\mathbf{M} = \sum_{k=1}^{N} \lambda_k \mathbf{q}_k \mathbf{q}_k^H \qquad (5.29a)$$

where λ_k and $\mathbf{q}_k$ ($k = 1, 2, \ldots, N$) are the eigenvalues and the eigenvectors of $\mathbf{M}$, respectively, and the symbol H stands for complex conjugate transpose operation. The N eigenvalues are real-positive, while the set of eigenvectors are orthonormal, i.e.,

$$\mathbf{q}_k^H \mathbf{q}_i = \begin{cases} 1, & k = i \\ 0, & k \neq i \end{cases} \qquad (5.29b)$$

The physical meaning of this decomposition can be understood by resorting to the Karhunen-Loeve expansion of the random vector $\mathbf{N}$ having the covariance matrix $\mathbf{M}$. As has been shown (Haykin, 1989, pp. 261–262), $\mathbf{N}$ can be expressed by the linear combination:

$$\mathbf{N} = \sum_{k=1}^{N} \alpha_k \mathbf{q}_k \qquad (5.30)$$

with the combination coefficients having the following property:

$$E\{|\alpha_k|^2\} = \lambda_k \qquad (5.31)$$

Also interesting to note is that the expected value of the energy in the random vector $\mathbf{N}$ is

$$E\{\mathbf{N}^H \mathbf{N}\} = \sum_{k=1}^{N} E\{|\alpha_k|^2\} = \sum_{k=1}^{N} \lambda_k \qquad (5.32)$$

In conclusion, the random vector $\mathbf{N}$ may be expressed as a linear combination of the orthonormal eigenvectors of its covariance matrix $\mathbf{M}$, the combination coefficients

---

[6]A complex-valued matrix is Hermitian if it is equal to its conjugate transpose.

have a mean square value equal to the homologous eigenvalues, and the total energy of the random vector is simply the sum of all the eigenvalues. Thus, the eigenanalysis is a powerful means to describe the intimate nature of the interference **N** received by the array antenna.

We note that, if the random vector **N** is just white Gaussian noise with power σ^2, the eigenvalues are all equal to σ^2. In the more general case of a correlated process, the eigenvalues have different numerical values. Figures 5.11 and 5.12 show, for several practical cases of interest, the eigenvalues, ranked by decreasing amplitude, *versus* the order number one to N. Figure 5.11 shows the eigenvalues for three situations, namely:

1. One monochromatic jammer at 45°, which corresponds to only one eigenvalue different from the other $(N - 1)$ noise eigenvalues, all equal to $\sigma^2 = 1$;
2. Two monochromatic jammers at $-5°$ and 45° that produce two equal eigenvalues with the remaining $(N - 2)$ associated with the noise;
3. Five monochromatic jammers giving five large eigenvalues (not equal) and five noise eigenvalues equal to σ^2.

In the three cases, the jammers have equal power-to-noise ratio value JNR = 30 dB.

Thus, looking at the ranked eigenvalues, in principle, we could count the number of monochromatic sources impinging on the array. When the jammer sources are not monochromatic, the number of eigenvalues that differ from the noise eigenvalues is larger than the jammer number. In fact, a wideband jammer can be considered as a cluster of spatially distributed, narrowband jammers. This is shown in Figure 5.12, which refers to the same arrangement of jammers as the previous figure, but the fractional bandwidth of the jammers is $B/f_0 = 0.1$.

The spectrum of the eigenvalues is also helpful to study the following two operational conditions:

1. The presence of correlated jammers;
2. The presence of very close jammers.

Figure 5.13 gives the eigenvalue spectrum of two monochromatic equal power jammers ($JNR_1 = JNR_2 = 40$ dB) with DOAs, $\theta_1 = 45°$ and $\theta_2 = -45°$, respectively, impinging on an array of $N = 10$ elements. The two jammers are correlated with correlation coefficient $|\rho|$, indicated as a parameter of the curves. When ρ is equal to zero, there are two principal eigenvalues of equal strength. As the correlation coefficient increases from zero to unity, the number of principal eigenvalues progressively reduces from two to one. This means that the array will not be able to distinguish highly correlated sources. We should expect poor performance against correlated sources with superresolution algorithms. This will be discussed in the course of this chapter. Figure 5.14 illustrates the spectrum of eigenvalues for two monochromatic, uncorrelated, equal-power jammers (JNR = 40 dB) received by an array of $N = 10$ elements. As their DOAs, θ_1 and θ_2, are progressively closer, the two

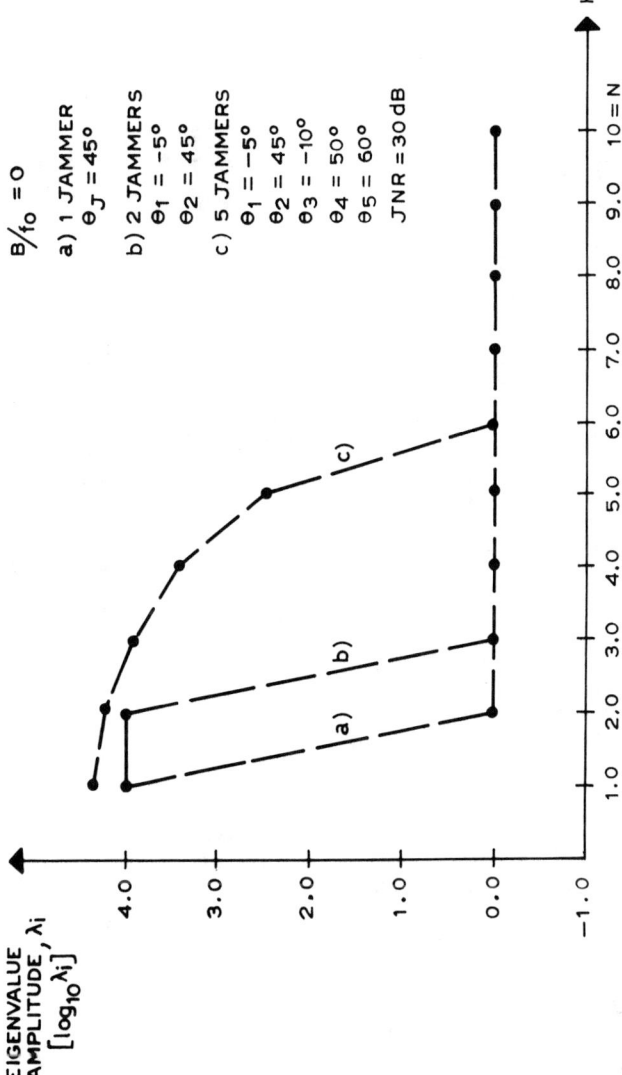

Figure 5.11 Ranked eigenvalues for equal-power monochromatic jammers; array element number, $N = 10$.

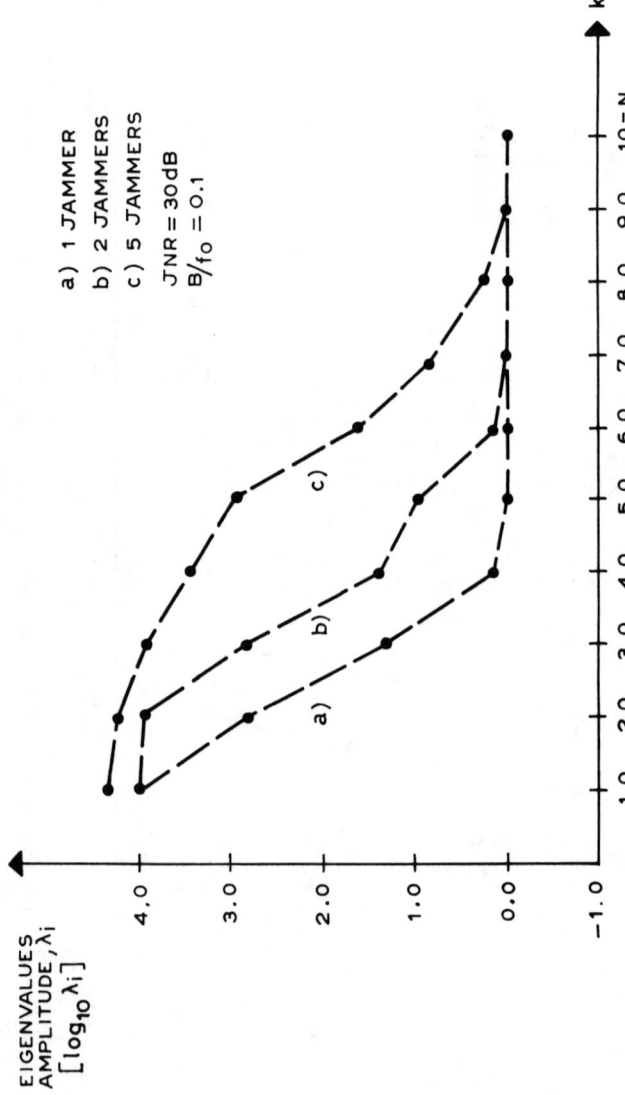

Figure 5.12 Ranked eigenvalues for equal-power broadband jammers; array element number, $N = 10$.

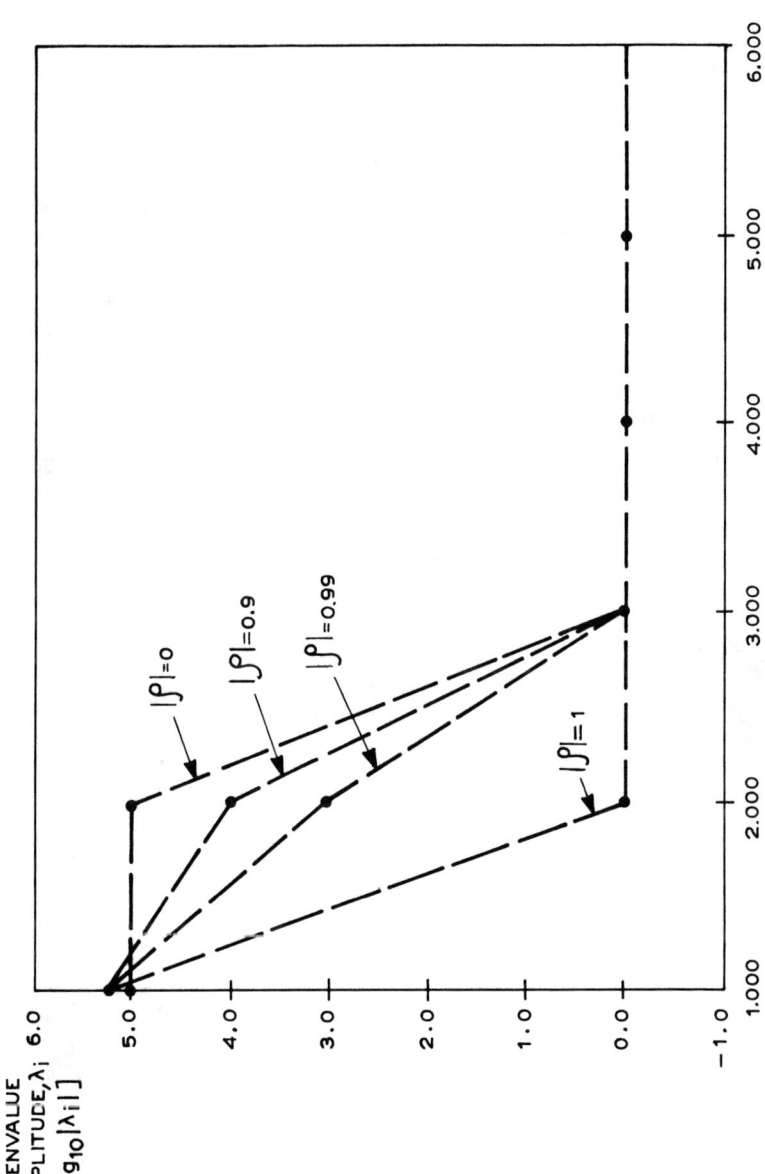

Figure 5.13 Ranked eigenvalues for two equal-power monochromatic but correlated jammers; $JNR_1 = JNR_2 = 40$ dB, $\theta_1 = 45°$, $\theta_2 = -45°$, array element number, $N = 10$.

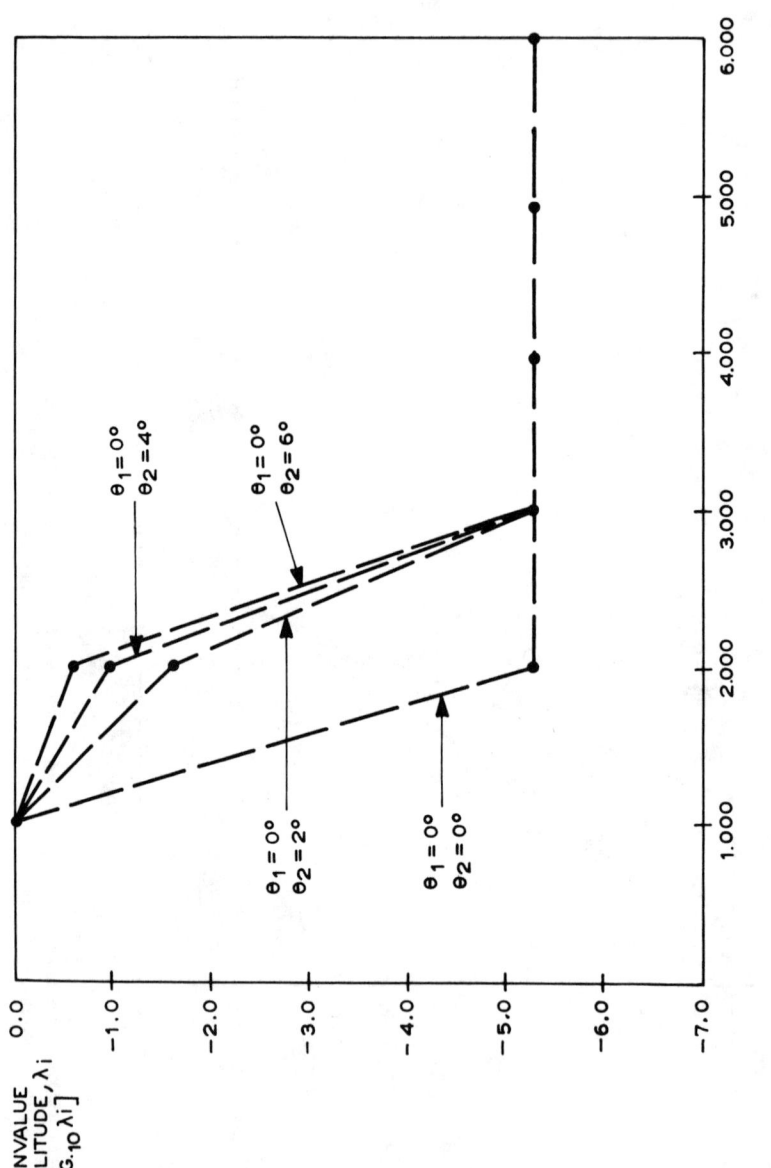

Figure 5.14 Ranked eigenvalues for two equal-power monochromatic and uncorrelated jammers having various incidence angles, θ_1 and θ_2; $JNR_1 = JNR_2 = 40$ dB; number of array elements, $N = 10$.

eigenvalues collapse into one eigenvalue. Note that the collapsing occurs when the two jammers are well contained within the main beam of the array, which is $\theta_{3dB} = 102/N \approx 10°$. This means that the capability of the adaptive array to resolve two sources is only generally related to the beamwidth steered in a given direction, while it is certainly dependent on the power values of the arriving wavefronts.

The eigenanalysis is a powerful tool for better understanding the nature of (5.23). The inverse of the covariance matrix $\mathbf{M}$, normalized by the receiver noise power σ^2 (equal to the minimum eigenvalue of $\mathbf{M}$), is (Gabriel, 1986):

$$\sigma^2 \mathbf{M}^{-1} = \sum_{k=1}^{N} \left(1 - \frac{\lambda_k - \sigma^2}{\lambda_k}\right) \mathbf{q}_k \mathbf{q}_k^H = \mathbf{I} - \sum_{k=1}^{q} \frac{\lambda_k - \sigma^2}{\lambda_k} \mathbf{q}_k \mathbf{q}_k^H \quad (5.33)$$

where q is the number of principal eigenvalues, i.e., those which differ from the noise eigenvalues σ^2.[7] By substituting this equation into (5.23), we obtain

$$\hat{\mathbf{W}} = \frac{\mu}{\sigma^2} \left\{ \mathbf{I} - \sum_{k=1}^{q} \frac{\lambda_k - \sigma^2}{\lambda_k} \mathbf{q}_k \mathbf{q}_k^H \right\} \mathbf{S}^* = \mu' \left\{ \mathbf{S}^* - \sum_{k=1}^{q} \frac{\lambda_k - \sigma^2}{\lambda_k} a_k \mathbf{q}_k \right\} \quad (5.34a)$$

where $\mu' = \mu/\sigma^2$, and $a_k = \mathbf{q}_k^H \mathbf{S}^*$. This new expression for the optimum weights illustrates the concept of *retrodirectivity*. In fact, the optimum weight $\hat{\mathbf{W}}$ is obtained by subtracting, by the quiescent weight $\mathbf{S}^*$ (which is optimum in the absence of jammers), the q principal eigenvectors associated with the q principal eigenvalues. The principal eigenvectors are related to the incident jammers. The concept is demonstrated by considering the components of a principal eigenvector as the weights of the adaptive array. Thus, the receiving antenna pattern is directed toward the jammer. Hence, the retrodirective principle for jammer cancellation is just the subtraction by the quiescent pattern of a number of beams focused on the incident jammers. The principle is illustrated by Figures 5.15 and 5.16, which refer to an array with $N = 10$ elements, receiving two monochromatic equal power jammers with JNR = 30 dB and with DOAs of $\theta_1 = -45°$ and $\theta_2 = +45°$, respectively. The desired look direction is $\theta_S = 0°$. Figure 5.15 shows the pattern obtained by the two principal eigenvectors, and Figure 5.16 depicts the resulting adapted pattern obtained subtracting the principal eigenvector patterns (eigenpatterns) by the quiescent pattern. The result is represented by two deep nulls along the jammer incident angles. The noise eigenvectors are orthogonal to the principal eigenvectors associated with the jammers. This means that, by utilizing the components of a noise eigenvector as weights in our adaptive array antenna, the resulting receiving pattern will have nulls

---

[7]Remember that:

$$\sum_{k=1}^{N} \mathbf{q}_k \mathbf{q}_k^H = \mathbf{I}$$

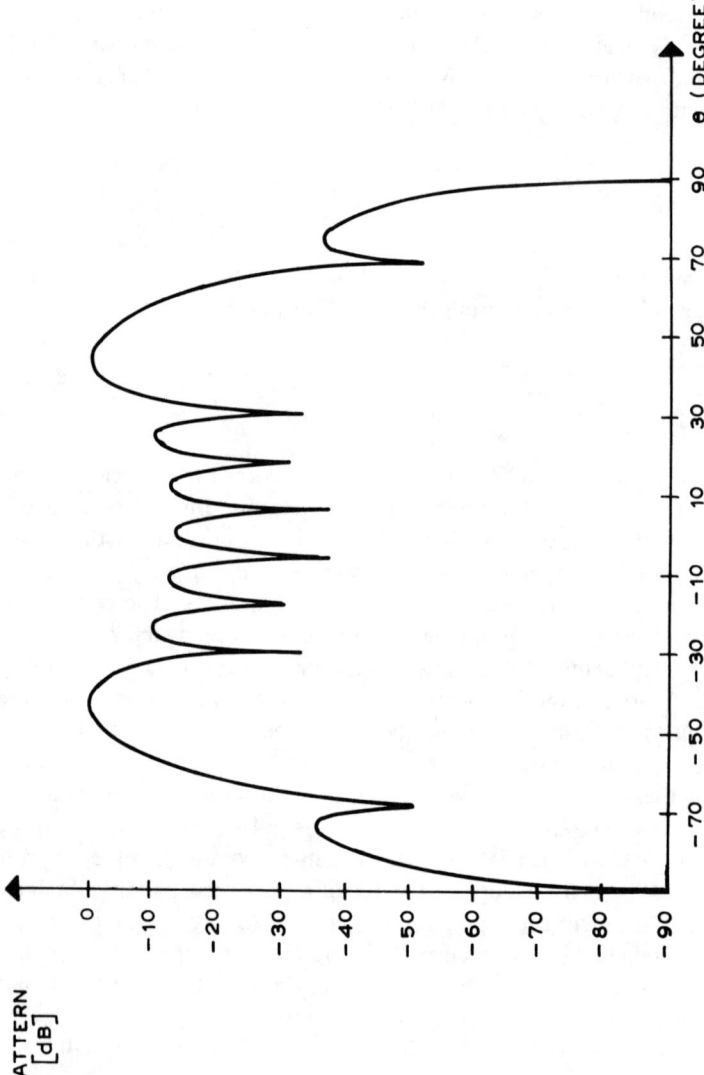

Figure 5.15 Pattern obtained by the two principal eigenvectors; two equal-power (JNR = 30 dB) monochromatic jammers at $\theta_1 = -45°$ and $\theta_2 = +45°$.

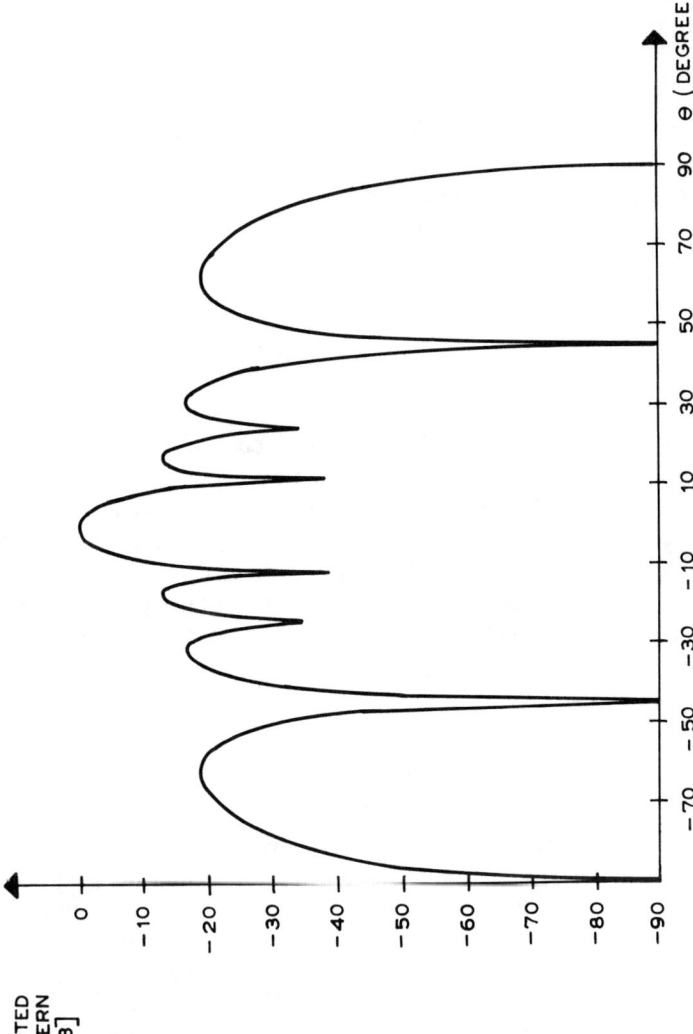

Figure 5.16 Adapted pattern; desired look direction, $\theta_s = 0°$, jammer incidence angles, $\theta_1 = -45°$ and $\theta_2 = +45°$.

in the jammer incident angles and be uniform for all the other angles. This pattern is illustrated by Figure 5.17, which refers to the same jammer scenario as described in Figures 5.15 and 5.16.

With some minor algebra, (5.34a) can be rewritten in another form to show explicitly the so-called "power inversion" law:

$$\hat{\mathbf{W}} = \frac{\mu}{\sigma^2}\left(\mathbf{I} - \sum_{k=1}^{q} \mathbf{q}_k\mathbf{q}_k^H\right)\mathbf{S}^* + \mu \sum_{k=1}^{q} \lambda_k^{-1} a_k \mathbf{q}_k \qquad (5.34b)$$

In this development, the optimum weight vector is the sum of two terms. The first is the projection of the desired weight set $\mathbf{S}^*$ on the complementary subspace of the jammers via the projection matrix

$$\mathbf{P} = \left(\mathbf{I} - \sum_{k=1}^{q} \mathbf{q}_k\mathbf{q}_k^H\right)$$

The second component is the sum of the jammer eigenvectors, each attenuated by their inverse eigenvalues times the projection of the eigenvector onto the desired steering vector $\mathbf{S}^*$. This approach naturally leads to the concept of eigenpatterns, already introduced in relation to Figure 5.15. Also, the power inversion rule (Compton, 1979) is evident from (5.34b). The term derives from the fact that the subspace associated with each jammer's degree of freedom is attenuated by its inverse power, i.e., its inverse eigenvalue. Finally, we note that the projection matrix can be also expressed by means of the noise eigenvectors as

$$\mathbf{P} = \sum_{k=q+1}^{N} \mathbf{q}_k\mathbf{q}_k^H$$

As an additional comment on the analysis of the optimum weight vector $\hat{\mathbf{W}}$ of (5.23), consider the normalized amplitude of the weight components applied at the array elements. Figure 5.18 illustrates the normalized amplitudes of the optimum weight components for one monochromatic jammer at 45° with JNR = 30. Then, we compare this situation with the case of a broadband jammer having $B/f_0 = 0.3$, and the corresponding normalized amplitudes of optimum weight components are depicted in Figure 5.19. We note that in this case, the optimum weight vector produces an amplitude tapering on the antenna aperture to obtain low sidelobes. This is needed to contrast the jammer action on a large portion of sidelobes. The corresponding adapted patterns for these two cases are illustrated in Figures 5.20 and 5.21, respectively. A similar behavior is obtained against five equal-power jammers with incident angles −20°, −40°, 30°, 40°, and 50°, respectively. The normalized

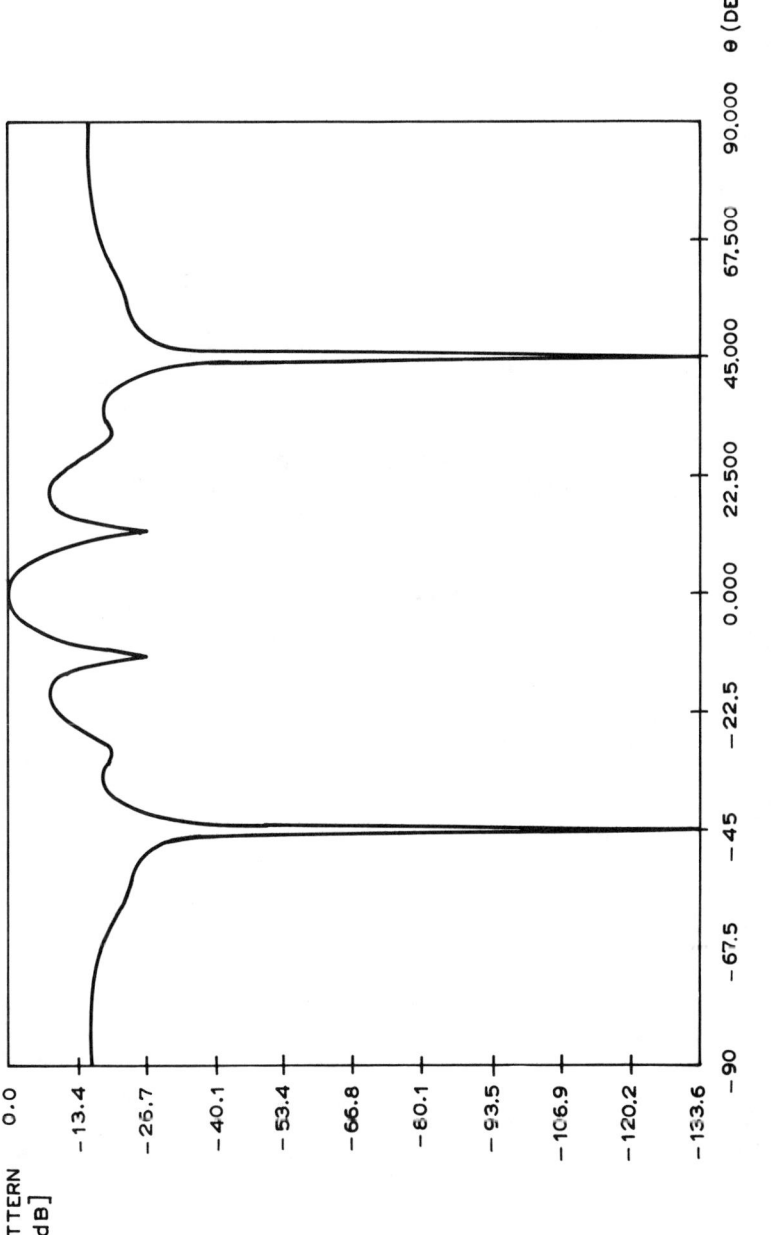

Figure 5.17 Receiving pattern by using the noise eigenvector as weight vector in the adaptive array; two equal-power (JNR = 30 dB) monochromatic jammers with incidence angles, $\theta_1 = -45°$ and $\theta_2 = +45°$; number of array elements, $N = 10$.

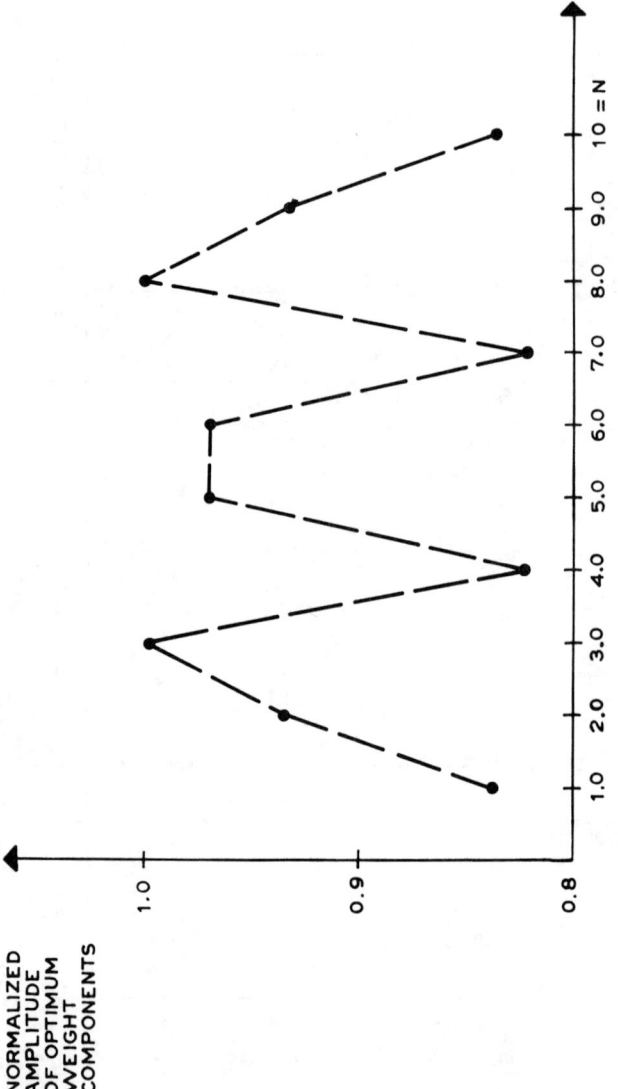

Figure 5.18 Normalized amplitude of optimum weight components for each receiving element of the array; one monochromatic jammer at 45° with JNR = 30 dB; desired look angle, $\theta_s = 0°$; number of array elements, $N = 10$.

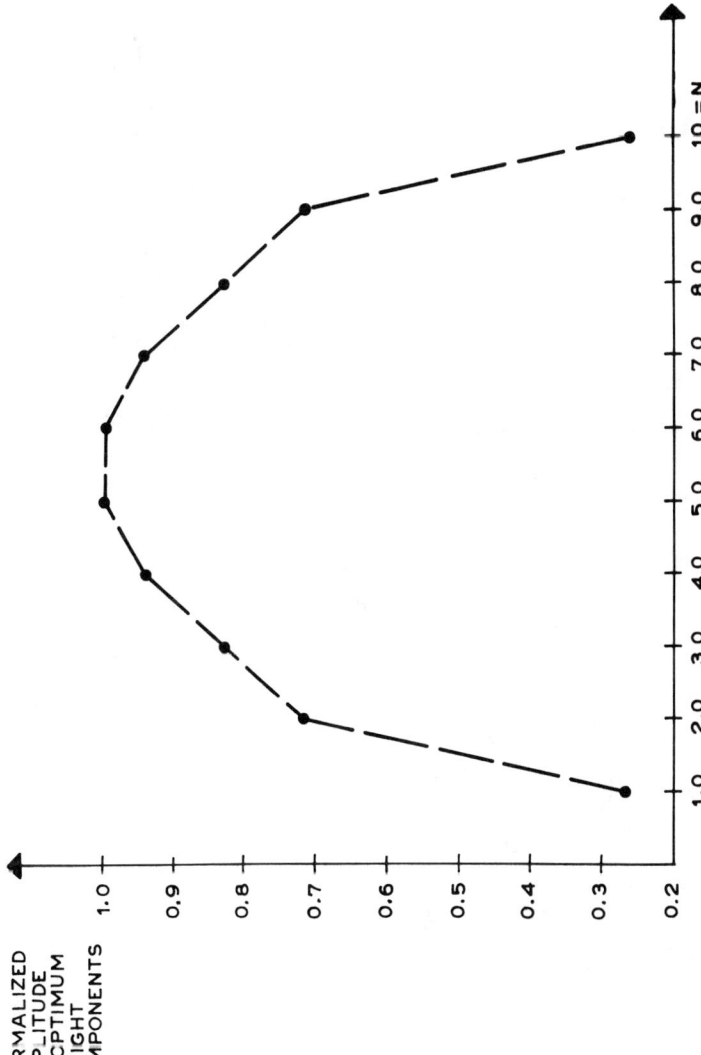

Figure 5.19 Normalized amplitude of optimum weight components for each receiving element of the array; one wideband jammer ($B/f_0 = 0.3$) at 45° with JNR = 30 dB; desired look angle, $\theta_s = 0°$; number of array elements, $N = 10$.

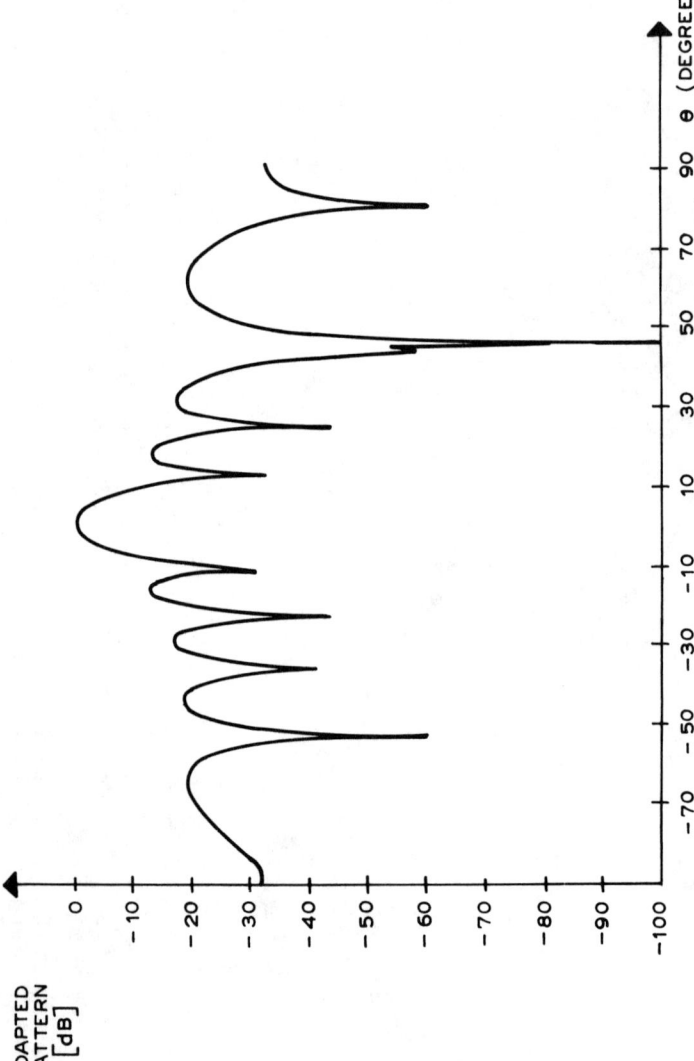

Figure 5.20 Adapted pattern for the operational conditions of Figure 5.18.

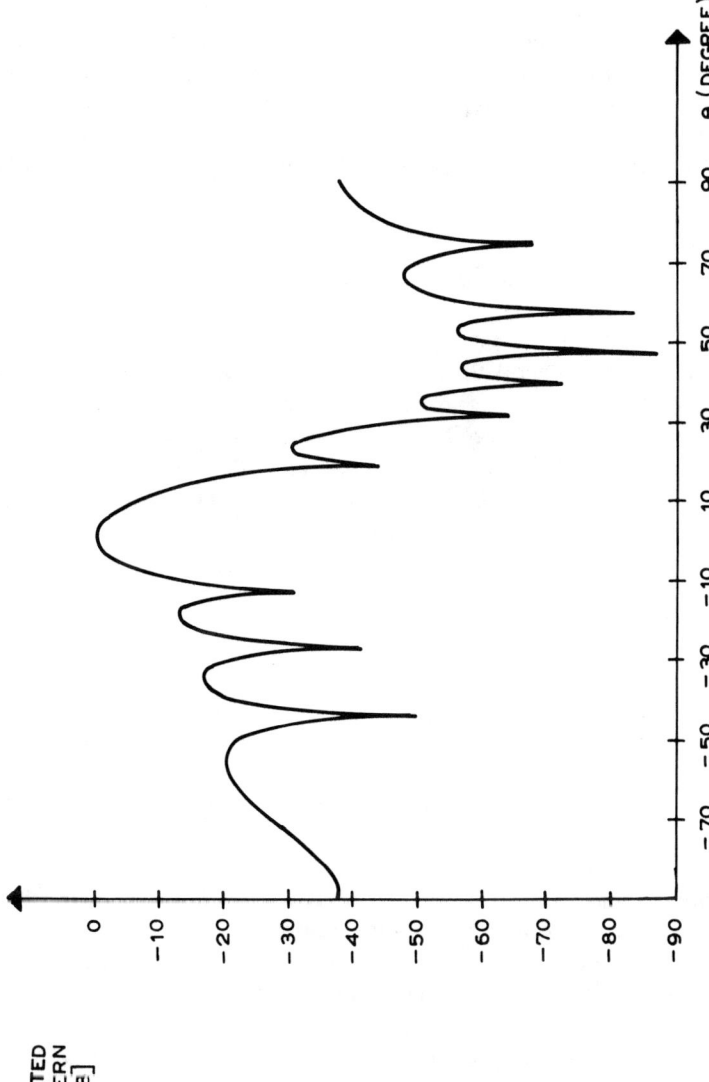

Figure 5.21 Acapted pattern for the operational conditions of Figure 5.19.

amplitudes of the optimum weight components and the adapted pattern are shown in Figures 5.22 and 5.23, respectively.

At this point, let us check if the jammer cancellation concept also works well with a quiescent weight **S** having an amplitude tapering to lower the sidelobes of the quiescent beam (see Section 2.3.2). Constraining the sidelobe levels during both quiescent and interference conditions is important for systems, such as airborne radars, where sidelobe clutter may represent a serious problem. We note that in the previous examples assumption was made of a uniform aperture tapering corresponding to a sin $(\cdot)/(\cdot)$ type of pattern. To this end, the following raised-cosine amplitude tapering is analyzed:

$$|\mathbf{S}_k| = 0.54 - 0.46 \cos\left[\frac{2\pi}{10}(k-1)\right], \quad k = 1, 2, \ldots, 10 \quad (5.35)$$

Figure 5.24 shows the quiescent pattern with a sidelobe level of -35 dB and a larger main beam than with uniform tapering. Figure 5.25 illustrates the pattern modification as a consequence of the cancellation of a monochromatic jammer at 44.5° of incident angle. Note that the sidelobe level is still around -35 dB, notwithstanding the adaptive jammer cancellation. The conclusion is that low-sidelobe and adaptive jammer cancellation are compatible techniques.

As a final remark on (5.34), assume JNR $\gg 1$, i.e., $\lambda_k/\sigma^2 \gg 1$ for $k = 1, 2, \ldots, q$; thus, (5.34) reduces to

$$\hat{\mathbf{W}} = \mu'\left(\mathbf{I} - \sum_{k=1}^{q} \mathbf{q}_k \mathbf{q}_k^H\right)\mathbf{S}^* \quad (5.36a)$$

$$= \mu'\left(\sum_{k=q+1}^{N} \mathbf{q}_k \mathbf{q}_k^H\right)\mathbf{S}^* \quad (5.36b)$$

Hence, the jammer suppression process is independent of the jammer power. We are only required to know the q principal eigenvectors or the $(N - q)$ noise eigenvectors. When one jammer impinges on the antenna, the corresponding output of the array ideally will be zero because the jammer is orthogonal to the noise eigenvectors incorporated in the optimum weight $\hat{\mathbf{W}}$ of (5.36b). The plane wave corresponding to the wished target will be reinforced at the array output. Equation (5.36) is the limiting case as the jammer powers approach infinity by virtue of the inverse power rule. Algorithms that are well known in the framework of numerical analysis can be used to calculate the principal eigenvectors; see, for instance (Marple, 1987, pp. 92, 109–110; Buhring, 1978a,b; Carhoun, 1991). Because the principal eigenvectors represent a good approximation of the jammer DOAs, another approach to

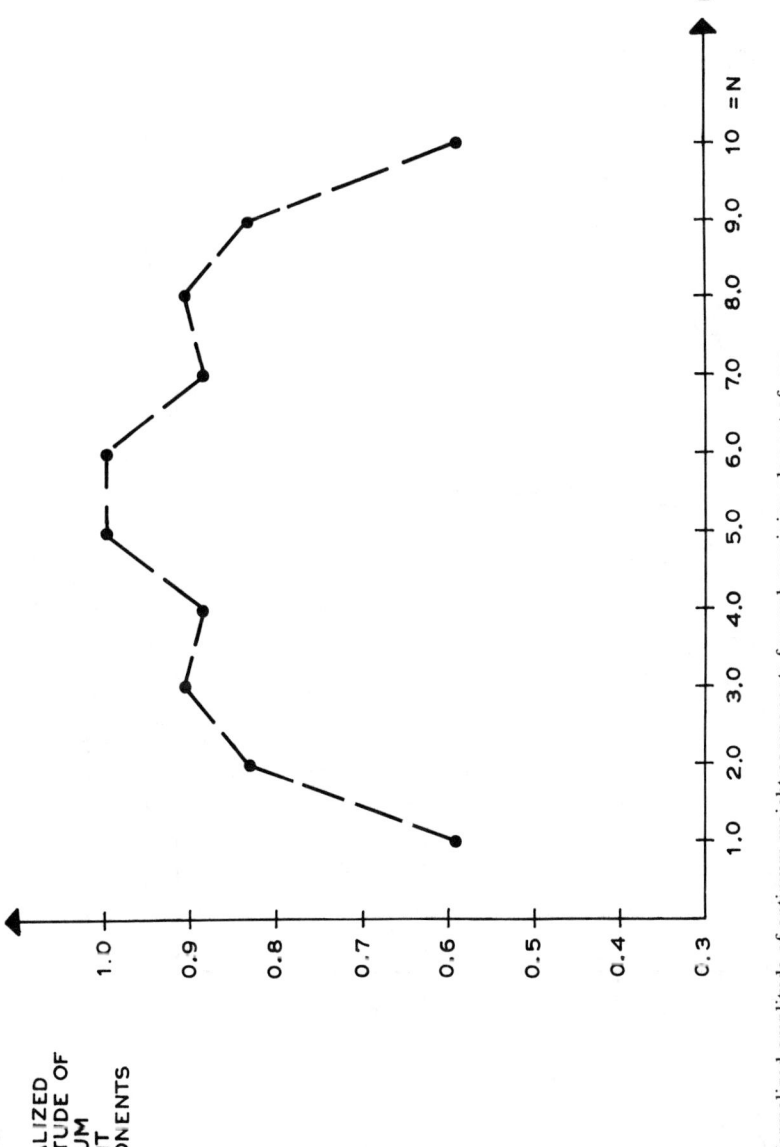

Figure 5.22 Normalized amplitude of optimum weight components for each receiving element of array; five monochromatic equal-power jammers (JNR = 30 dB) at incidence angles, $-20°$, $-40°$, $+30°$, $+40°$, and $50°$, respectively; desired look angle, $\theta_s = 0°$; number of array elements, $N = 10$.

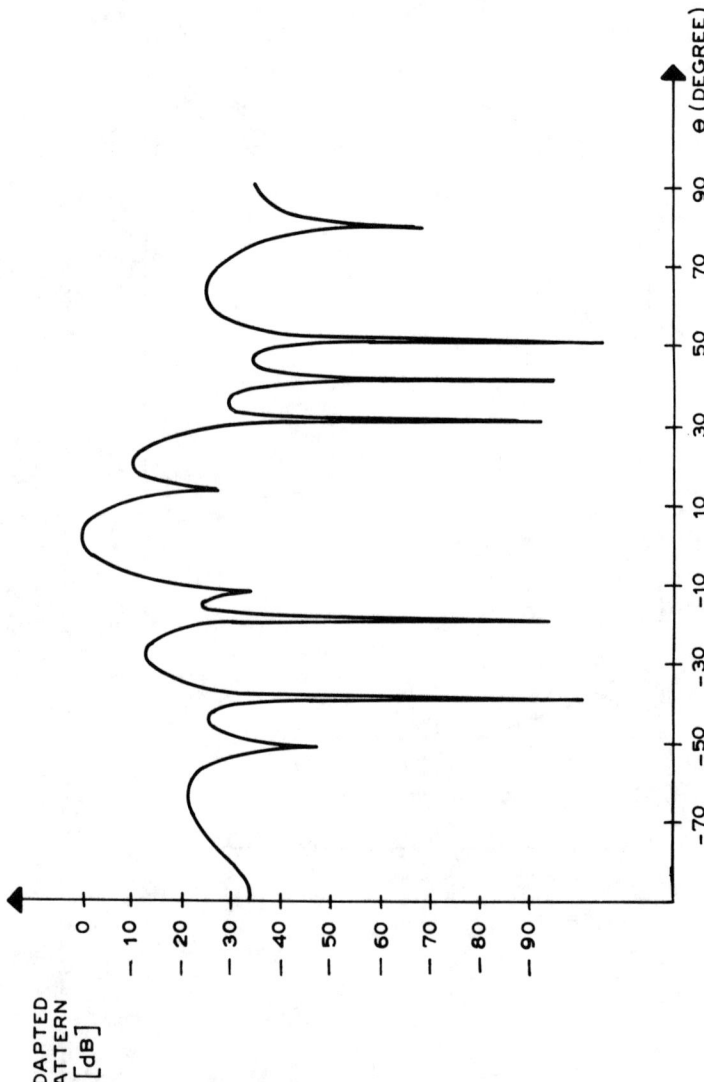

Figure 5.23 Adapted pattern for the operational conditions of Figure 5.22.

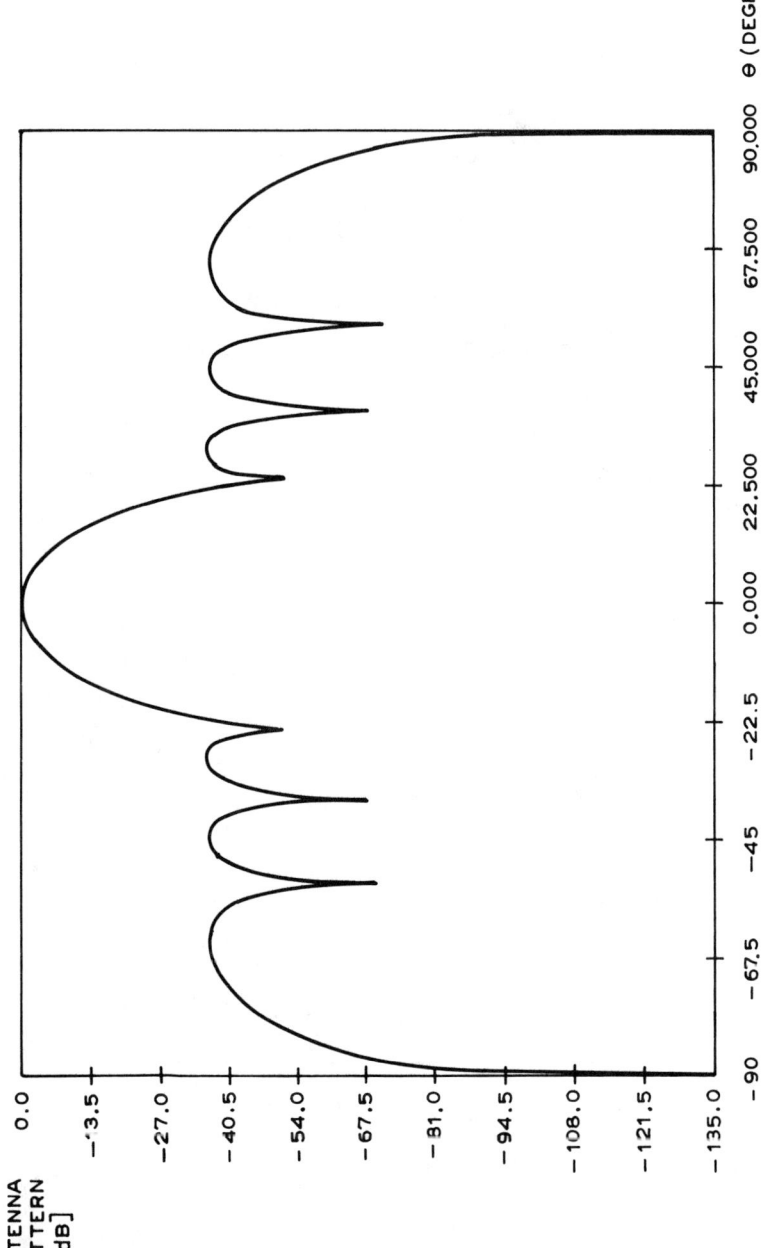

Figure 5.24 Quiescent antenna pattern with raised cosine; number of array elements, $N = 10$.

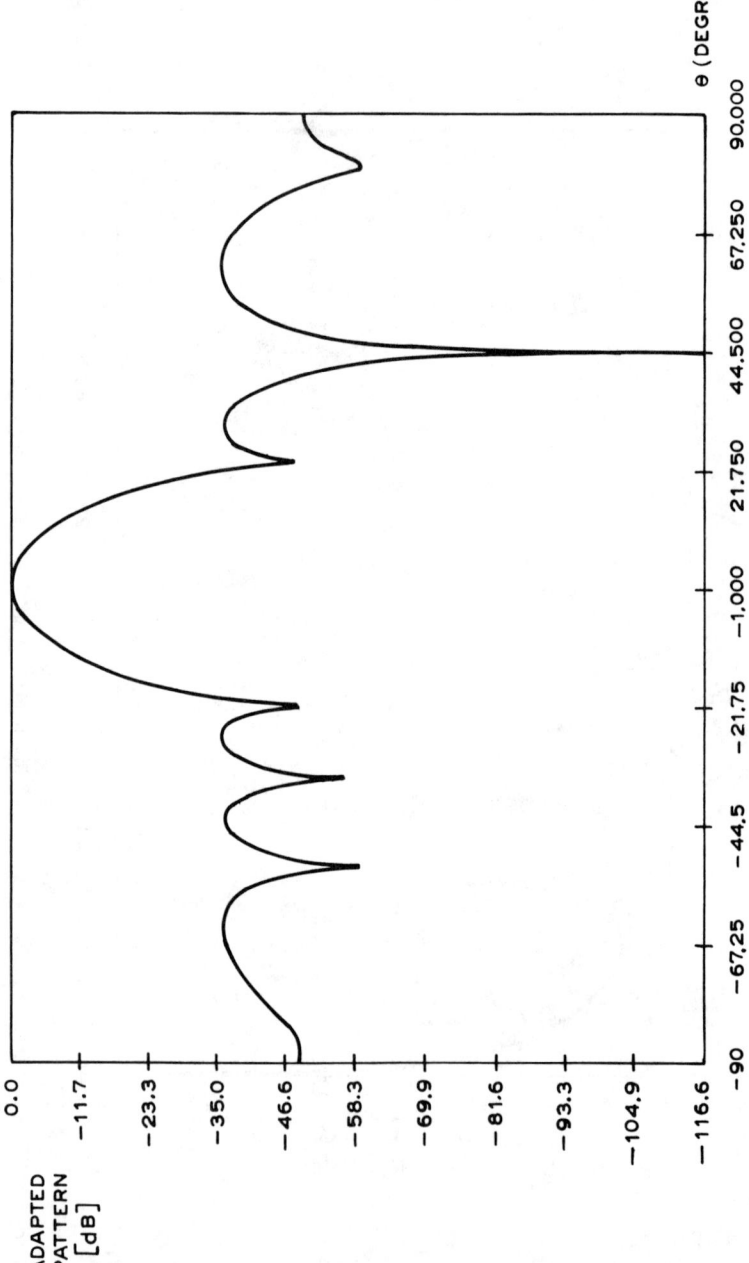

Figure 5.25 Adapted pattern with quiescent raised cosine against a monochromatic jammer with incident angle, 44.5°.

the jammer nulling problem is to estimate first the jammer DOAs and subsequently use this information to place nulls on the jammer incident angle; see, for instance (Friedlander and Porat, 1989). The estimation of the DOA of sources is considered in Section 5.7.

5.3.3 Performance Limitations due to Mismatching of Receiving Channels

As for the SLC, there are several effects of hardware errors that produce degradation on adaptive array performance. These include fixed amplitude and phase errors, array element position errors, unknown mutual coupling errors, mismatching, as a function of frequency, between the receiving channels downstream of each array element, limited number of bits in the ADCs and signal processor, and so on. The fixed errors are compensated by the adaptive cancellation algorithm, and do not affect the jammer cancellation ratio. However, they do affect the $(SINR)_o$ because they represent errors in the mathematically defined steering vector (Kelly, 1989; Zunich and Griffiths, 1991). In the following, the results of computer simulation quantitatively show the performance reduction due to mismatching between the receiving channels. This exercise is a generalization of work for the SLC (Subsection 4.3.2.2).

The model for the analysis of the mismatching in the receiving channels of the adaptive array is shown in Figure 5.26. The fractional bandwidth of each receiving channel is B/f_0. The amplitude response of the channels is assumed to differ from

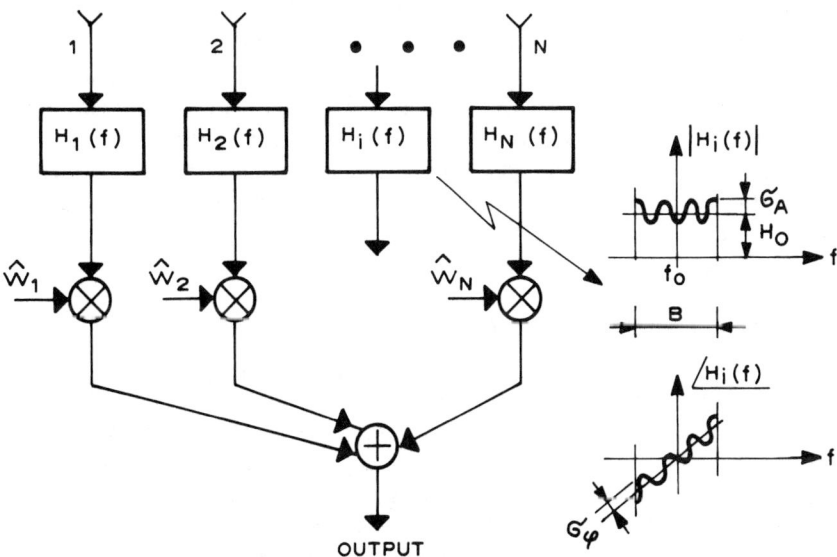

Figure 5.26 Random amplitude and phase mismatching in the receiving channels of an adaptive array antenna.

the ideal constant value $|H_0|$ by a factor $e_{\text{AMP},i}(f)$, which is a random variable having zero mean value and a standard deviation σ_A equal for all the channels. The random variables $e_{\text{AMP},i}(f)$ are independent from channel to channel and from frequency to frequency. Similarly, the phase response of the channels differs from the ideal linear term, Kf, as a function of frequency by a factor $e_{\text{PHASE},i}(f)$, which is a zero-mean random variable having a standard deviation σ_ϕ equal for all the channels. The random variables $e_{\text{PHASE},i}(f)$ are independent from channel to channel and from frequency to frequency. In mathematical terms, the model of the ith channel, i.e., the amplitude and phase of the ith channel transfer function, is

$$|H_i(f)| = |H_0| + e_{\text{AMP},i}(f)$$

$$\angle H_i(f) = Kf + e_{\text{PHASE},i}(f), \quad f \in \left(-\frac{B}{2}, +\frac{B}{2}\right) \qquad (5.37)$$

The probability density functions of the random variables $e_{\text{AMP},i}(f)$ and $e_{\text{PHASE},i}(f)$ have been hypothesized as Gaussian. Following a similar mathematical procedure, explained in Subsection 4.3.2.2, we are able to evaluate the IF of the array under mismatching conditions. Figure 5.27 depicts the IF values as a function of the number N of antennas for couples of σ_A/H_0 and σ_ϕ having a fractional bandwidth B/f_0 = 10%. The situation refers to a desired look angle of 0° and a wideband jammer having a JNR = 40 dB and a DOA = 45°. Three curves are shown: curve (a), which refers to the absence of channel mismatching (see also Figure 5.7); curve (b), related to σ_A = 0.085 H_0 and σ_ϕ = 2.5°; and curve (c) with mismatching parameters σ_A = 0.17 H_0 and σ_ϕ = 5°. Of course, means to compensate for the mismatching are recommended for a high-performance array. Deterministic sources of mismatching can be discovered by injecting test signals into the channels. Random sources are compensated by using a channel equalizer, based on the tapped delay-line technique, described in Subsection 4.3.2.4.

5.3.4 Effects of Correlated Sources

Correlation may exist between the desired signal and the interference and among the interference sources themselves. The correlation coefficient between two sources is indicated by $\rho_{12} = |\rho_{12}| \exp(j\phi_{12})$. When $|\rho_{12}| = 1$, the two sources are said to be coherent, i.e., one is a scaled and delayed replica of the other. Two uncorrelated sources, which do not have fixed phase and amplitude relationships, are characterized by $\rho_{12} = 0$. The purpose of this section is to discuss the operational cases where the correlation may be present and their effect, if any, on the performance of the adaptive arrays.

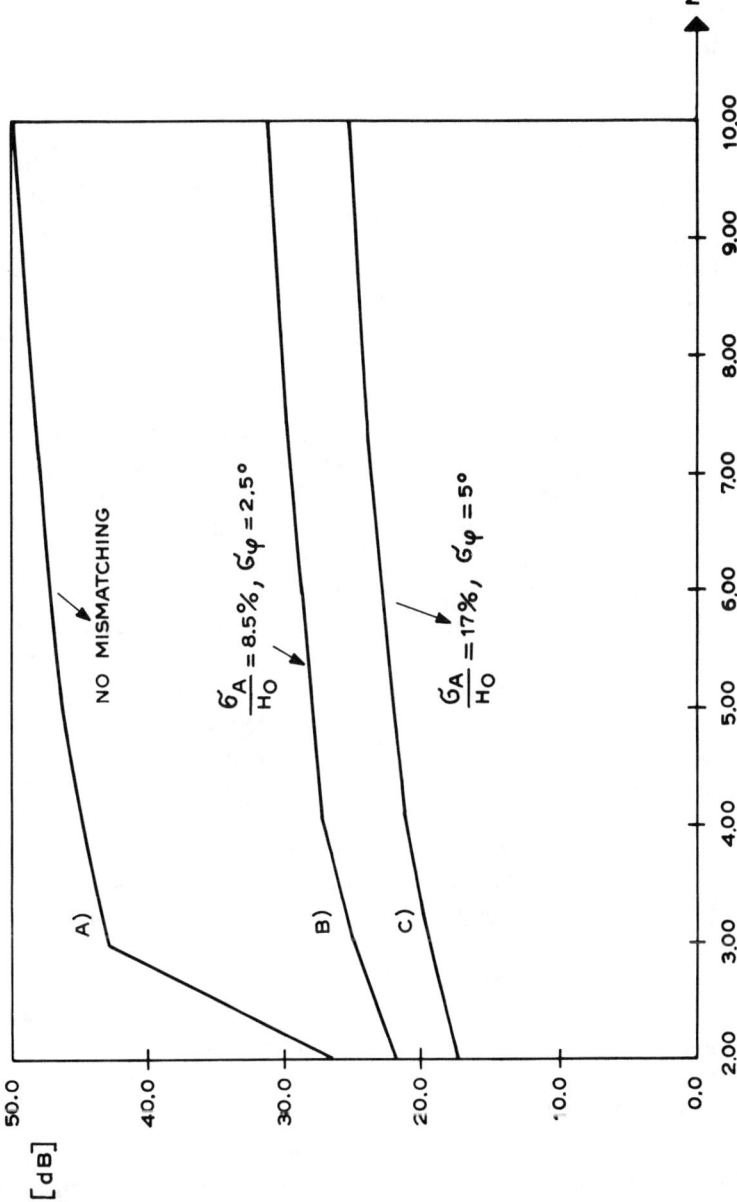

Figure 5.27 Improvement factor *versus* the number of array elements in presence of receiving channel mismatching; $B/f_0 = 10\%$, $\theta_s = 0°$, $\theta_J = 45°$, JNR = 40 dB.

1. The correlation may occur between the target echoes and an interference source when a "smart" jammer deliberately retrodirects the radar energy to the array. However, the jamming signal must be in the same range cell of the target to which it has cohered in order to contribute a correlated wavefront snapshot to the covariance process. If this is so, the jammer is most likely in a track-breaking mode (see Section 1.2) and at a target location where adaptive nulling is ineffective. Other methods are required to counter such a treat. If it is not located very close to the target, it is unlikely to generate a pulse that coincides with the target signal due to different path delays.
2. Another situation is observed when tracking a target at a certain elevation angle. The direct echo path is disturbed by the reflected path, which is correlated with the desired echo. In reality, target multipath is *not* an ECCM issue; nevertheless, adaptive nulling could be helpful in solving this problem. Of course, adaptive nulling in azimuth is not effective, while the target multipath problem can be handled with arrays having adaptive degrees of freedom in elevation.
3. An additional operational situation of interest results when a *stand-off jammer* (SOJ) deliberately directs the jamming beam toward the ground to produce a reflected jammer that is highly correlated with the direct-path jammer. Two situations may occur, concerning the direct and ground-scattered jamming signals. First, if all signals arrive in the sidelobe region, they can be adequately cancelled by adaptive nulling, as will be shown later. Second, if the scattered jamming signal arrives within the main-beamwidth, it is diffuse in angle and cannot be cancelled by array nulling alone (otherwise, the whole main beam is nulled). Superresolution (Section 5.7) could help in this case.

    There is an additional situation for jamming multipath, i.e., near-field or close-in reflections, such as an aircraft wing for an airborne radar or a nearby building for a ground-based radar. However, the real problem is the lack of correlation between the direct and reflected paths into the main beam due to doppler shifts and the fact that the path delay may exceed the size of a range cell in the ground-based example. Space-time adaptive processing (Brennan and Reed, 1973) could be adopted to counteract the doppler shift in the aircraft case, and near-field nulling (Fenn and Tsandoulas, 1989) could be a possible technique to alleviate the problem in the ground-based case.
4. A final example of correlated sources occurs when we wish to form the adapted beam with only one data snapshot. In this case, although the sources may be uncorrelated, they produce a received signal having source contributions with fixed relative phase (Gabriel, 1980a, p. 664).

In the following, we analyze Case 3 of two correlated jammers into the sidelobes and Cases 2 and 4 of correlation between desired target signal and interferences, which is instructive from a mathematical viewpoint. The reader is cautioned that this area of research has not reached a definite consensus.

We start analyzing the case of two correlated jammers received into the sidelobes of an array. We consider the usual array of $N = 10$ receiving elements with uniform weighting amplitude for the quiescent condition. Two monochromatic jammers, having equal jammer-to-noise power values of JNR = 40 dB, impinge on the array with DOAs of $\theta_1 = 15°$ and $\theta_2 = 20°$, respectively. Figure 5.28(a) shows the ranked eigenvalues for three values of correlation coefficient: $|\rho| = 0$, $|\rho| = 0.9$, $|\rho| = 1$. For the case of $|\rho| = 0$, the two principal eigenvalues are not equal because the two DOAs are very close. Well separated DOAs would produce two equal eigenvalues, as already shown in Figure 5.11. Only one main eigenvalue remains when $|\rho|$ tends toward one. Figure 5.28(b) presents the two patterns related to the two principal eigenvectors for $|\rho| = 0$. We note that, owing to the small angular separation of the two sources, the two eigenpatterns are not precisely directed toward the two DOAs. Nevertheless, the net result is the formation of two deep nulls along the two jammer DOAs, as shown in the adapted array pattern of Figure 5.28(c). The patterns related to the two principal eigenvectors for the case of $|\rho| = 0.9$ are depicted in Figure 5.28(d). These patterns are approximately equal to those of Figure 5.28(b) as far as the main-beam portions are concerned. They differ in the sidelobe regions, specifically, the level is higher for the case of $|\rho| = 0.9$. Also, in this case, the cancellation is successful, as illustrated by Figure 5.28(c).

Finally, we consider the case of $|\rho| = 1$; only one eigenpattern is identical to the eigenpattern (a) of Figure 5.28(d). The corresponding pattern of the adapted array, shown again in Figure 5.28(c), does not exhibit a deep null. However, this fact does not mean that the two correlated jammer interferences are not cancelled. This is proved by the two following arguments. (1) The signal-to-interference-plus-noise power ratio, $(SINR)_O$, at the output of the array is approximately ten for all three cases, corresponding to the three values of ρ; therefore, the two interference sources have been perfectly cancelled and the only remaining disturbance is the receiver noise; the useful signal is then spatially integrated by the ten elements of the array. (2) By calculating the value of the two products $(\hat{\mathbf{W}}^T \mathbf{J}_1)$ and $(\hat{\mathbf{W}}^T \mathbf{J}_2)$, where $\hat{\mathbf{W}}$ is the optimum weight vector, and $\mathbf{J}_1$ and $\mathbf{J}_2$ are the direction vectors of the two interferences, we note that, when $\rho = 0$, the two products are separately equal to zero, i.e., $\hat{\mathbf{W}}^T \mathbf{J}_1 = 0$, $\hat{\mathbf{W}}^T \mathbf{J}_2 = 0$, which reflects the two deep nulls in the adapted pattern; when $\rho = 1$, $\hat{\mathbf{W}}^T \mathbf{J}_1$ and $\hat{\mathbf{W}}^T \mathbf{J}_2$ are different from zero, but $\hat{\mathbf{W}}^T (\mathbf{J}_1 + \mathbf{J}_2) = 0$. Note that this situation corresponds to "virtual nulling," as described by Bressler et al. (1988, p. 835). The interference sources combine destructively so that the resulting antenna gain pattern need not exhibit nulls in the individual directions of the interferers. An essential point to note is that probing the weigh set with single far-field wavefronts to display an adapted pattern is meaningless insofar as the correlated wavefront is concerned. However, if we probe the weight set with the same correlated wavefront that defined the weight set, we then discover the cancellation null, i.e., $\hat{\mathbf{W}}^T (\mathbf{J}_1 + \mathbf{J}_2) = 0$. The only caveat here is that the cancellation suffers as soon as the amplitude-phase relationship between the two correlated wavefronts

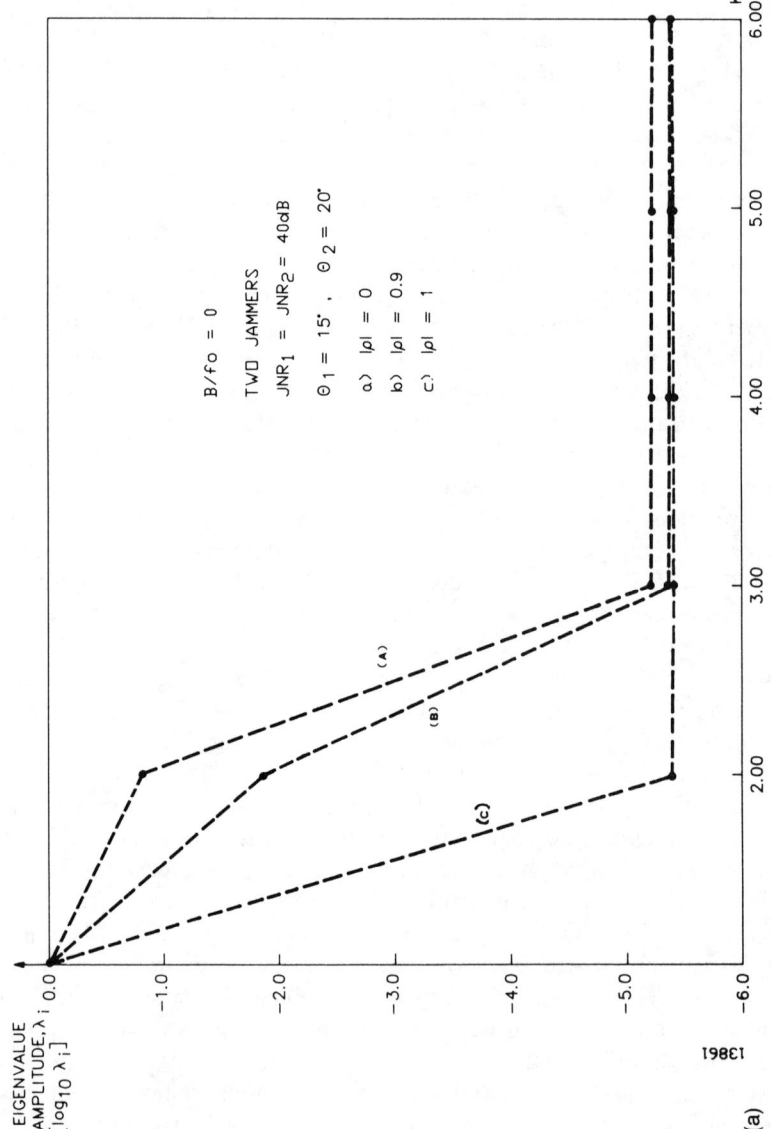

Figure 5.28 (a) Ranked eigenvalues for two monochromatic correlated jammers; array element number, $N = 10$;

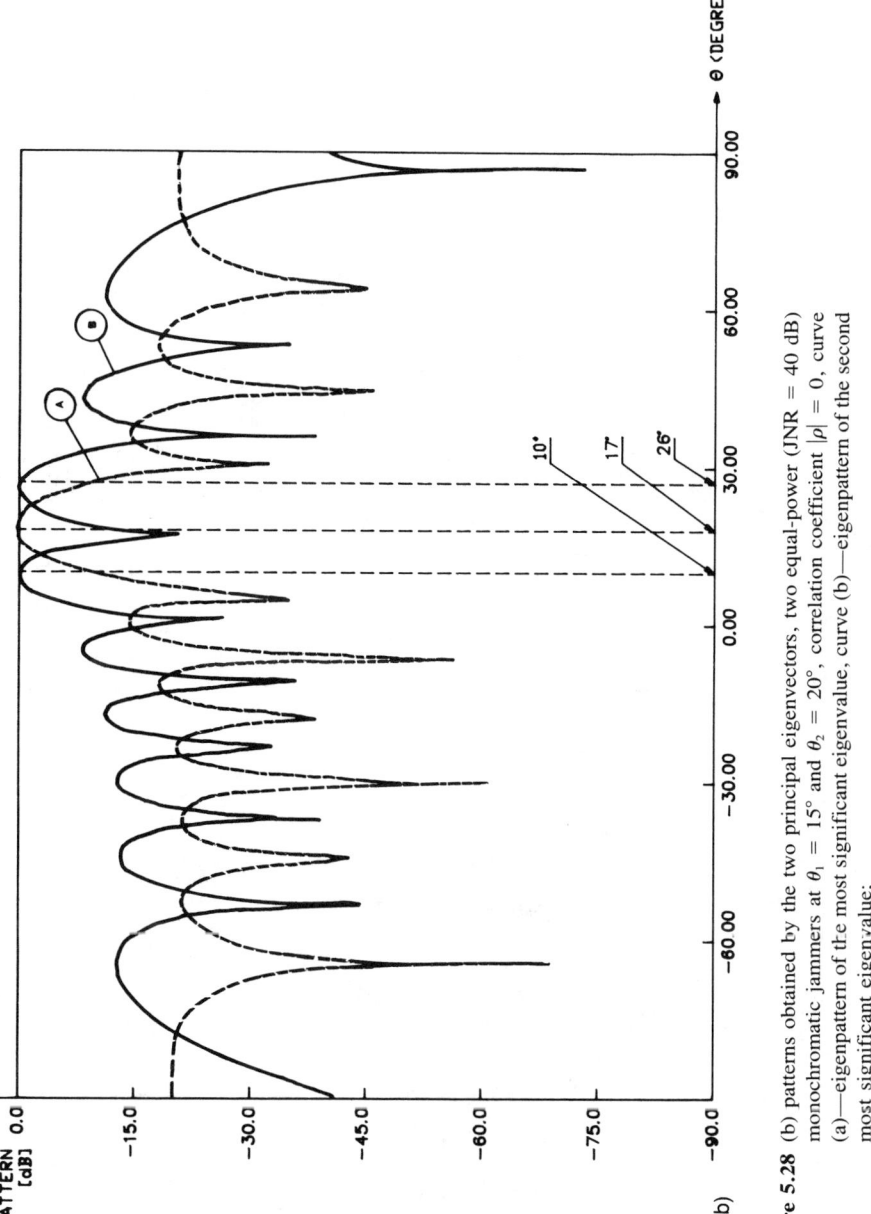

Figure 5.28 (b) patterns obtained by the two principal eigenvectors, two equal-power (JNR = 40 dB) monochromatic jammers at $\theta_1 = 15°$ and $\theta_2 = 20°$, correlation coefficient $|\rho| = 0$, curve (a)—eigenpattern of the most significant eigenvalue, curve (b)—eigenpattern of the second most significant eigenvalue;

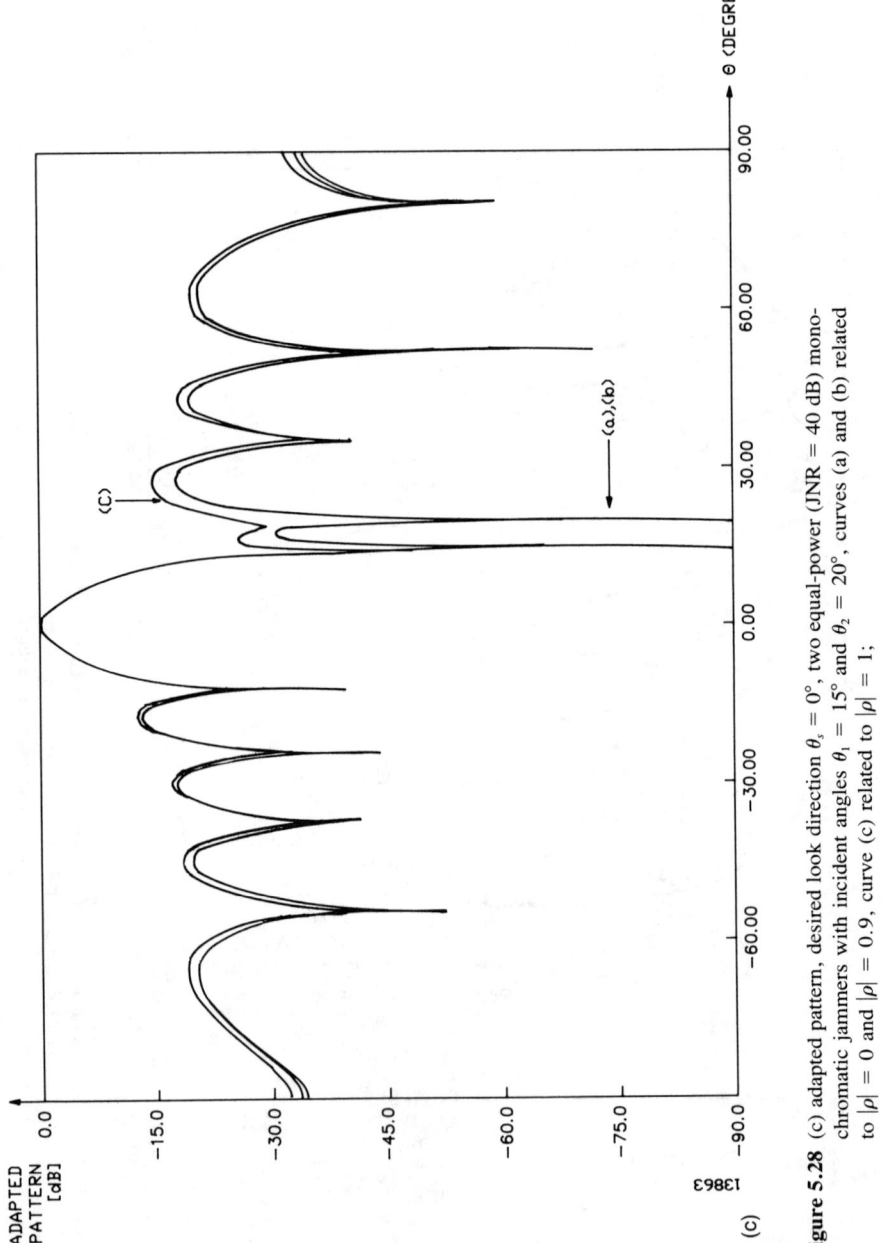

Figure 5.28 (c) adapted pattern, desired look direction $\theta_s = 0°$, two equal-power (JNR = 40 dB) monochromatic jammers with incident angles $\theta_1 = 15°$ and $\theta_2 = 20°$, curves (a) and (b) related to $|\rho| = 0$ and $|\rho| = 0.9$, curve (c) related to $|\rho| = 1$;

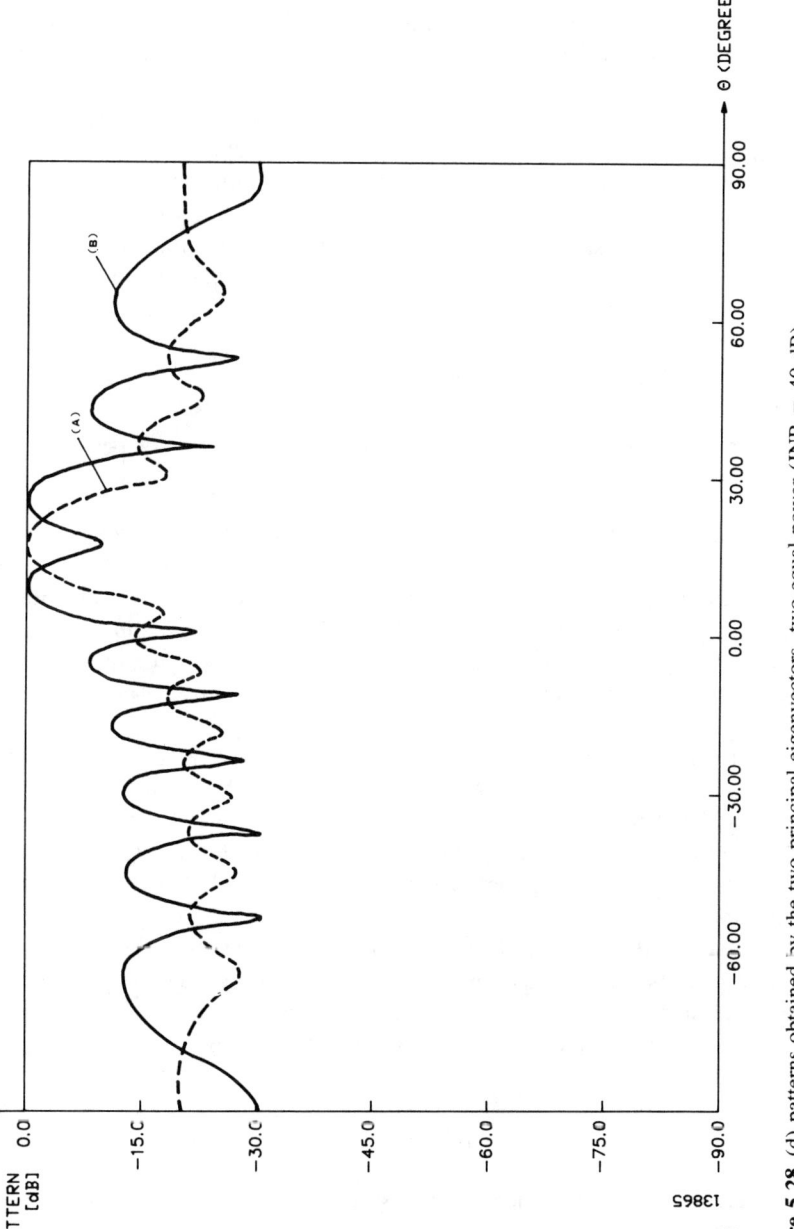

Figure 5.28 (d) patterns obtained by the two principal eigenvectors, two equal-power (JNR = 40 dB) monochromatic jammers at $\theta_1 = 15°$ and $\theta_2 = 20°$, correlation coefficient $|\rho| = 0.9$, curve (a)—eigenpattern of the most significant eigenvalue, curve (b)—eigenpattern of the second most significant eigervalue.

changes. Thus, the rate of adaptive updating may be affected by the dynamics of the environment. However, if the statistical averaging is sufficiently long in a dynamic environment to decorrelate the two wavefronts, the adaptive process will assign two degrees of freedom and separately cancel each wavefront (see also Section 5.7.6).

Another way to look at the problem of cancellation of two correlated jammers is by resorting to the theory of linear subspaces, which represents a much more fundamental approach to this problem in contrast to the retrodirective beam method. Thus, if we consider subspace, S_1, to be the jammer signal subspace of dimension one (as there is only one associated eigenvalue) and the complementary subspace, S_N, to be the system-noise subspace (which includes as much projection as possible in the main-beam direction), the monochromatic jammer can be eliminated by simply deleting its eigenvector from the eigenvector expansion of the inverse covariance matrix in the weight set solution equations, as follows:

$$\hat{\mathbf{W}} = \mathbf{M}^{-1}\mathbf{S}^* = \mathbf{E}\mathbf{\Lambda}^{-1}\mathbf{E}^H\mathbf{S}^* = [\mathbf{E}_J \mathbf{E}_N] \begin{bmatrix} \mathbf{\Lambda}_J^{-1} & \\ & \mathbf{\Lambda}_N^{-1} \end{bmatrix} \begin{bmatrix} \mathbf{E}_J^H \\ \mathbf{E}_N^H \end{bmatrix} \mathbf{S}^* \quad (5.38)$$

where $\mathbf{E}$ is the matrix of eigenvectors, $\mathbf{\Lambda}$ is the diagonal matrix of eigenvalues, $\mathbf{E}_J$ is the jammer eigenvector, and $\mathbf{E}_N$ is the matrix containing the $(N - 1)$ noise eigenvectors. We immediately see that if $\mathbf{E}_J$ is eliminated, there is no projection of the correlated wavefront onto the weight set, thus eliminating the jammer response. Of course, in the optimal solution, the $\mathbf{E}_J$ is not eliminated, but just reduced by a quantity proportional to the inverse of the jammer power (the inverse power rule) as discussed in relation to (5.34b).

The effect on the adaptive array performance of the correlation between the desired signal and an interference source is also of interest. We shall show that the adaptive array will fail in its mission in two ways. In fact, there is a cancellation of the desired signal and a reduction of the null depth along the jammer DOA. The obvious motivation is that no distinction exists, or is limited, between the useful signal and the interference.

First, we note that the optimum weight vector of (5.23) has been derived by implicitly assuming that no correlation exists between the desired signal and the disturbance. The introduction of this correlation modifies the optimum weight expression as follows (Reddy et al., 1987), for the case of two sources impinging on the array:

$$\hat{\mathbf{W}} = \mu(\sigma_1^2 \mathbf{V}_1^* \mathbf{V}_1^T + \sigma_2^2 \mathbf{V}_2^* \mathbf{V}_2^T + \sigma_1 \sigma_2 \rho \, \mathbf{V}_2^* \mathbf{V}_1^T + \sigma_1 \sigma_2 \rho^* \mathbf{V}_1^* \mathbf{V}_2^T + \sigma^2 \mathbf{I})^{-1} \mathbf{V}_1^* \quad (5.39)$$

where σ_1^2, σ_2^2, and σ^2 are the variances of the useful signal, disturbance, and receiver noise, respectively; $\mathbf{V}_1$ and $\mathbf{V}_2$ are the directional vectors relative to the useful signal and the interference. The constant μ is selected to have, for instance, unity gain along the look direction and ρ is the correlation coefficient between the two sources. Equation (5.39) is easily interpreted as a generalization of the classical expression (5.23). The new matrix to be inverted indeed contains the covariance matrix $\sigma_1^2 \mathbf{V}_1^* \mathbf{V}_1^T$ of the useful signal (now considered as a statistic process with zero-mean value), the covariance matrix $\sigma_2^2 \mathbf{V}_2^* \mathbf{V}_2^T$ of the interference, the two cross-covariance matrices $\sigma_1 \sigma_2 \rho^* \mathbf{V}_1^* \mathbf{V}_2^T$ and $\sigma_1 \sigma_2 \rho \mathbf{V}_2^* \mathbf{V}_1^T$, and the noise term $\sigma^2 \mathbf{I}$, where $\mathbf{I}$ is the N-dimensional identity matrix. Equation (5.39) reduces to (5.23) when $\rho = 0$. In fact,

$$\hat{\mathbf{W}} = \mu(\sigma_1^2 \mathbf{V}_1^* \mathbf{V}_1^T + \sigma_2^2 \mathbf{V}_2^* \mathbf{V}_2^T + \sigma^2 \mathbf{I})^{-1} \mathbf{V}_1^* = \mu(\sigma_1^2 \mathbf{V}_1^* \mathbf{V}_1^T + \mathbf{M})^{-1} \mathbf{V}_1^* \quad (5.40)$$

which, when applying a well-known matrix identity, reduces to (Shan and Kailath, 1985; Monzingo and Miller, 1980, p. 103):

$$\hat{\mathbf{W}} = \mu \left(\mathbf{M}^{-1} \mathbf{V}_1^* - \sigma_1^2 \frac{\mathbf{M}^{-1} \mathbf{V}_1^* \mathbf{V}_1^T \mathbf{M}^{-1} \mathbf{V}_1^*}{1 + \sigma_1^2 \mathbf{V}_1^T \mathbf{M}^{-1} \mathbf{V}_1^*} \right) = \mu_0 \mathbf{M}^{-1} \mathbf{V}_1^* \quad (5.41a)$$

with

$$\mathbf{M} = \sigma_2^2 \mathbf{V}_2^* \mathbf{V}_2^T + \sigma^2 \mathbf{I} \quad (5.41b)$$

Returning to the more complete (5.39), we are interested in calculating:
1. The output power of the optimum beamformer when the signal and interference are correlated;
2. The response of the optimum beamformer in the interference direction.

The output power is evaluated as follows:

$$P_{\text{out}} = E\{|\mathbf{W}^T \mathbf{Z}|^2\} = \mathbf{W}^H (\mathbf{M}_s + \mathbf{M}) \mathbf{W} \quad (5.42)$$

where $\mathbf{Z}$ is the received signal by the array and $\mathbf{M}_s$ is the covariance matrix of the desired signal:

$$\mathbf{M}_s = \sigma_1^2 \mathbf{V}_1^* \mathbf{V}_1^T \quad (5.43)$$

When the receiver noise is zero (i.e., $\sigma^2 = 0$), P_{out} is (Reddy et al., 1987):

$$P_{\text{out}} = \sigma_1^2 (1 - |\rho|^2) \quad (5.44)$$

The response of the optimum beamformer along the interference direction is calculated as:

$$G(\theta_2) = \mathbf{W}^T\mathbf{V}_2 \tag{5.45}$$

which when $\sigma^2 = 0$ becomes (Reddy et al., 1987):

$$G(\theta_2) = -\frac{\sigma_1}{\sigma_2}\rho \tag{5.46}$$

This means that in the absence of noise, the beamformer response in the interference direction is different from zero and equal to the correlation coefficient multiplied by σ_1/σ_2. The output power (5.44) is maximum only when the signal and interference are uncorrelated. For any finite correlation, some signal cancellation is present and the jammer cancellation is imperfect. When $\rho = 1$, in the ideal case of $\sigma^2 = 0$, the output power vanishes and the beamformer fails to form a deep null in the jammer DOA. In the more practical situation of moderate receiver noise power ($0 < \sigma^2 \ll \sigma_1^2, \sigma_2^2$) the output power does not decrease to zero, even with perfect coherence (i.e., $\rho = 1$), and there is a contribution to the output power from the signal, interference, and noise. However, looking at the expression of $G(\theta_2)$, in presence of receiver noise, fully uncorrelated sources do not guarantee perfect interference rejection (Reddy et al., 1987). Figure 5.29 shows the gain pattern of an adaptive array with $N = 10$ elements. The adaptive weights are those given by (5.39). Two correlated sources are impinging on the array: (1) the desired source with $\theta_1 = 0°$ and $\sigma_1 = 1$, and (2) the interference with $\theta_2 = 20°$ and $\sigma_2 = 1$. Negligible receiver noise is assumed, i.e., $\sigma = 0.00001$. The parameters of different curves are characterized by the correlation coefficient, ρ, between the sources. Only when $\rho = 0$ is there a null at 20°. As ρ increases toward unity, the array shows a beam at 20°, which is equal to the main beam at 0°. The signal and interference combine destructively in the array, producing zero power at the output, as stated by (5.44).

A usual means to avoid cancellation of the desired signal and restore the jammer cancellation has been envisaged in (Gabriel, 1980b; Lee and Wu, 1989; Reddy et al., 1987; Shan and Kailath, 1985; Shan et al., 1985) among other relevant papers on this important topic. This technique is called *subaperture sampling* or *spatial smoothing*. The basic idea is to restore the full rank of the matrix $\mathbf{M} - \sigma^2\mathbf{I} \equiv \sigma_1^2\mathbf{V}_1^*\mathbf{V}_1^T + \sigma_2^2\mathbf{V}_2^*\mathbf{V}_2^T$. In fact, when the desired signal and the interference are coherent, the number of eigenvalues different from zero for the matrix $(\mathbf{M} - \sigma^2\mathbf{I})$ is one, i.e., the matrix does not have full rank. (Because the matrix $\sigma^2\mathbf{V}*\mathbf{V}^T$ is a diad, its rank is one; thus, the full rank of $\mathbf{M} - \sigma^2\mathbf{I}$ is two in this specific example.) The processing scheme is based on combining the measurements from overlapping subarrays, as shown in Figure 5.30. Given a snapshot of N sensor outputs at the tth time instant:

$$\mathbf{Z}(t) = [Z_1(t)Z_2(t)\ldots Z_N(t)]^T \tag{5.47}$$

defined, for the general case of k sources, p subsets, as follows:

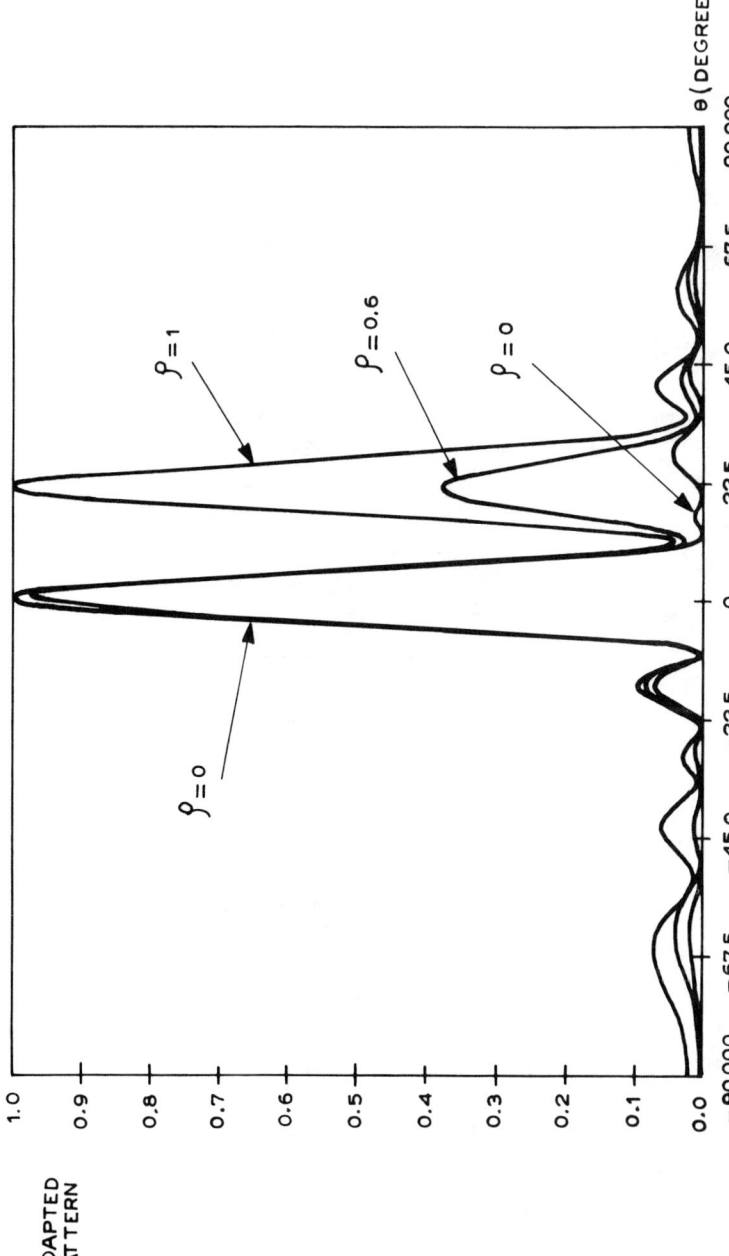

Figure 5.29 Pattern of the adaptive array for two correlated sources; number of array elements, $N = 10$; $\theta_1 = 0°$, $\theta_2 = 20°$, $\sigma_1 = 1$, $\sigma_2 = 1$, and $\sigma = 0.00001$.

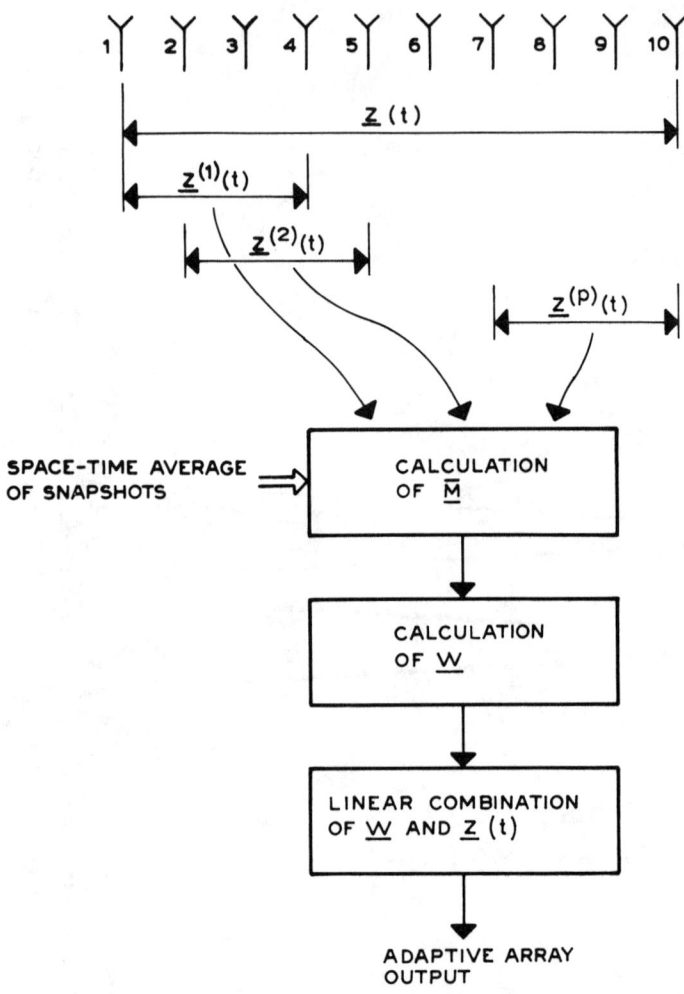

Figure 5.30 Spatial smoothing processing scheme ($N = 10$, $k = 3$, $p = 7$).

$$\mathbf{Z}^{(1)}(t) = [Z_1(t) \ldots Z_{k+1}(t)]^T$$
$$\mathbf{Z}^{(2)}(t) = [Z_2(t) \ldots Z_{k+2}(t)]^T$$
$$\vdots$$
$$\mathbf{Z}^{(p)}(t) = [Z_p(t) \ldots Z_{k+p}(t)]^T \tag{5.48}$$

Next define the spatially smoothed correlation matrix:

$$\bar{\mathbf{M}} = \frac{1}{p}\sum_{l=1}^{p} E\{[\mathbf{Z}^{(l)}(t)]^*[\mathbf{Z}^{(l)}(t)]^T\}$$
$$= \frac{1}{p}\sum_{l=1}^{p}\left\{\frac{1}{T}\sum_{t=1}^{T_a}[\mathbf{Z}^{(l)}(t)]^*[\mathbf{Z}^{(l)}(t)]^T\right\} \quad (5.49)$$

where $T_a \geq 1$ is the number of available time snapshots.

As has been shown (Shan and Kailath, 1985), the full rank of $(\mathbf{M} - \sigma^2\mathbf{I})$ is restored by the spatial smoothing if and only if $p \geq k$. The price for this result derives from the condition that $k + p = N$ (for construction of the subarrays), which, with the constraint on p, requires that $N \geq 2k$. Therefore, the spatial smoothing works if the array elements are at least twice the number of sources impinging on the array. This condition is stronger than usual for uncorrelated sources, i.e., with N elements, $N - 1$ jammers can be suppressed. The spatial smoothing technique is helpful in working with one data snapshot. The spatial smoothing technique will be considered again in Section 5.7.6, where the problem to solve will be the angular resolution of correlated sources.

A completely new approach to the problem of optimum beam-forming for correlated signal and interference is described by Bresler *et al.* (1988), which is now briefly recalled. The novelty of the approach is in the definition of new optimality criteria and in the conception of computationally efficient algorithms. Actually, three criteria are presented by the authors. The first criterion, called *optimum combiner*, is for coherently combining the correlated useful signal and interference to maximize the output signal-to-receiver-noise ratio, thus avoiding the problem of mutual cancellation between the signal and interference as in earlier beamformers. This type of beamformer finds useful application in multipath situations where the direct path from the look direction may have power lower than that of the multipath. In the second beamformer, all interferers, whether correlated or uncorrelated, are considered undesired, and the weight vector is developed to maximize the output signal-to-interference-plus-noise ratio. When two or more correlated interferences impinge on the array, this technique mixes these signals in a destructive manner, instead of forcing deep nulls in their direction. The technique is then applicable to those situations where an SOJ produces two correlated disturbances by pointing the beam toward the ground to produce a multipath. The third beamformer places nulls along the direction of interferers, irrespective of its coherence or incoherence with the desired signal. This beamformer helps deal with smart jammers that imitate the desired signal. The paper also shows that when the useful signal and interference sources are uncorrelated, the three beamformers reduce to the classical relation (5.23).

5.4 SUPPRESSION OF JAMMERS BY MULTIPLE-BEAM SIGNAL PROCESSING

The optimum configuration for adaptive nulling is the fully adaptive array (Subsection 5.3.1), which, by definition, suppresses the interference before beam-forming by applying some matrix operation to all array elemental outputs. For the near future, this is not realizable with an array of thousands of elements for reasons of cost. Additionally, we know that if N is the number of degrees of freedom of the adaptive array, the system must spend many iterations ($2N - 5N$) before convergence, and the computational burden for the inverse of the interference covariance matrix (N^3) may be prohibitive. So, the number of channels for processing must be reduced. The problem has been considered in many references (Chapman, 1976; Fante, 1980; Krucker, 1986; Morgan, 1978; Klemm, 1975; Wirth, 1976, 1977; Takao and Uchida, 1989; Gerlach, 1990c; Brookner and Howell, 1986, 1987). The purpose is, in general, achieved by three types of channel reduction strategies (Nickel, 1989b):

1. Post-beam-forming jammer suppression, in which a sum beam (Σ) plus auxiliary beams are formed from the same array, as illustrated in Figure 5.31;
2. Jammer suppression after partial beam-forming, in which the reduced channels consist of subarrays of approximately the same size, and the adaptive weights are applied to the subarray outputs (see Figure 5.32);

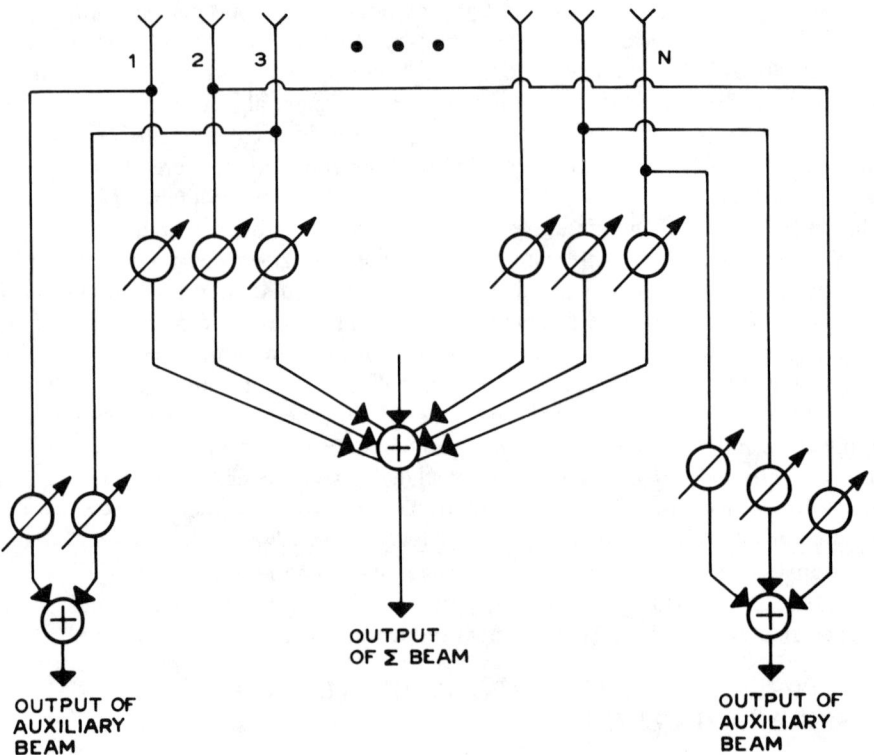

Figure 5.31 Sidelobe canceler configuration forming auxiliary beams from the same main array.

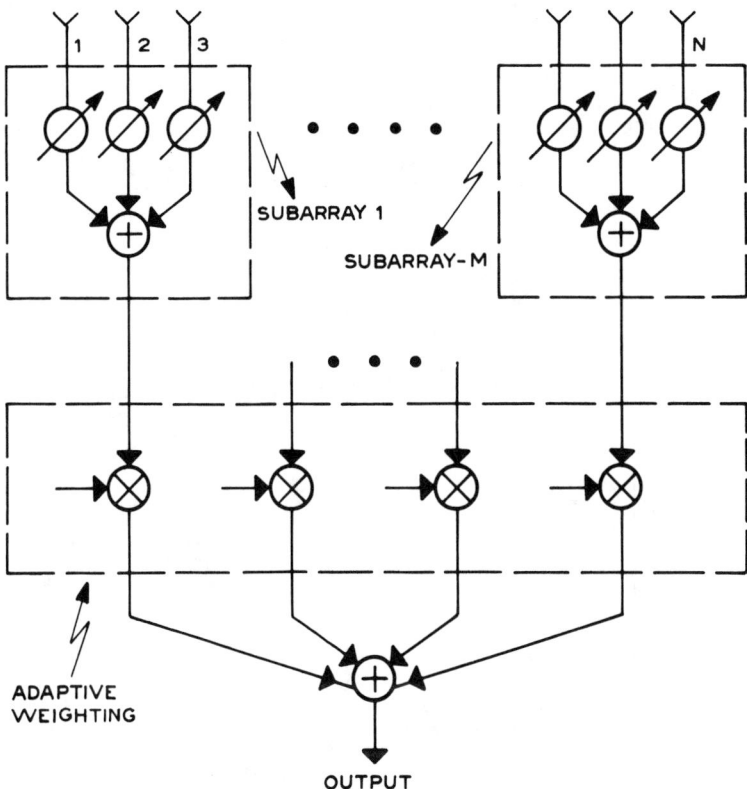

Figure 5.32 Interference suppression at subarray level.

3. Multibeam antenna concept, where a number of steerable beams are formed in the direction of the useful signal and jammers, and the output beams are adaptively combined to suppress the jammers (see Figure 5.33).

The first array configuration is quite conventional, and has the advantage of using homogeneous hardware for the main and auxiliary antennas. The auxiliary antennas may have a directional radiation pattern. This approach may suffer for some practical problems, as discussed in Nickel (1989b).

The second method for reducing the number of channels is best suited to large phased arrays. After phase-shifting, subarrays are formed at the element level. The outputs of the subarrays are digitized and constitute a superarray, consisting of elements with patterns corresponding to the subarray patterns. The arrangement preserves the homogeneous antenna front-end hardware, which is desirable for the application of highly integrated modules. Unfortunately, grating lobes may be present

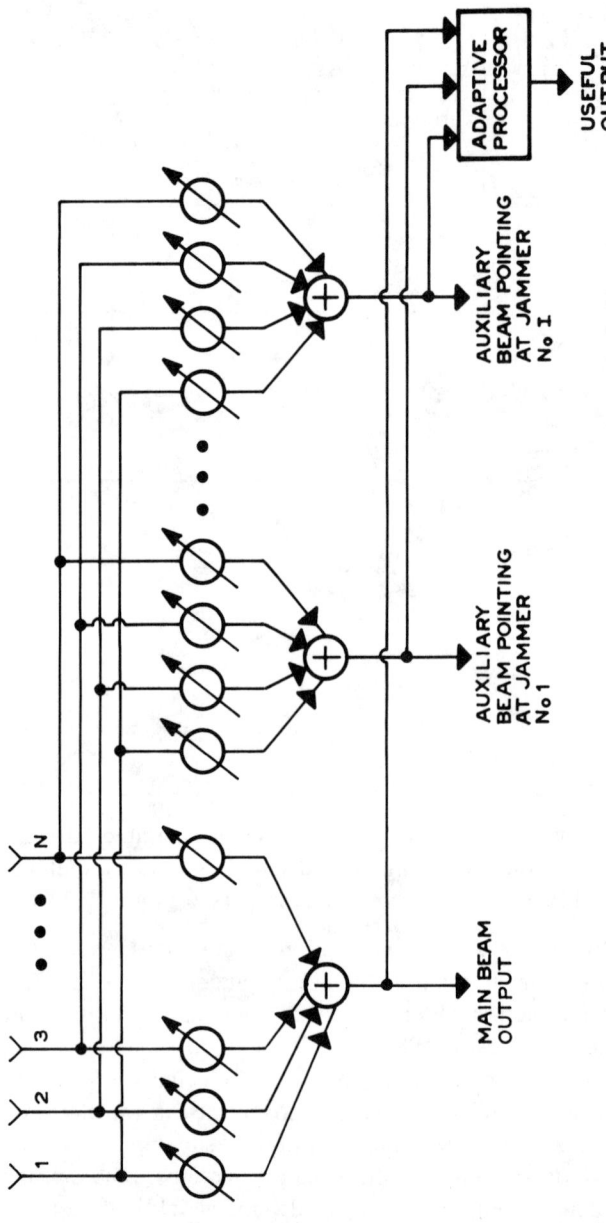

Figure 5.33 Interference cancellation with multiple-beam array antenna. (Derived from Brookner and Howell, IEE Intl. Radar Conf., Radar 1987, London, 19–21 October 1987, pp. 257–263.)

due to the spacing of the elements of the superarray, creating spurious notches, resulting in blind directions for the antenna. For multiple jammers, the number of blind directions soon becomes intolerably high. Grating lobes can be avoided by irregular spacing of the superarray, thereby forming irregular subarrays (White, 1983). Design rules for subarray configurations may be taken from the theory of thinned arrays. Nickel (1989b) has found a design rule for the arrangement of subarrays in a regular array, and has recommended that the centers of the subarrays have distances without a common divisor. The great advantage of beam-space nulling (a name often used in reference to this second technique), as compared with element-space nulling, is that the dimensionality of the adaptive control problem can thereby be reduced from the number of array elements to the number of beams, which equates to the number of interferers. For a large array, the number of interference sources, of course, will be far fewer than the number of elements.

The third processing configuration (Figure 5.33) has the potential of preserving the low-sidelobe level of the unadapted antenna pattern, as discussed in Gabriel (1986) and by Brookner and Howell (1986; 1987). This advantage and the reduced number of computations to form and invert the covariance matrix are obtained at expense of hardware complexity. The technique involves the transformation of a large array of N elements into $(I + 1)$ arrays, where I is the number of interferers to be nulled. Each array uses all the elements of the main aperture. The technique requires the estimation of the jammer DOAs, which can be done by resorting to one of the methods described in Section 5.7. Once the number and locations of the jammers have been determined, beams are formed in the direction of the jammers, and the corresponding outputs are combined with the main array output in an adaptive fashion to improve the signal-to-interference-plus-noise ratio. The advantages of reducing the number of degrees of freedom from N (some thousands) to $I + 1$ (around ten) are related to the reduction of required computations and the increase of the adaptation speed of the processor.

We believe that the three techniques have advantages and disadvantages which should be carefully analyzed for the practical applications at hand. A basic problem still remains unsolved, represented by the lack of a theory for specifying an effective transformation matrix from the array-element domain to the beam-element domain. Some preliminary mathematical effort has been made in this direction; see, for instance (Takao and Uchida, 1989; Gerlach, 1990).

5.5 DETERMINISTIC SPATIAL FILTERING

5.5.1 Principles

In addition to the adaptive methods for interference suppression, a fixed reduction of the sidelobes of the antenna pattern in those directions where the interference is expected to be present may be useful. This technique, called *deterministic spatial filtering*, because we know *a priori* the direction of arrival of the disturbance, is now described (Groger, 1989; Wirth, 1977; Drane and McIlvenna, 1970; Kurth, 1974; Steyskal, 1982; Prasad and Charan, 1984). The basis of deterministic spatial filtering assumes a fictional interference situation, distributed over a certain angular sector where we wish to have low sidelobes. The corresponding weight vector of the array is calculated by means of the classical equation (5.23). The covariance matrix of the fictional interference is calculated as follows for the case of a planar array.

Suppose we have a number of independent interference sources distributed over a certain angular sector. At the array elements, we receive the sum of all interferences. If we let the number of sources tend to infinity, we obtain an angular interference power density $p(u,v)$, where u and v are the directional cosines, and the sum of all interferences becomes an integral. We thus write the limiting form of the (ik)th element of the interference covariance matrix $\mathbf{C}$ as

$$C_{ik} = \int_{u^2+v^2 \leq 1} p(u,v)a_i(u,v)a_k^*(u,v) du\, dv \tag{5.50}$$

where $a_i(u,v)$ denotes the complex gain of the ith array element in the direction (u,v). A similar expression applies to the linear array case:

$$C_{ik} = \int_{-1}^{1} p(u)a_i(u)a_k^*(u) du \tag{5.51}$$

We have a covariance matrix $\mathbf{C}$ and a beam-steering vector $\mathbf{a}(u_0,v_0)$ for a target in the direction (u_0,v_0). The components of $\mathbf{a}(u_0,v_0)$ are the complex gain of the array elements for a target in the direction u_0,v_0. We obtain the weighting for the array elements as follows:

$$\mathbf{W}_0 = \mathbf{C}^{-1}\mathbf{a}^*(u_0,v_0) \tag{5.52}$$

This equation corresponds to the optimum solution of the adaptive method (5.23). The assumed spatial power distribution $p(\cdot)$ determines the kind of sidelobe reduction.

General Reduction of Sidelobes

In the case of $p(\cdot)$ = constant, the corresponding covariance matrix **C** becomes the identity matrix, **I**, if we have a linear antenna at $\lambda/2$-spacing and the complex gains are described by pure phase shifts corresponding to the time delay between the elements. The optimum weighting is then equal to that for a plane wavefront. If a reduction of the sidelobes in the entire visible region is desired, a broadening of the main beam must be accepted. This means that, in the main-beam area, the assumed power density $p(\cdot)$ must be very small compared to the other regions. This formulation includes all the techniques of low sidelobe antennas and has been already considered in Chapter 2.

Reduction of Sidelobes in a Limited Solid Angle

The fictional power distribution $p(\cdot)$ is made higher in the region Ω_1, where low sidelobes are desired (e.g., the horizon region), as compared to the other regions, $\Omega_2 = \Omega - \Omega_1$, where sidelobes will be allowed to increase (Ω indicates the antenna field of view). Then, by computing a covariance matrix based on the fictional interference and computing the adaptive weights to null this fictional interference, low sidelobes are formed in the desired region. The ratio of the powers for Ω_1 and Ω_2 determines approximately the sidelobe reduction within Ω_1 against Ω_2. The sidelobe reduction is achieved by a redistribution of the sidelobes; a reduction within Ω_1 leads to an increase of the sidelobes within Ω_2. The sidelobe level increase in the region Ω_2 is related to the extent of Ω_1 and the desired sidelobe reduction in this region. Interesting experimental results have been reported by Groger (1989). For instance, -50 dB sidelobes can be achieved in three angular sectors with the experimental phased-array system, ELRA, operating at S-band. An interesting aspect of this experiment is the capability of shaping the pattern at the subarray level, rather than the array level, a point relevant for large phased arrays.

5.5.2 Applications

The region with interferences is likely to be the horizon (see Figure 5.34) for ground-based or long-range jammers. For a stationary radar in a stationary environment, the complex gain values $a_i(\cdot)$ ($i = 1, \ldots, N$) can be considered known and stored in the computer. The situation is different for a mobile radar system because the near-field scattering and, consequently, the parameters $\{a_i(\cdot), i = 1, \ldots, N\}$ change from place to place. The application of deterministic spatial filtering seems to be difficult in this case. A second application relates to the observation of targets over sea. A reduction of the sidelobes in a certain region below the main lobe could eliminate

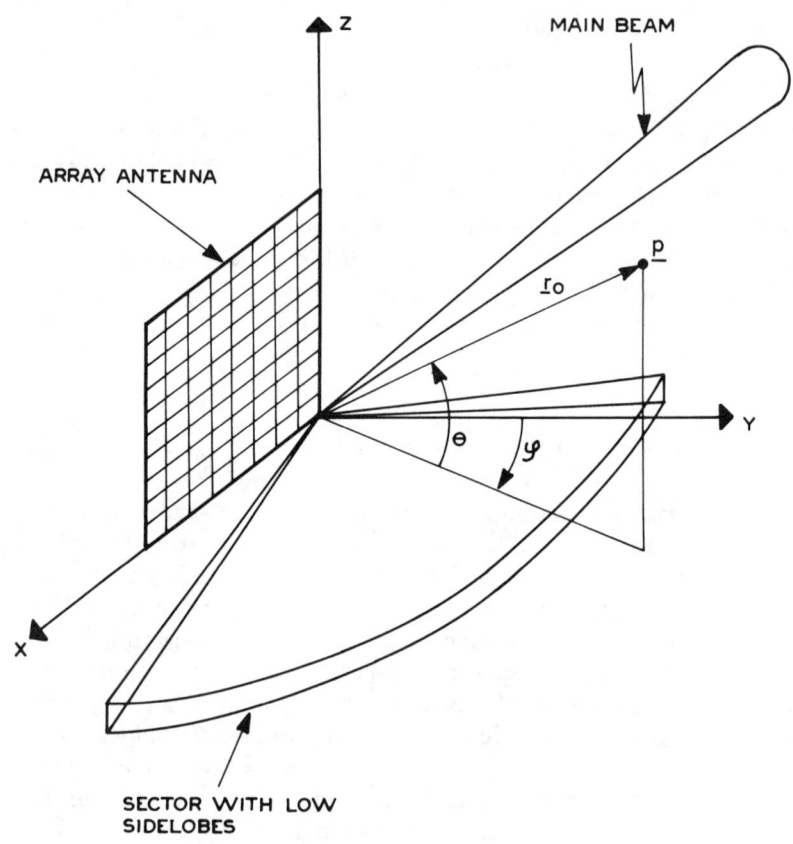

Figure 5.34 Phased-array antenna with a sector having low sidelobes.

multipath. This has been demonstrated in Pearson *et al.* (1982) by an eight-element vertical array with digital beam-forming.

For these applications, an accurate measurement of the antenna in the real environment is necessary, i.e., the complex gains $a_i(u,v)$ ($i = 1, \ldots , N$) must be known as a function of the direction (u,v). A theoretical approximation for these complex gains will not be sufficient for the given task. If we have a regular array, these measurements can be reduced to a single element, as all elements behave equally in this case. The transition to other elements is then described by the position-dependent phase shift. With such measured complex gains, we can calculate the elemental weighting for a deterministic sidelobe reduction in given solid angles. The required accuracy for the complex weights corresponds to the general requirements for low-sidelobe antennas.

5.5.3 Examples

As a first example, consider the reduction of sidelobes for a linear array (Farina et al., 1980). We wish to put an extended null around the direction $\hat{u} = \sin\hat{\theta}$. The fictional interference power density $p(u)$ has the following mathematical form:

$$p(u) = \sigma_J^2 \, \text{rect}_\delta(u - \hat{u}) \tag{5.53}$$

with

$$\text{rect}_\delta(u - \hat{u}) = \begin{cases} 1, & \hat{u} - \delta \le u \le \hat{u} + \delta \\ 0, & \text{otherwise} \end{cases} \tag{5.54}$$

The $(i - k)$th element of the covariance matrix $\mathbf{C}$ to be used in (5.52) is

$$C_{ik} = \sigma_J^2 \int_{\hat{u}-\delta}^{\hat{u}+\delta} \exp\left[-j\frac{2\pi}{\lambda}(x_i - x_k)u\right] du + \sigma^2 \delta(i,k)$$

$$= 2\sigma_J^2 \frac{\sin\left[\dfrac{2\pi}{\lambda}(x_i - x_k)\delta\right]}{\dfrac{2\pi}{\lambda}(x_i - x_k)} \exp\left[-j\frac{2\pi}{\lambda}(x_k - x_i)\hat{u}\right] + \sigma^2 \delta(i,k) \tag{5.55}$$

where x_i is the spatial coordinate of the ith element of the array, σ^2 is the power of a white fictional noise, and $\delta(i,k)$ is the Kronecker operator. Figure 5.35 shows the gain pattern for an array of 64 elements spaced 0.5λ with the main beam directed at $2.5°$ and an extended null of $4°$ width centered around $10°$. We assume $\sigma_J = 100\,\sigma$. For comparison, the classical $\text{sinc}(\cdot)$ or $(\sin x)/x$ type of gain pattern also is shown.

The next exercise refers to a planar array. We are interested in forcing low sidelobes in a wide sector, as shown in Figure 5.34. To this end, the fictional power distribution is assumed to be a constant value σ_J^2 in the region S shown in the direction cosine plane (u,v) of Figure 5.36, while, in the remaining part of the unit circle, the power is set to σ^2. The $(i - k)$th element of the covariance matrix $\mathbf{C}$ is

$$C_{ik} = \sigma_J^2 \int\int_S \exp\left[-j\frac{2\pi}{\lambda}(x_i - x_k)u - j\frac{2\pi}{\lambda}(z_i - z_k)v\right] du\, dv + \sigma^2 \delta(i,k) \tag{5.56}$$

where (x_i, z_i) are the spatial coordinates of the ith element of the array. The antenna gain pattern has been calculated for the following parameters, characterizing the re-

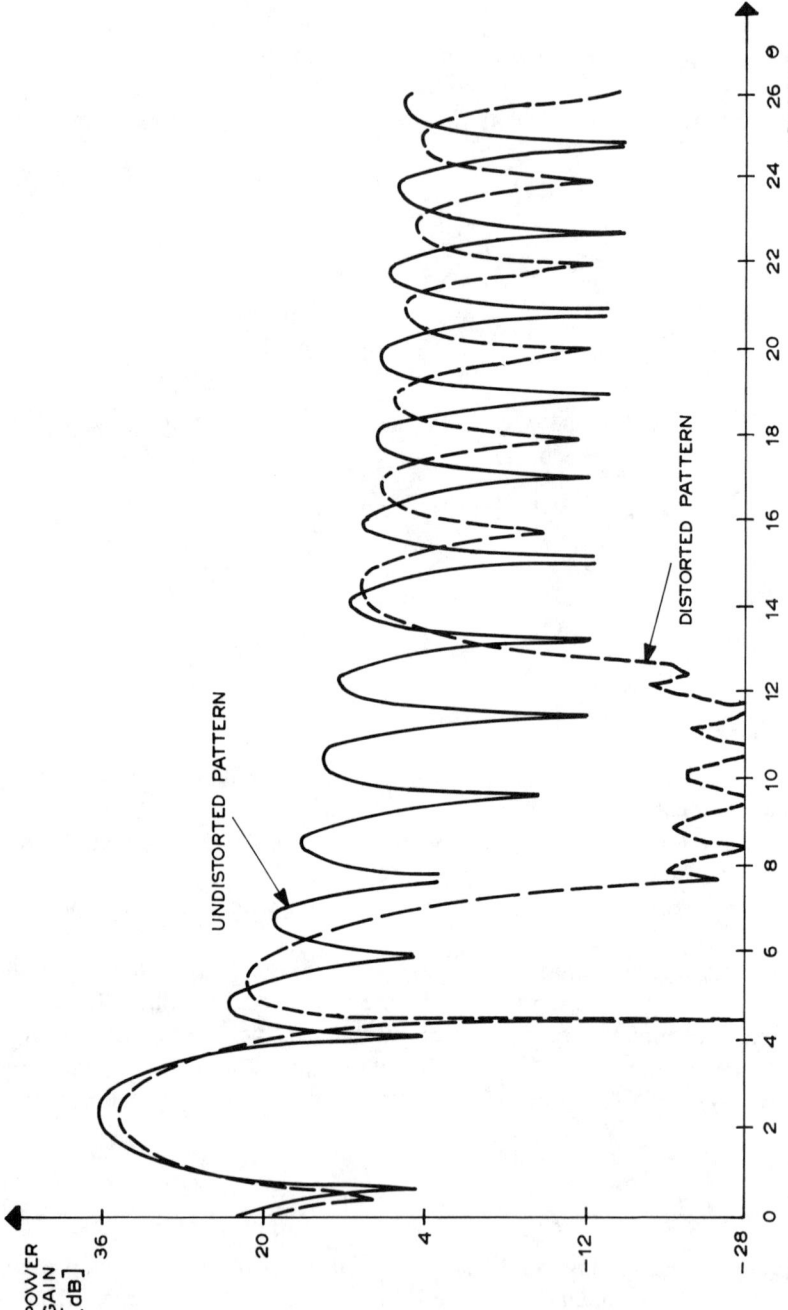

Figure 5.35 Gain pattern, with low sidelobes in one sector, of a linear array.

gion S of Figure 5.36: $q = 0.292$, $p = 1$, and w taken to give a 2° wide null. The ratio of the fictional jammer power to noise power is $\sigma_j^2/\sigma^2 = 100$. A planar antenna of 10 × 10 elements, regularly spaced by 0.5λ, was taken for the exercise. The classical undistorted pattern is shown in Figure 5.37, while the distorted pattern with the required wide null is depicted in Figure 5.38.

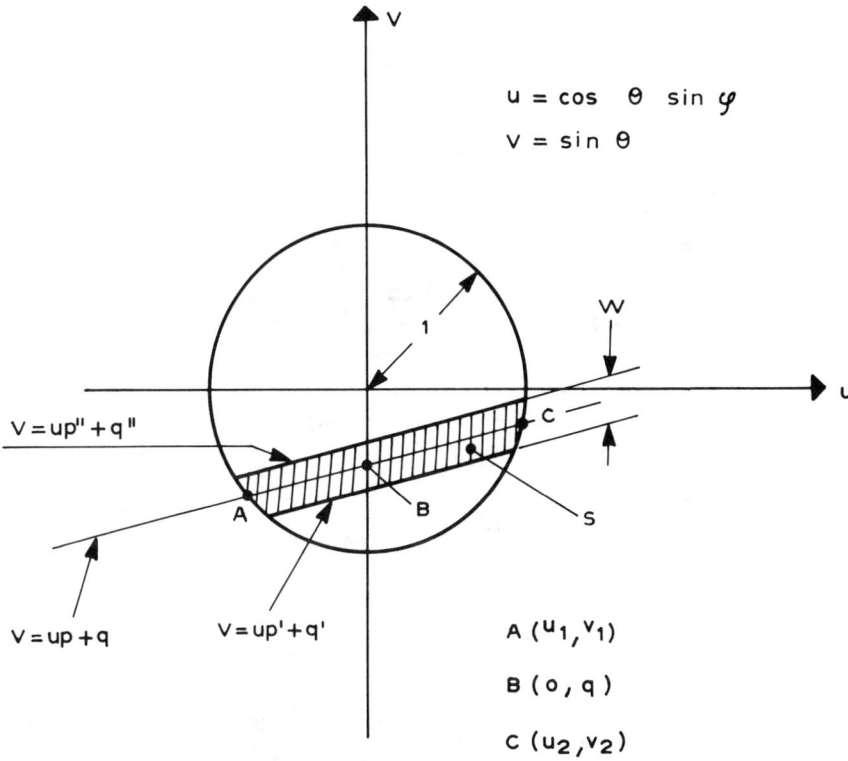

Figure 5.36 Definition of the sector S in which the sidelobes are low for a 10 × 10 planar array.

5.6 PHASE-ONLY NULLING

The idea of phase-only nulling in phased-array antennas is appealing because the phase shifters are already available as part of the beam-steering system. Hence, if the same phase shifters can be employed for the dual purposes of beam-steering and adaptive nulling of unwanted interference, costly retrofitting becomes unnecessary.

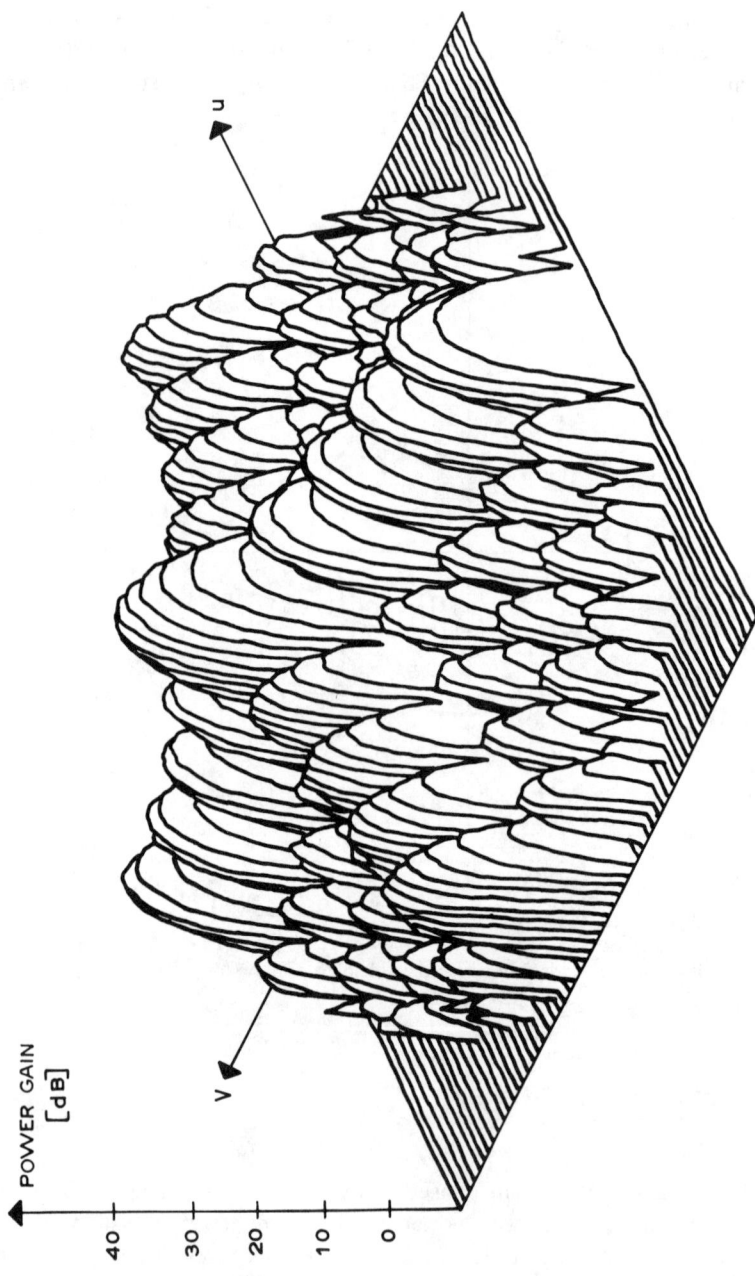

Figure 5.37 Undistorted gain pattern of a 10 × 10 planar array.

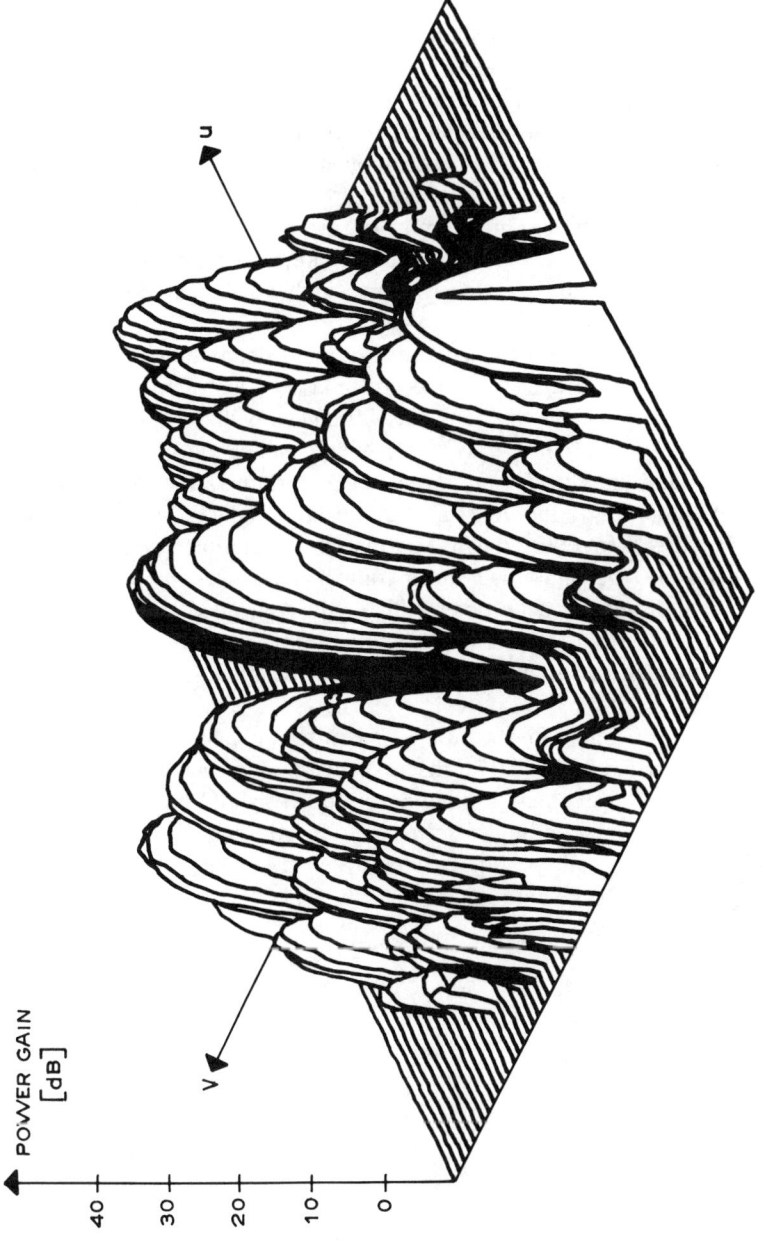

Figure 5.38 Distorted gain pattern of a 10 × 10 planar array.

5.6.1 Null Placement in *a priori*-Known Directions

Let us begin our review of phase-only nulling by considering the deterministic methods of null placement. These methods can be applied when the direction of incidence of the jammer or other disturbances such as an extended clutter is known *a priori*. In the latter case, an extended null can be obtained by generating a series of nulls in an angular sector. This concept has been applied in an operational three-dimensional phased-array radar, where several nulls are synthesized in a small angular sector in elevation of an array pattern (Giusto and DeVincenti, 1983). The nulls are aimed toward the horizon to cancel the ground clutter. Wide nulling, by adjusting the phase, can also be obtained by resorting to the pattern synthesis method described in Dufort (1989).

Phase-only null synthesis presents analytic and computational difficulties that are not present when both the amplitude and phase of the element weights can be freely perturbed. The principal source of the difficulties is the restriction of the weight perturbations to be of the phase-only type, which makes the nulling problem non-linear in general and analytically unsolvable. The minimization of the mean-square error must now incorporate the constraint that each weight W_i must have unit magnitude, i.e., $W_i^* W_i = 1$ for every i. This contrasts with combined phase and amplitude control, in which case the pattern is a linear combination of the complex array weights. The number of nulls that can be deterministically controlled with phase-only synthesis is considerably smaller than with combined phase and amplitude control, and the resulting distortion of the adapted pattern is more pronounced, but there is still sufficient flexibility for most purposes.

The problem of generating a null by adapting only the phase of the weights can be described in mathematical terms as follows. Indicate with $\mathbf{Z}$ the set of N samples received by the jammers impinging on the array and the receiver noise, and with $y = \mathbf{W}^T \mathbf{Z}$ the corresponding array output. In addition, denote with $s = \mathbf{W}^T \mathbf{S}$ the desired signal at the array output, where $\mathbf{S}$ is the expected signal vector, and $e = y - s$ is the error to be minimized. The weights $\mathbf{W}$ are usually calculated by minimizing the mean-square error:

$$J(\mathbf{W}) = E(|e|^2) \tag{5.57}$$

In the phase-only nulling technique, the minimization of the functional $J(\mathbf{W})$ is subject to the constraints:

$$W_i^* W_i = 1, \quad i = 1, 2, \ldots, N \tag{5.58}$$

Therefore, the new function to be minimized without constraint is

$$J'(\mathbf{W}) = J(\mathbf{W}) + \sum_{i=1}^{N} e_i(1 - W_i^* W_i) \qquad (5.59)$$

where e_i is a Langrage multiplier. In principle, the minimization of (5.59) requires the solution of $2N$ nonlinear algebraic equations, obtained by equating to zero the derivatives of J' with respect to W_i and e_i ($i = 1, 2, \ldots, N$). This is a quite cumbersome approach for large values of N. For instance, a planar phased array of 64 × 64 radiating elements brings to $N = 4096$. A more efficient approach has been suggested by Baird and Rassweiler (1978). Their method is based on a modification of the eigenvalue analysis of the optimum weights $\hat{\mathbf{W}} = \mu \mathbf{M}^{-1} \mathbf{S}^*$ for unconstrained weights. From (5.34), we remember that the optimum unconstrained weights are obtained through subtracting by the quiescent weights $\mathbf{S}^*$ the direction vectors $\mathbf{V}_k$ ($k = 1, \ldots, I$) associated with the I jammers to be nulled:

$$\mathbf{W} = \mu \left[\mathbf{S}^* - \sum_{k=1}^{I} \alpha_k \mathbf{V}_k \right] \qquad (5.60)$$

Baird and Rassweiler (1978) have shown that the constrained weights, i.e., the solution of (5.57) and (5.58), are as follows:

$$W_i = \exp\left[j\, \text{phase}\left(\mathbf{S}^* - \sum_{k=1}^{I} \gamma_k \mathbf{V}_k \right) \right], \quad i = 1, 2, \ldots, N \qquad (5.61)$$

where phase $(\cdot)$ stands for the phase value of the complex number in parentheses. We note that the unknowns to be calculated now are the γ_k ($k = 1, \ldots, I$), which are considerably less in number than the $2N$ unknowns in the previous minimization of (5.59). The unknowns γ_k are evaluated by resorting to an iterative minimization algorithm such as the random search or simplex method. This numerical procedure is iterated for each weight for the array. Other numerical methods also are possible, such as the branch and bound algorithm (Mendelovicz and Oestreich, 1979).

The algorithm has been tested by simulating a linear equispaced array of $N = 64$ elements (Farina et al., 1980). The requirement is to reduce the level of the sixth sidelobe by forcing three nulls, closely spaced with the following DOAs: $\theta_1 = 10.5°$, $\theta_2 = 11°$, and $\theta_2 = 12°$. The resulting gain pattern is shown in Figure 5.39, which also contains for comparison the undistorted pattern. A reduction of 15 to 20 dB has been obtained at the expense of the increase of sidelobes in other regions of no interest. To test the robustness of the technique to coding the weights with a digital word having a limited number of bits (note that Figure 5.39 refers to an infinite number of bits), refer to Figures 5.40 and 5.41. Figure 5.40 relates to the coding of the phases of the weights with four bits, while Figure 5.41 depicts the gain pattern

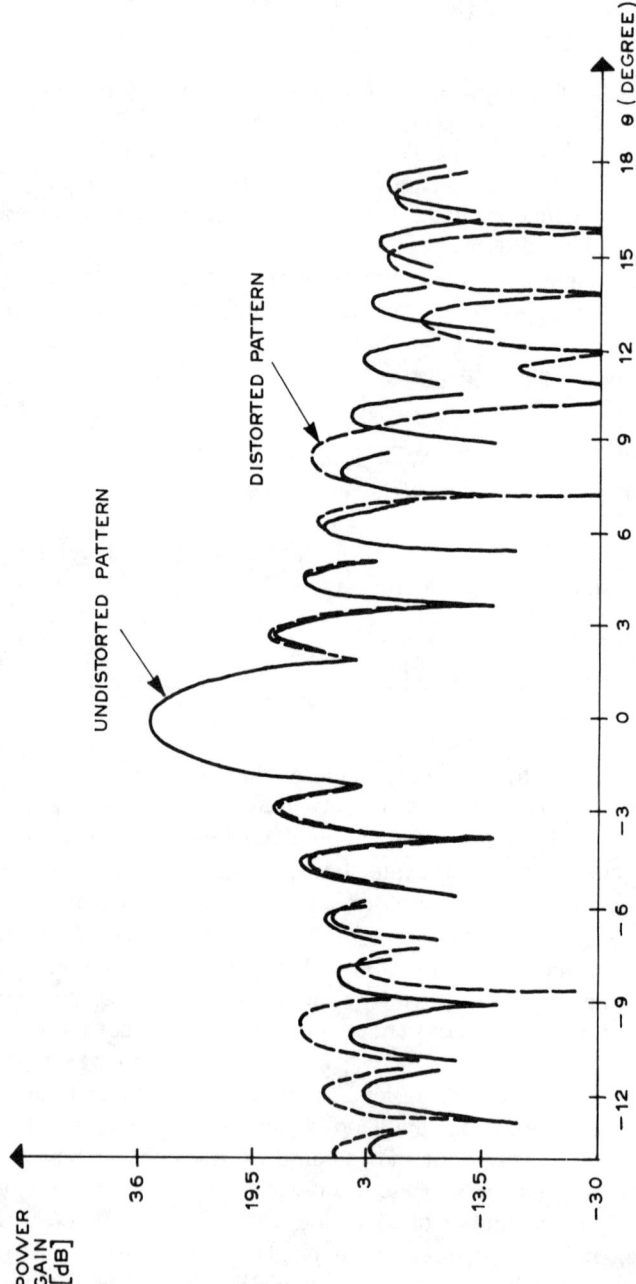

Figure 5.39 Extended null by phase-only nulling technique (infinite number of bits for phase coding).

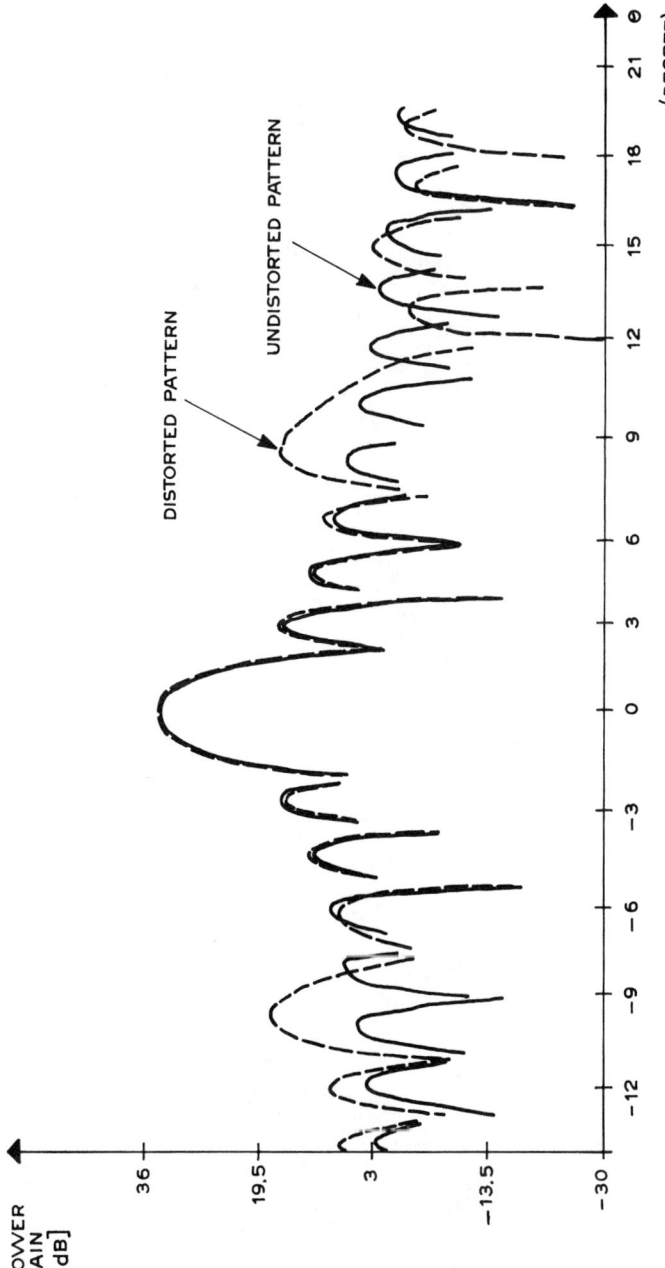

Figure 5.40 As in Figure 5.39, but using 4 bits to code the weights.

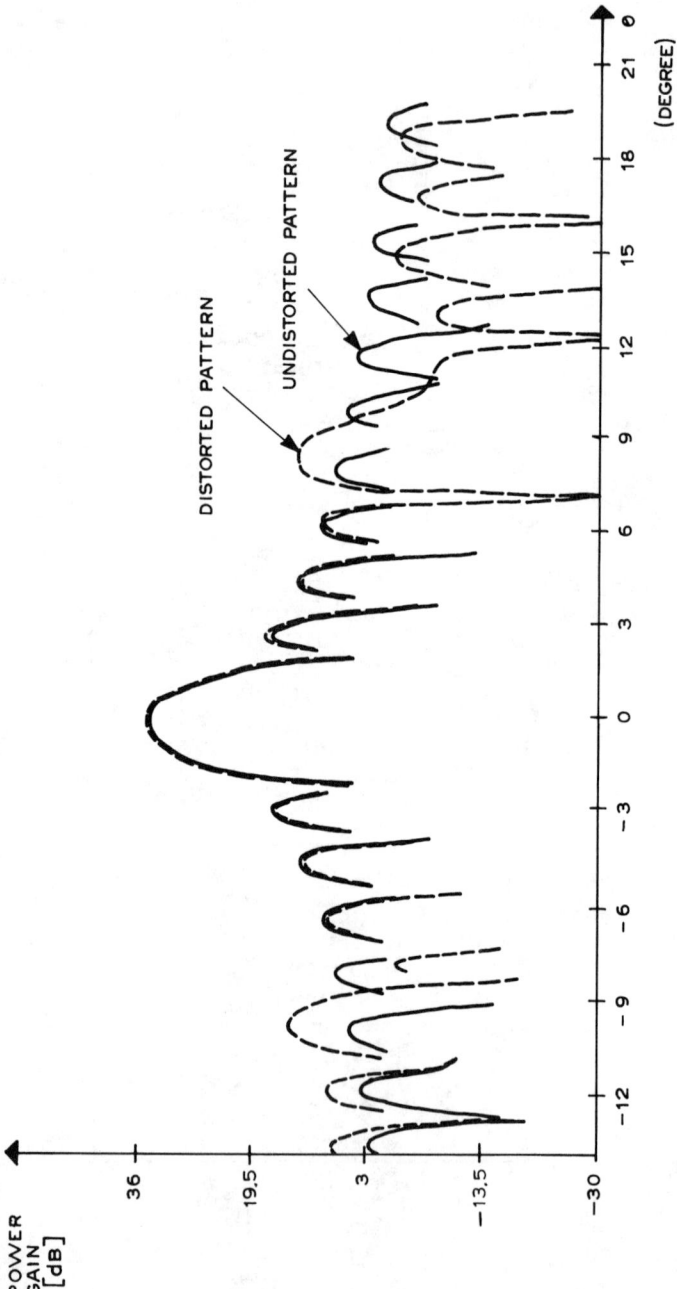

Figure 5.41 As in Figure 5.39, but using 5 bits to code the weights.

when we use five bits for coding the weights. The numerical results indicate that more than five bits must be used to code the weights for an adequate reduction (e.g., 20 dB) of sidelobe level.

A practical application of this technique has been tested in a three-dimensional (3D) phased-array radar (Giusto and DeVincenti, 1983). The objective was to verify the feasibility of putting a wide deterministic null (3° width) in the antenna elevation pattern to filter the ground clutter from the horizon. An average value of -35 dB for the null depth is practical, even in the presence of manufacturing inaccuracies (i.e., the phase and amplitude errors of the array excitation have rms values of 5° and 0.5 dB, respectively).

5.6.2 Adaptive Methods

Let us now turn our attention to the adaptive methods of phase-only sidelobe nulling. Baird and Rassweiler (1978) discussed the application of the beam-space decomposition to null steering by adaptive controlling the phase values of the beam-space coefficients (5.61). They employed a digital random search optimization algorithm to evaluate the weights, and performed computer simulations using their search technique, applying it successfully to a 16-element linear array, including cases of a uniformly illuminated array without errors and a tapered array with amplitude and phase errors. Both single and multiple interferers were cancelled. The effect of phase quantization was not included in their simulation. In addition to their computer simulation, Baird and Rassweiler constructed a 16-element breadboard, null-steering, C-band linear array with a uniform amplitude distribution and high-precision (9-bit) phase shifters. They were successful in forming deep nulls with a single interference source.

Turner *et al.* (1977; 1978) have described an adaptive S-band phased-array, consisting of a space-fed lens of 295 elements, each having a 3-bit PIN diode phase shifter. Adaptive nulling is accomplished by control of the element-steering phase. Their approach is oriented toward arrays with coarsely quantized phase shifters; thus, it is of considerable practical interest. In the technique implemented, a subarray beam is directed toward the interference and progressively spoiled by adjusting the steering phases of the subarray elements until the subarray field is equal in amplitude and in phase opposition to the main array field. When multiple interference sources are present, a subarray must be assigned to each interferer. The subarray beams then interact through their sidelobes, thus making creation of deep nulls more difficult as adjusting a subarray to reduce the interference from one source reintroduces interference from the other sources via the subarray sidelobes. An iterative random search type of adaptive algorithm was developed, whereby the interactive residues continuously decrease with each adjustment of the subarrays. Computer simulations of a simplified model of the array demonstrated nulling with up to five interference sources, and successful preliminary experimental results were also reported with a cancellation of 25 dB against a single CW interference.

Adaptive phase-only nulling investigations have also been described by Haupt et al. (1984). The experiments were performed on a precision linear array of 80 S-band, H-plane sectoral horn radiators with a Taylor amplitude illumination taper for the sum pattern. Phase control of each element was accomplished with an 8-bit digitally controlled phase shifter, making the *least significant bit* (LSB) about 1.4°. Although not designed to be adaptive, the antenna was made adaptive by linking control of the phase shifters to a computer for which the adaptive algorithms were programmed. Adaptive cancellation was performed with two different algorithms, designed to minimize the total output power of the array: an element-space gradient search algorithm, and a beam-space algorithm. Note that the beam-space algorithm, unlike the element-space method, requires a knowledge of the number and direction of the interferers. The adaptive performance of the antenna was evaluated and the gradient algorithm was tested with a fully adaptive array and a variety of partially adaptive configurations, using a single CW interference signal. The partially adaptive configurations, with from four to ten adaptive elements, resulted in considerably shorter adaptation times than the fully adaptive array. All but one of the configurations was effective in placing a pattern null of better than 20 dB below the quiescent pattern value. The one exception was a configuration with all adaptive elements placed on one side of the array center. Other than this case, there was little difference in the null depth achieved by the various configurations of adaptive elements, the patterns being lowered to essentially the noise level at the location of the interference source. Sidelobe distortion was greatest for the four-element configurations, but did not exceed 3.5 dB. There was virtually no main-beam distortion.

In conclusion, despite the theoretical and experimental research done, some doubts remain about the practical details of adaptive phase-only nulling technique. The quantization issue is especially important in large phased arrays because of the high cost of high-precision phase shifters. While there are some indications that coarsely quantized phase shifters can be used to give deep nulls, the price may be an unacceptable degree of pattern distortion elsewhere. Beam-space or subarray nulling methods appear to offer the most promise for large planar arrays because of the reduction in the number of adaptive controls that they offer, and they should be emphasized in future work.

5.7 ADAPTIVE ARRAYS WITH ANGULAR SUPERRESOLUTION

The resolution of a conventional antenna is limited by the well known Rayleigh criterion, which states that two equal-amplitude radiating sources can be resolved if they are separated in angle by roughly λ/L, in radians, where λ is the wavelength and L is the aperture length. An improvement of resolution could, in principle, be achieved by increasing the radar carrier frequency and antenna aperture length. Unfortunately, this is not always the case because the selection of the proper carrier frequency is the result of a trade-off analysis involving other operational requirements (e.g., search and track capabilities), limitations from the environment (e.g., atmospheric attenuation), and technological considerations. Increasing the antenna length is limited by physical and cost constraints.

There is, however, a solution to the problem of improving the angular resolution if the radar is equipped with a phased-array antenna and has an adequate signal processing capability. To understand how this can be done, consider again the scheme of Figure 5.1(a), showing that a wavefront incident on the array produces a sinusoid-like snapshot having a frequency related to the wavefront DOA. At this point, the problem of estimating the DOA of a number of sources impinging on the array can be recast in the framework of the classical problem of estimating the frequencies of multiple sinusoids embedded in noise from a limited number of data samples. If the sinusoids in the observed process are well separated in frequency, discrete Fourier transform processing of the samples is efficient and effective for estimation of the frequencies. However, as the frequencies are more closely spaced, DFT processing is ineffective and so-called "high-resolution" techniques are required.

In summary, the proposed system for improved angular resolution beyond the Rayleigh limit comprises an array antenna having a receiver and analog-to-digital conversion facilities downstream from each element (or a group of them, in the subarray approach). The digitized signals are sent to a digital signal processor, where modern spectral estimation algorithms sort closely spaced frequencies to determine the DOAs of either closely spaced radiating sources (e.g., jammers) or closely spaced targets. Additional information can be had, such as the source strengths, their cross-correlation, the number of sources, and, more generally, a high-quality (i.e., without sidelobes) spatial spectrum of radiating or backscattered energy from the environment.

When evaluating the performance of estimation methods operating on a certain set of noisy data, in addition to the resolution capability, we also should consider the bias and standard deviation affecting the angle estimates. The bias represents the systematic error of an estimation method, while the standard deviation is a measure of the nonsystematic error.

The purpose of this and the following sections are to illustrate the rationale of the processing algorithms, to give a quantitative feeling of the resolution achievable, to collect information concerning experimental systems described in the open literature, and to address the implementation problems, emphasizing the need for adequate parallel processing architecture and device technology to obtain real-time automated operation.

The achievement of a spatial spectrum pattern is of great interest for ECCM purposes and other radar applications. First, the estimated angles of the jammers can be used to form beams in the jammer directions and to use them as auxiliary beams for adaptive interference suppression as suggested by Brookner and Howell (1986;

1987) and by Gabriel (1986). See also Friedlander (1988), Friedlander and Porat (1989), and Hudson (1985b). Additional information, such as the source strengths, number of sources, and their coherence can be used to catalog the interference sources, to sort those of interest, to track them as their power values and DOAs change with time, and to react properly to them. Another application refers to tracking an interesting target in the presence of interference received through the antenna main beam. Possible solutions to this intriguing problem make use of high-resolution angular techniques, as shown in Gabriel (1986) and Nickel (1987), and illustrated in the computer experiment described in Section 5.7.5. Additional applications of angular superresolution are the following (Nickel, 1988):

1. *Resolution of a cluster of targets.* This is only of interest for targets on track. This application requires the angular resolution in azimuth and elevation, however, within a limited angular steerable sector. A fast reaction time for the radar system may be required.
2. *Emitter passive location for triangulation.* Here, the achievement of azimuthal resolution is sufficient. Because of the missing range resolution, "ghosts" may occur, which, however, can be discriminated from true sources on the basis of power-level estimation. In this application, the flow of data is less than in the previous case, and thus more time is available for signal processing.
3. *Reduction of multipath error at low elevation over sea* (Haykin et al., 1985; Theil, 1989 1990). The problem is again one-dimensional. In fact, we are interested in the angular resolution in elevation. In an advanced approach, to discriminate the direct path from the reflected path would be advisable looking also at their doppler frequency values, which are indeed very close to each other (Picardi and Spina, 1985). This would produce, however, a two-dimensional resolution problem. As only error reduction is of interest, no complete resolution of all reflecting points constituting the target's fine structure and its image is required. Indeed a two-target model is generally sufficient for error reduction. These two targets can be highly correlated, a situation which raises problems for some of the algorithms, as will be discussed in Section 5.7.6. In this application, a fast reaction time may be needed.
4. *Reduction of glint errors.* This is mainly a problem for complex targets of an angular extent comparable to the antenna beamwidth. Good rate of error reduction is achieved by working with a multiple-point target model and improving the resolution in azimuth and elevation.

At this point, the reader is warned that the high-resolution methods cannot substitute for the more conventional processing technique based on the use of sum-and-difference beams. Indeed, if there is only one target present, the monopulse method asymptotically provides (i.e., for large number of array elements or large number of processed time echoes) unbiased and minimum variance estimation of target angles. Additionally, if more than one target is present, separated by several beamwidths, and the antenna has a low sidelobe level, the monopulse performance is still reasonable. This means that high-resolution techniques should be considered as a second processing step, which applies after a first processing step based on the conventional monopulse techniques, the latter having attractive features of robustness against measurement errors. Downstream of the coarse estimation phase, the fine structure of the spatial (angular) spectrum pattern is analyzed by resorting to high-resolution methods. However, high-resolution techniques require high-precision, calibrated, multichannel receivers downstream from the radiating elements of the array antenna.

5.7.1 Classical Beam-Forming Methods

Since the 1970s, numerous high-resolution algorithms have been conceived. The interested reader should search the specialized literature and, in particular, the following remarkable references (Kay and Marple, 1981; Kay, 1988; Marple, 1985, 1987; Childers, 1978; IEEE, 1982; Gabriel, 1980a). Here we intend to present a classification of the methods illustrating the rationale of the main ones.

We start our description by recalling the definition of the power spectral density (in the frequency analysis of time series) or angular spectrum (in the wavenumber analysis for array antenna applications) of a sequence $\{y_k, k = 1, \ldots, N\}$. The psd is obtained as the DFT of the autocorrelation function of the sequence (the well-known Wiener-Khintchin theorem). Two classical methods of psd estimation are based on this theorem (Marple, 1985; 1987). Namely,

1. The *indirect approach,* in which an estimation of the correlation sequence (or correlogram) is first made on the basis of the sequence $\{y_k, k = 1, \ldots, N\}$ of available data, followed by a DFT;
2. The *direct approach* (or periodograms), in which the psd is obtained by taking the squared magnitude of the DFT of the data sequence with appropriate statistical averaging.

These methods, when employing the FFT algorithm, are computationally efficient and, in general, produce good results. One limitation is that of resolution, which is limited by the available data window. This is expressed by the familiar [(time interval over which the data have been measured) × (frequency resolution achievable)] product equal to unity in the analysis of time series of data, or by the already mentioned Rayleigh limit in the analysis of spatial series of data collected by an array of antennas. Bear in mind that this limit applies for a deterministic sequence of data. With a stochastic sequence, the limitation should be expressed as the stability-time-bandwidth product, $QTB,$ equal to unity (Marple, 1985, p. 145; 1987). The statistical stability, $Q,$ of the spectral estimate is defined as the ratio of the variance of the estimated psd, divided by the squared value of the expectation of the estimated psd. Small values of Q correspond to smooth psd estimates. This new limiting equation says that, for a given available record length T, we must trade off the statistical stability of the psd estimate with the achievable resolution. Another limitation produced by the windowing of data is leakage in the spectrum, i.e., energy in the main lobe migrates into the sidelobes, thus masking and distorting other spectral responses that are present. These two performance limitations are particularly relevant when analyzing short data records, as in the case of data gathered by an array of antennas.

The classical methods for the psd estimate in time-series analysis find their counterpart in spatial-series analysis with conventional beam-forming procedures. The common way to determine the waves' direction of arrival is shown in Figure 5.42(a), where a scanning beam explores the antenna field of view. The signals received by the array elements are linearly combined with weights:

$$W_k = \exp(-j\omega x_k), \quad k = 1, 2, \ldots, N \tag{5.62}$$

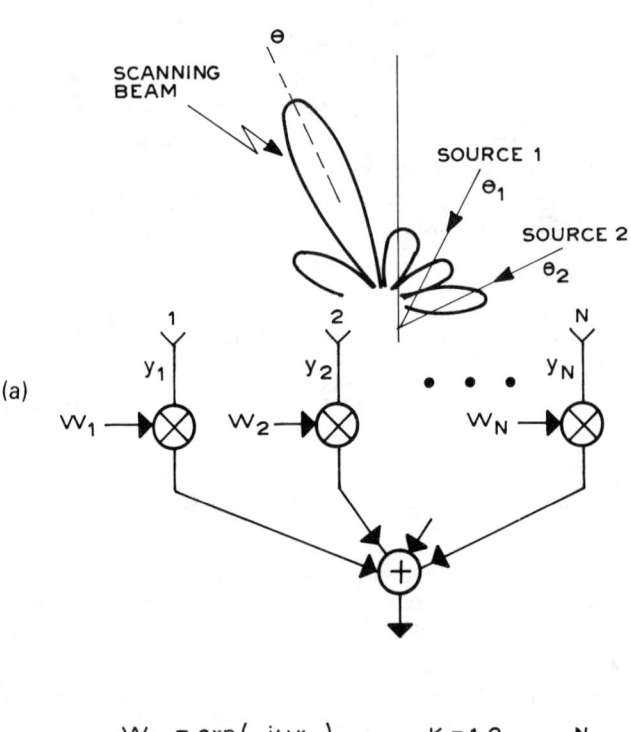

Figure 5.42 (a) Processing scheme for the classical beam-forming method; (b) power spectral density obtained by the scheme of Figure 5.42(a); two sources present in the antenna field of view, $\theta_1 = -45°$, $\theta_2 = +45°$, $JNR_1 = 30$ dB, $JNR_2 = 40$ dB, $N = 10$.

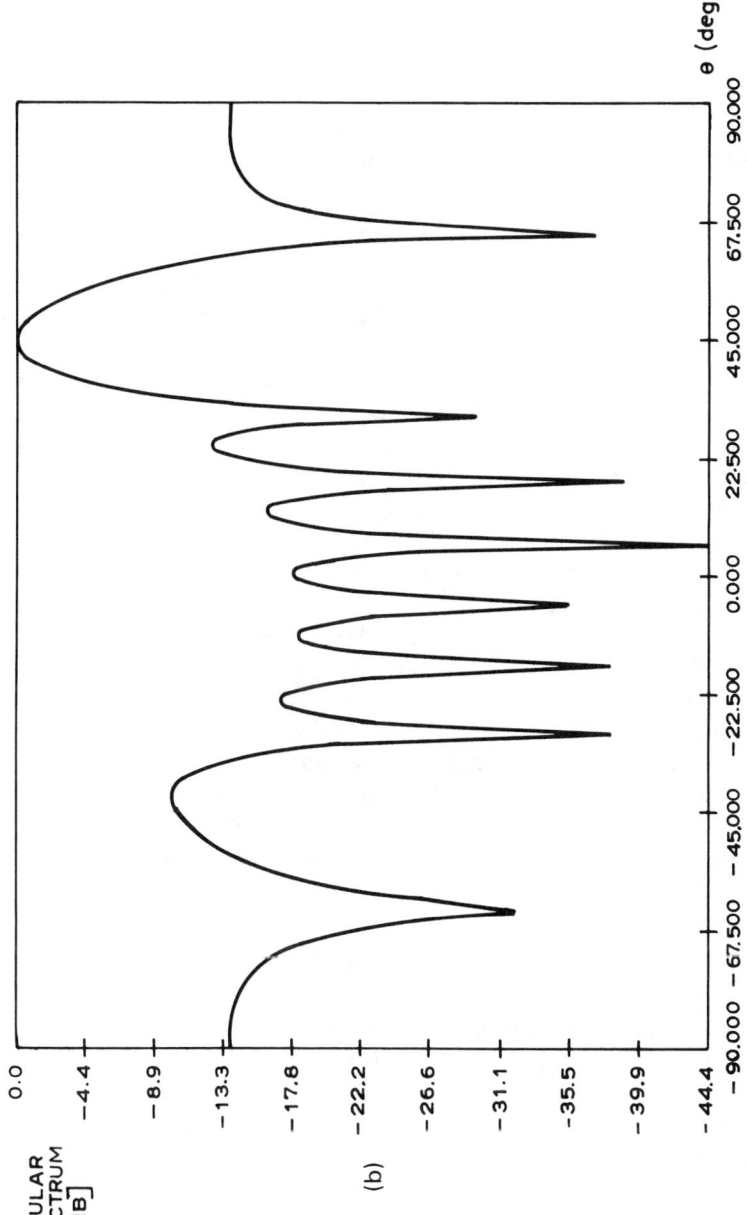

Figure 5.42 Continued

with $\omega = 2\pi \sin\theta$, where θ is the antenna look direction and x_k being the array element coordinate measured in wavelengths. When θ coincides with θ_l, i.e., the DOA of a source, the received signals are phase-compensated and summed coherently. Thus, the signal propagating in the look direction, θ_l, is reinforced, while other signals, such as noise, are not enhanced in strength. The mathematical expression of the angular spectrum, $P(\theta)$, is given by the power received in each look direction as the steering vector $\mathbf{W}(\theta)$ is varying with the angle θ:

$$P(\theta) = E\{|\mathbf{W}^T(\theta)\mathbf{Y}|^2\} = \mathbf{W}^H(\theta)\mathbf{M}\mathbf{W}(\theta) \qquad (5.63)$$

where $\mathbf{M}$ is the correlation matrix of the received signals $\mathbf{Y}^T = [y_1, y_2, \ldots, y_N]$.

Figure 5.42(b) shows the angular spectrum when there are two sources having incident angles $\theta_1 = -45°$ and $\theta_2 = +45°$ from boresight. The power levels of the sources with respect to the receiver noise power are $JNR_1 = 30$ dB and $JNR_2 = 40$ dB, respectively. We assume that $\mathbf{M}$ is exactly known *a priori*. The peaks of the two beams formed along θ_1 and θ_2 are linearly proportional to the corresponding values of JNR (Marple, 1987, p. 202). The beamwidths are inversely proportional to the number N of elements in the array ($N = 10$ in the example). The DOAs of the two sources are well estimated in this case.

As we know from the theory developed in Section 5.3.1, the weight vector $\mathbf{W} = \mathbf{S}^*$ represents the optimum solution to enhance the SNR at the output of the array. The optimality is related to the conditions of having only one monochromatic source and spatially white noise. However, if the assumptions do not hold, this beam-forming procedure will produce inaccurate estimates of the amplitude and DOA of the source. As in the time-series analysis, the advantages of the classical beam-forming method are their computational efficiency and proportionality of the angular spectrum to the power of the incoming plane wave. The disadvantages are related to the angular resolution being limited to the reciprocal of the array length, distortion of the spectrum due to the sidelobes, and suppression of a weak signal's main-lobe response by strong signal sidelobes.

5.7.2 Linear Prediction Applied to DOA Estimate

In search of new approaches that are able to overcome the limitations of classical methods, we consider the so-called *parametric methods,* initially conceived for the analysis of temporal sequences of data (Marple, 1985; 1987). The rationale to obtain a better psd estimate is to use a model for the process that generates data samples and to determine the model parameters from the same available data. The corresponding performances are related to the capability of the assumed model to match the process under analysis. The efficiency of the method is based on the capability of limiting the number of parameters to be estimated. Many deterministic and stochastic discrete-time processes are well approximated by a rational transfer function model. In this model, the input driving sequence n_k and output sequence y_k are related by the system transfer function:

$$H(z) = B(z)/A(z) \qquad (5.64)$$

The model is termed *autoregressive–moving average* (ARMA). $A(z)$ is the z-transform of the *autoregressive* (AR) branch, $B(z)$ is the z-transform of the *moving average* (MA) branch. The psd of the ARMA output process, $\{y_k\}$, driven by a white noise process, $\{n_k\}$, is

$$P_y(f) = \sigma^2 \left| \frac{B(f)}{A(f)} \right|^2 \qquad (5.65)$$

and $A(f) \triangleq A(z = e^{j2\pi f \Delta t})$, $B(f) \triangleq B(z = e^{j2\pi f \Delta t})$, where Δt is the sampling interval and σ^2 is the variance of the input noise sequence.

A very interesting case derives from the use of an AR model. With this model, the present value of the process is expressed as a weighted sum of p past values plus a noise term having variance σ^2. The psd in this case is

$$P_{AR}(f) = \frac{\sigma^2}{|A(f)|^2} = \frac{\sigma^2}{\left| 1 + \sum_{k=1}^{p} a_k \exp(-j2\pi f k \Delta t) \right|^2} \qquad (5.66)$$

To estimate the psd, we need to estimate the parameters $\{a_1, a_2, \ldots, a_p, \sigma^2\}$. If we know the $p + 1$ autocorrelation lags $R_{yy}(0)$, $R_{yy}(1)$, $\ldots$, $R_{yy}(p)$ of the sequence y_k, the $p + 1$ unknowns may be calculated by solving the Yule-Walker normal equations. A computationally efficient algorithm, known as Levinson recursion (see also Strobach, 1991), provides the solution with order p^2 operations. The AR-psd preserves the known autocorrelation lags and recursively calculates the autocorrelation for the lags beyond the window of known lags. The AR spectra do not exhibit the traditional sidelobes due to windowing. The extrapolation of the correlation is responsible for the high-resolution property of the AR-psd estimate. An alternative interpretation for the autocorrelation extrapolation is due to Burg (1968), who argued that the autocorrelation extrapolation should be selected to yield maximum entropy (i.e., the process with such an autocorrelation sequence would be the most random one possible), but would still preserve the autocorrelation lag values from 0 to p. This is because when the prediction error has been minimized; its power spectrum is equivalent to that of white noise. Thus, the uncertainty in the error sequence has been maximized. This is the celebrated *maximum entropy method (MEM)*. It recalls the thermodynamic concepts, and a generalization recently has been proposed, based on the minimization of free energy (Silverstein and Pimbley, 1988).

In most practical situations, we have data samples rather than autocorrelation values. We could estimate from the data the autocorrelation values for a number of lags, and then apply the Levinson recursion. However, a more efficient and better performing method does not require the estimation of the autocorrelation. It is based on the linear least-squares prediction of the nth sample of a data sequence $y_1, y_2, \ldots, y_N$ as a linear combination with coefficients $a_1, a_2, \ldots, a_p$ of the p data samples. The linear prediction error is defined as

$$e_{p,n} = y_n - \hat{y}_n = y_n - \sum_{k=1}^{p} a_k y_{n-k} \tag{5.67}$$

Roughly, the algorithm, also due to Burg, finds the prediction coefficients $a_1, \ldots, a_p$ that minimize the squared prediction error as the prediction filter moves along the data stream. Also note that N should be at least two times the value of p.

One problem with the AR-psd estimate is in the choice of AR order, which is not known *a priori*. Too low a guess for the AR order results in a highly smoothed psd. Too high an order introduces spurious details into the psd estimate. Statistical criteria have been conceived to find estimates of p. They are based on a function of p, which is the sum of two terms, the first being the prediction error that decreases with p, and the second, called the *penalty function*, which increases with p to prevent too high an order selection. The sum of the two terms has a minimum, indicating the best value of p. Among the order criteria, we note the *final prediction error* (FPE), the *Akaike information criteria* (AIC), and the *criterion autoregressive transfer* (CAT) (Marple, 1987, pp. 230, 374; Haykin, 1989, pp. 335–336).

A measure of the resolution of equal-power sinusoids has been determined in the case of a known autocorrelation function (Marple, 1982):

$$F = \frac{1.03}{Tp[\text{SNR}\,(p+1)]^{0.31}} \tag{5.68}$$

where F is the resolution in Hz, T the sampling interval in seconds, p the highest lag of the autocorrelation sequence, and SNR is the signal-to-noise ratio of a single sinusoid, expressed in linear (rather than in dB) units.

When the autoregressive psd estimation is used to indicate the presence of sinusoidal components in the data, the power associated with components in the AR-psd estimate is correctly computed by integrating the area under the power spectral density. As an alternative, we are tempted to use the height of each spectral peak as an indicator of sinusoidal power. The height is a reliable indicator of relative power with classical spectral estimators because their peaks are linearly proportional to the sinusoidal power when the input consists of sinusoids in white noise. This will not work with AR spectral estimators. Marple has shown (1987, p. 202) that, for a single sinusoid in white noise, the peak is proportional to the square of the power for high SNR. The same reasoning applies for multiple, well separated peaks.

The linear prediction spectral estimate that is commonly used in time-series analysis can also be applied in the angular resolution problem with an array of antennas, (Gabriel, 1979; 1980a; Borgiotti, 1978). This method is based on estimating the output of one array element by a weighted combination of the output of the other elements. The optimum weights are found by minimizing the mean squared prediction error. An additional effect is to whiten the spectrum of the output error, thus maximizing the uncertainty (the concept of maximum entropy spectral estimation). A conceptual scheme of the system is shown in Figure 5.43. We note that the adaptive array is similar to the well known sidelobe canceler in which the main beam is simply an array element. We know from the theory developed in Chapter 4 and Section 5.3 that the system operates placing nulls in the DOA of the jammers. The adapted pattern is obtained by subtracting the summed array pattern from the main-beam pattern. We also remember that the summed array pattern consists of properly weighted eigenvector beams that are formed by the finite array elements, and therefore cannot violate the conventional aperture resolution limit of the array (see, for instance, Figure 5.15). The desired spatial spectrum pattern is then obtained as the inverse of the adapted pattern. As Gabriel (1979; 1980a) indicates, such inverse patterns are not true antenna patterns, and linear superposition does not hold because the processing is nonlinear. These spatial spectra indicate the array pattern null points, and the nulls may be located arbitrarily close in terms of array beamwidth. The formation of the nulls utilizes as many array degrees of freedom as needed to cancel

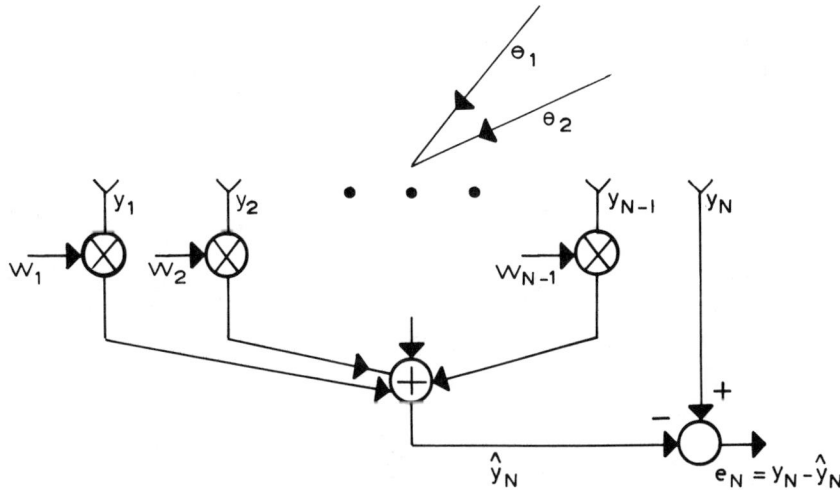

Figure 5.43 Linear prediction spatial filter for superresolution.

the sources, but conventional beam-forming utilizes the degrees of freedom at its disposal in the same manner as for the different scanning angles, rather than adaptively. We reiterate that the adaptivity of the weights to the sources gives rise to a nonlinear processing array system for which the Rayleigh diffraction limit is invalid; this resolution limit applies only for antennas operating as linear devices. An important question is whether the filter null points can accurately represent the DOA of the sources, which is one of our next points to discuss.

The linear prediction method can mathematically be described by the minimization of the output power of the adaptive array of Figure 5.43 under the constraint that one element in the weight vector is equal to unity. The function to be minimized is

$$\mathbf{W}^H\mathbf{M}\mathbf{W} - \alpha(\mathbf{U}_p^T\mathbf{W} - 1) \tag{5.69}$$

where $\mathbf{U}_p$ is a column vector having its pth element equal to one and the other elements equal to zero (in Figure 5.43, $p = N$), and α is the Lagrange multiplier. By standard calculations, the optimum weight, i.e., that which minimizes (5.69), is found to be (Huizing and DeValk, 1987; Theil, 1989):

$$\mathbf{W} = \frac{\mathbf{M}^{-1}\mathbf{U}_p}{\mathbf{U}_p^T\mathbf{M}^{-1}\mathbf{U}_p} \tag{5.70}$$

We note that the optimum weight $\mathbf{W}$ is the pth column of the matrix $\mathbf{M}^{-1}$ divided by the scalar $\mathbf{U}_p^T\mathbf{M}^{-1}\mathbf{U}_p$, which is the element (p,p) of the same matrix $\mathbf{M}^{-1}$. A way to define the linear prediction angular spectrum derives from the observation that the optimum weights produce nulls in the jammers' directions. Thus, the reciprocal of the adapted pattern shows peaks in the jammers' DOAs. Consequently, the linear prediction angular spectrum is

$$P(\theta) = \frac{\cos t}{|\mathbf{W}^H\mathbf{S}(\theta)|^2} \tag{5.71}$$

where $\mathbf{W}$ is the optimum weight of (5.70) and $\mathbf{S}(\theta)$ denotes the steering vector for the look angle, θ. The explicit expression of the angular spectrum is given by (5.71), taking the output power $|\mathbf{W}^H\mathbf{M}\mathbf{W}|$ as the constant of the array. The angular spectrum becomes

$$P(\theta) = \frac{\mathbf{W}^H\mathbf{M}\mathbf{W}}{|\mathbf{W}^H\mathbf{S}(\theta)|^2} \tag{5.72}$$

which, by substitution of (5.70), can be rewritten as

$$P(\theta) = \frac{\mathbf{U}_p^T \mathbf{M}^{-1} \mathbf{U}_p}{|\mathbf{U}_p^T \mathbf{M}^{-1} \mathbf{S}(\theta)|^2} \qquad (5.73)$$

Usually, the value of p is selected as $p = 1$ or $p = N$. By choosing $p = 1$, the previous equation becomes

$$P(\theta) = \frac{(\mathbf{M}^{-1})_{1,1}}{|(\mathbf{M}^{-1})_1 \mathbf{S}(\theta)|^2} \qquad (5.74)$$

where $(\mathbf{M}^{-1})_{1,1}$ and $(\mathbf{M}^{-1})_1$ are the element (1,1) and the first row of $\mathbf{M}^{-1}$, respectively.

To show how critical the DOA estimate is, consider the following figures. Figure 5.44 shows the angular pattern of the adaptive array *versus* the beam-scanning angle, θ, when two sources impinge on the array with DOAs, $\theta_1 = 45°$ and $\theta_2 = -45°$, and values of $JNR_1 = 40$ dB and $JNR_2 = 30$ dB, respectively. The array antenna has $N = 10$ elements, equally spaced at $0.5\ \lambda_0$. The receivers downstream of the array elements have a fractional bandwidth B/f_0, with respect to the carrier frequency, f_0, which is the parameter. The incoming sources have been assumed to have a flat spectrum in the channel bandwidth, B. The receiver noise is white Gaussian and independent among the channels. The figure shows that the resolution capability of the system decreases as the receivers' bandwidth increases. This is particularly true for off-boresight sources, whereas, for near broadside directions, the time dispersion $d \sin\theta/c$ is mitigated by the small value of $\sin\theta$. Also confirmed is that the peak values of the angular spectrum are proportional to $(NJNR)^2$ (see Marple, 1987).

At this point, we estimate the angular resolution capacity of the linear prediction method. To this end, consider two equal-power, monochromatic, uncorrelated sources impinging on the array from two directions that are symmetric with respect to the array boresight. By progressively decreasing the angular separation of the two sources, we reach a situation in which the two peaks of the angular spectrum will not be distinguishable. Consequently, the resolving capability of an estimator can be defined as the smallest difference in angle at which two equistrength targets can be still distinguished. For the purpose of the exercises in this section, we have decided to consider the two equal-power sources to be at the limit of resolution in angle if their beams cross at a level which is 3 dB down from their peak values. Curve (a) of Figure 5.45 shows the angular resolution limit of the linear prediction method. The limit curve is depicted on a chart having as x-axis the minimum angular separation $\Delta\theta$ of the method referred to the Rayleigh limit θ_{RAY} of the classical beamformer, and as y-axis the power values of the two sources. Note that, while the classical method of beam-forming (Section 5.7.1) has a resolution capability independent of the source power levels, the superresolution methods achieve a better

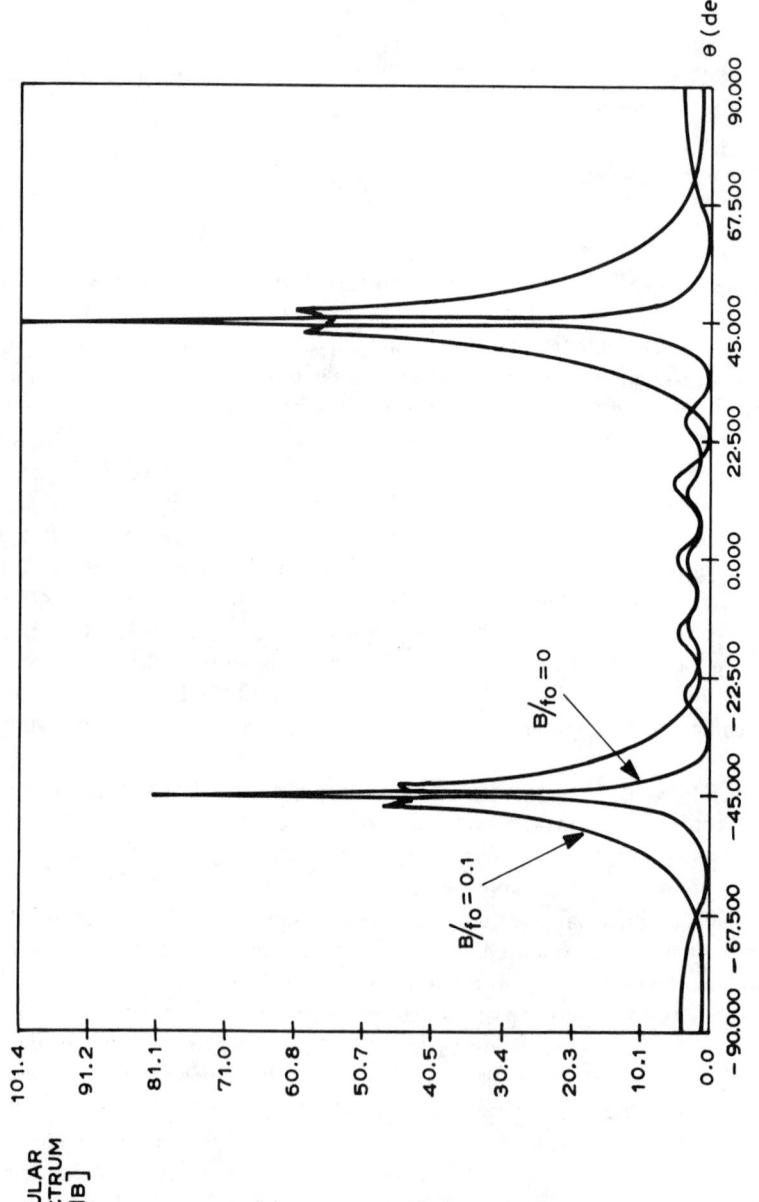

Figure 5.44 Angular spectrum of linear prediction method; two sources at $\theta_1 = 45°$ and $\theta_2 = -45°$ with $JNR_1 = 40$ dB and $JNR_2 = 30$ dB; B/f_0 is the fractional bandwidth of the receivers downstream from the array elements, number of array elements, $N = 10$.

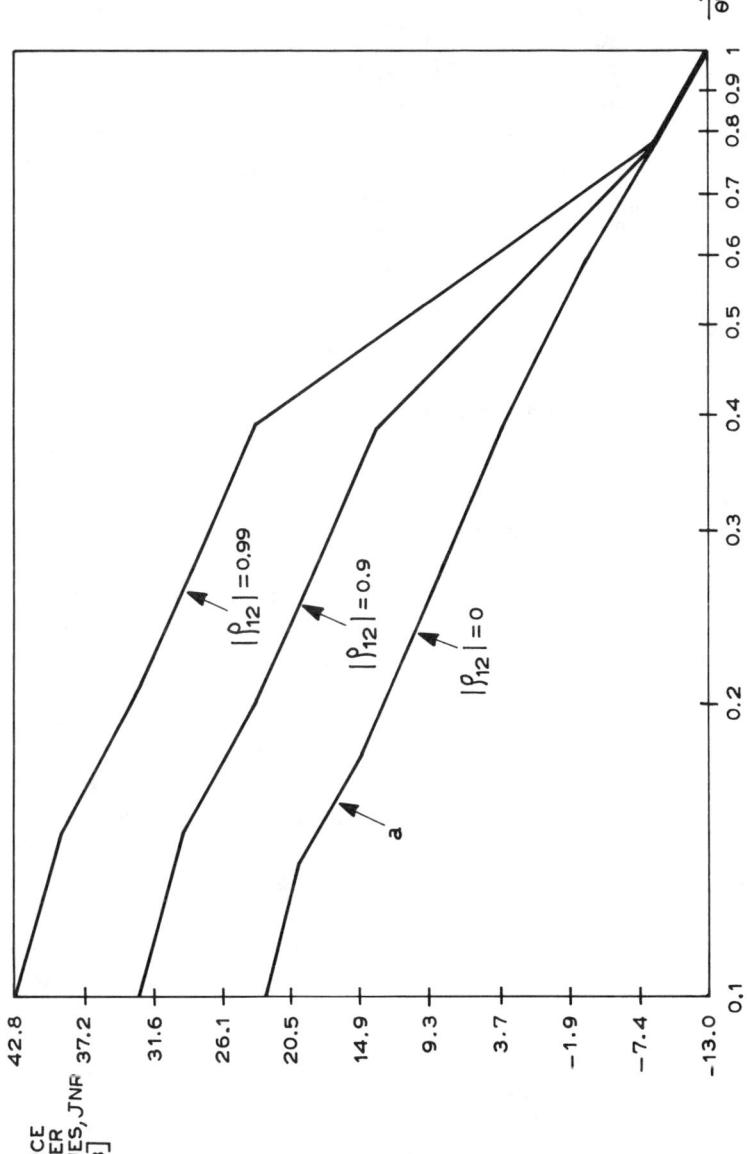

Figure 5.45 Angular resolution limit of the linear prediction method; two equal-power monochromatic sources with correlation coefficient, ρ_{12}.

resolution if enough source-to-noise power ratio is available. Figure 5.45 also shows two additional curves, which refer to two correlated sources with correlation coefficient ρ_{12}. The resolution capability decreases as the correlation coefficient tends to unity. The motivations of this phenomenon and possible remedies will be discussed in Section 5.7.6. The curves of Figure 5.45 have been derived by assuming that the covariance matrix **M** of the data received by the array is *a priori* known.

Curve (a) of Figure 5.45 is similar to one found by Gabriel (1980a). He concludes that the resolution limit of the maximum entropy method (which is equivalent to the linear prediction solution for a linear array with equidistant receiving elements) for two equal-strength, incoherent sources, embedded in white Gaussian noise, is a function of the JNR of each source at the array elements and the number N of the array elements. The source separation with the linear prediction (i.e., MEM), superresolution method is related to the 3 dB beamwidth of the Rayleigh criterion by the following equation:

$$(\text{source separation})_{\text{MEM}} = \frac{\theta_{\text{Ray}}}{(\text{JNR}N)^{1/3.26}} \qquad (5.75)$$

This relationship applies for a linear array having $\lambda/2$ elemental spacing, monochromatic incident signals, and no errors (e.g., mismatching in the receiving channels) in the array.

Therefore, if enough JNR or numbers of array elements are available, the two sources may be separated, even if their angular distance is less than the Rayleigh limit. For example, JNR = 10 dB and $N = 10$, the separation is 0.24 of the 3 dB beamwidth (see also Figure 5.45, curve (a)).

A further limitation on the resolution capability of the method is given by the mismatching between the receiving channels downstream from the array elements. Figure 5.46 illustrates the influence of the mismatching between the receiving channels. The mismatching model of the array is that shown in Figure 5.26 and (5.37). The fractional bandwidth of each receiving channel is $B/f_0 = 0.05$; the sources impinging on the array are assumed to have a flat spectrum in this bandwidth. The curves of Figure 5.46 have been derived by assuming that the covariance matrix **M** of the signals received by the array is *a priori* known. The channel mismatch greatly affects the resolution performance of the method. In fact, large source-to-noise power values are not always helpful.

5.7.3 Minimum Variance Applied to DOA Estimate

This method is due to Capon (1969) and can be applied to time-series analysis and to array processing for angular resolution. In time-series analysis, the spectral estimate is defined by a filter designed to pass the power in a narrow band around the

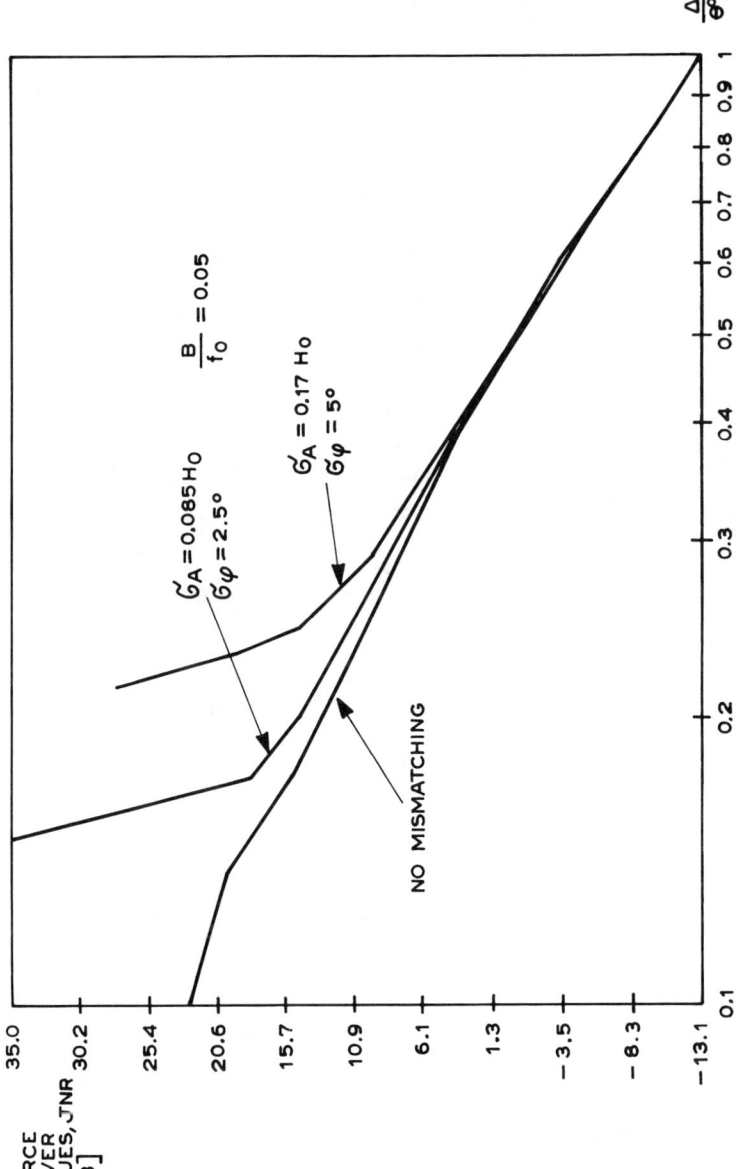

Figure 5.46 Angular resolution limit of the linear prediction method; two equal-power nonmonochromatic ($B/f_0 = 0.05$) sources impinging on an array of $N = 10$ receiving elements; amplitude mismatching, σ_A/H_0, and phase mismatching, σ_ϕ, of the receiving channels as parameters of the curves.

frequency of interest and to minimize all other frequency components in an optimal manner. Observe that no model parameters are explicitly computed. The difference between the minimum variance method and the classical periodogram method is that in the former the shape of the narrowband filters is generally different for each frequency, whereas the shape of the filters in the periodogram method does not change. The filters are *finite impulse response* (FIR) with weights selected so that the frequency response of the filter is unity at the frequency, f_0, under consideration (i.e., an input sinusoid at that frequency would pass unchanged) and the variance of the output process is minimized (i.e., the spectral components that are not near f_0 are rejected).

The minimum variance method derives an angular spectrum from the received data by minimizing the output power or variance of an adaptive array for each look direction, under the constraint that the gain in the look direction is equal to unity. The minimum variance method therefore reduces the contributions to the variance from targets and noise not propagating in the look direction. As a result, this method eliminates problems due to the presence of multiple targets associated with the conventional beam-forming procedure. This is identical to the use of a zero-order mainbeam directional gain constraint in adaptive arrays (see Montingo and Miller, 1980), where the spatial spectrum is estimated as the output residual power, $P(\theta)$:

$$P(\theta) = \hat{\mathbf{W}}^H(\theta)\mathbf{M}\hat{\mathbf{W}}(\theta) \tag{5.76}$$

with

$$\hat{\mathbf{W}}(\theta) = \mu \mathbf{M}^{-1}\mathbf{S}^*(\theta) \tag{5.77}$$

where $\mathbf{S}^*(\theta)$ is the usual weight vector for steering the main beam at the angle θ, and μ is a constant. The zero-order gain constraint implies the following:

$$\mathbf{S}^T\hat{\mathbf{W}} = 1 \tag{5.78}$$

which is equivalent to selecting the following value for the constant:

$$\mu = [\mathbf{S}^T\mathbf{M}^{-1}\mathbf{S}^*]^{-1} \tag{5.79}$$

By substitution of the constant into (5.77) and (5.76), the following angular spectrum is obtained:

$$P(\theta) = \frac{1}{[\mathbf{S}^T(\theta)\mathbf{M}^{-1}\mathbf{S}^*(\theta)]} \tag{5.80}$$

Upon sweeping the steering vector $\mathbf{S}^*(\theta)$ for a given inverse of the covariance matrix,

$P(\theta)$ provides the required angular spectrum. The basic configuration of the system is shown in Figure 5.47. The difference between conventional and adaptive beam-forming is quite evident from Figure 5.47. In conventional beamforming, the beam is formed and steered through all angles simply by setting $\mathbf{W} = \mathbf{S}^*(\theta)$. The spatial pattern is then obtained by measuring the received power at each angle. In minimum variance adaptive beam-forming, for each desired look direction, the weights of the adaptive array are evaluated to minimize the output power subject to a look-direction constraint. The spatial pattern is obtained by measuring the output power after convergence of the adaptive algorithm for each explored angle.

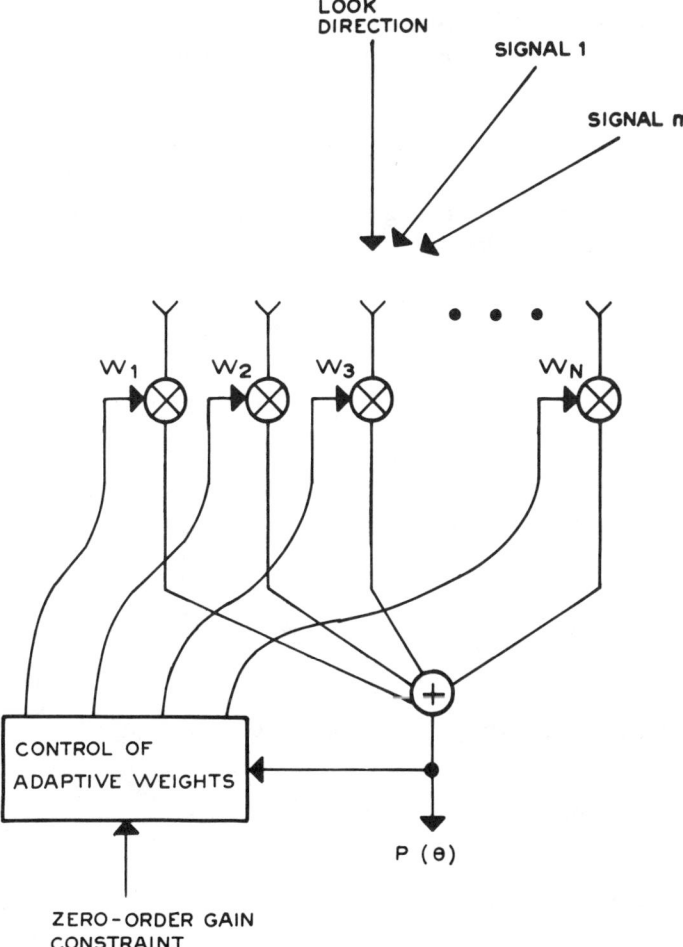

Figure 5.47 The adaptive array for high angular resolution with the minimum variance (Capon) method.

As noted by Marple (1987, p. 354) the minimum variance spectral estimator is not a true psd function because the area under the estimate does not represent the total power of the process, a characteristic of the true psd. In addition, the inverse transform of the minimum variance estimate does not match the autocorrelation sequence, which is used to create the minimum variance estimate. Instead, these two features are achieved with the AR-psd methods.

As an application example, consider the minimum variance angular estimate of only white spatial noise. By substitution of $\mathbf{M} = \sigma^2\mathbf{I}$ into (5.80), we obtain the minimum variance spatial pattern of white noise equal to σ^2/N, where σ^2 is the true spatial pattern of the noise and N is the number of array elements. As another example, consider the case of a monochromatic plane wave impinging on the array from angle θ_0. The covariance matrix $\mathbf{M}$ of the received signal plus thermal noise is

$$\mathbf{M} = \sigma_0^2 \mathbf{S}^*(\theta_0)\mathbf{S}^\mathrm{T}(\theta_0) + \sigma^2\mathbf{I} \tag{5.81}$$

where σ_0^2 is the power of the incoming plane wave and σ^2 is the thermal-noise variance. The inverse of this covariance matrix is

$$\mathbf{M}^{-1} = \frac{1}{\sigma^2}\left[\mathbf{I} - \frac{\frac{\sigma_0^2}{\sigma^2}\mathbf{S}^*(\theta_0)\mathbf{S}^\mathrm{T}(\theta_0)}{1 + \frac{\sigma_0^2}{\sigma^2}\mathbf{S}^\mathrm{T}(\theta_0)\mathbf{S}^*(\theta_0)}\right] \tag{5.82}$$

Note that $\mathbf{S}^\mathrm{T}(\theta_0)\mathbf{S}^*(\theta_0)$ is equal to the number N of array elements. The equation thus becomes

$$\mathbf{M}^{-1} = \frac{1}{\sigma^2}\left[\mathbf{I} - \frac{\sigma_0^2 \mathbf{S}^*(\theta_0)\mathbf{S}^\mathrm{T}(\theta_0)}{\sigma^2 + \sigma_0^2 N}\right] \tag{5.83}$$

which, substituted into (5.80), gives the following spatial pattern:

$$P(\theta) = \left\{\frac{1}{\sigma^2}\mathbf{S}^\mathrm{T}(\theta)\left[\mathbf{I} - \frac{\sigma_0^2 \mathbf{S}^*(\theta_0)\mathbf{S}^\mathrm{T}(\theta_0)}{\sigma^2 + \sigma_0^2 N}\right]\mathbf{S}^*(\theta)\right\}^{-1} \tag{5.84}$$

which, when evaluated in θ_0, takes the value:

$$P(\theta_0) = \sigma^2\left(\frac{1}{N} + \frac{\sigma_0^2}{\sigma^2}\right) \tag{5.85}$$

As a consequence, only when the SNR is high and the number N of antenna is large, the peak value of the minimum variance angular estimate is equal to the power σ_0^2 of the incoming monochromatic plane wave. In conclusion, the minimum variance method is a good indicator of the DOA of monochromatic plane waves, with the peak values of the function $P(\theta)$ related to the power of the sinusoids in the received samples.

As shown by Burg (Childers, 1978, p. 132), the Capon pattern is just the average of several maximum entropy patterns of different orders. See also (Nickel, 1988a) and (Marple, 1987, pp. 354–356). This averaging operation explains the much greater fluctuations of the maximum entropy patterns as compared to the minimum variance patterns.

In the following, a number of figures illustrate the performance of the minimum variance method in several cases of practical interest. In all the examples considered, we assume *a priori* knowledge of the covariance matrix **M** of the received data from an array of $N = 10$ elements, spaced $0.5\ \lambda_0$ apart.

Figure 5.48 shows the angular patterns for two sources having DOA values of $\theta_1 = -45°$ and $\theta_2 = +45°$, and $JNR_1 = 30$ dB and $JNR_2 = 40$ dB, respectively. The pattern has been found for three different values of the receiving channel bandwidth, B. The angular resolution deteriorates as the receiver bandwidth increases. We assume that the sources are random processes with flat spectrum in the channel bandwidth of the receivers. The angular resolution would not change as much as a function of the bandwidth, for sources having DOAs close to the array boresight. From the figure, we note that the peak values of the spectrum are proportional to (NJNR), i.e., to the product of the number of antennas in the array by the JNR of the source. Curve (a) of Figure 5.49 illustrates the angular resolution limit of the minimum variance method for two equal-power noncoherent sources. The relative bandwidth (B/f_0) of the receiving channel has been set equal to zero. The ordinate axis of the figure indicates the JNR value. The abscissa shows the angular resolution, $\Delta\theta$, of the method with respect to the Rayleigh limit resolution, θ_{Ray}. As in Section 5.7.2, we consider the two equal-power sources still resolvable in angle if their beams cross at a level that is 3 dB down from their peak values. The other two curves illustrate the influence of the correlation between the two monochromatic sources on the angular resolution capabilities of the Capon method. The correlation coefficient is $|\rho_{12}|$ (real value, i.e., $\phi_{12} = 0$). Note that we need more powerful sources to achieve the angular resolution values related to the absence of correlation.

Figure 5.50 illustrates the angular resolution limits when the fractional bandwidth B/f_0 of the receiving channels is equal to 0.05. In particular, the curve labeled with zero refers to the case of perfectly matched receiving channels. The other two curves show the degradation of angular resolution when there is a mismatch between the receiving channels downstream from each element of the array. The model of Section 5.3.3 has been adopted here. The amount of amplitude and phase mismatching is shown in the figure. Note that the presence of channel mismatching greatly

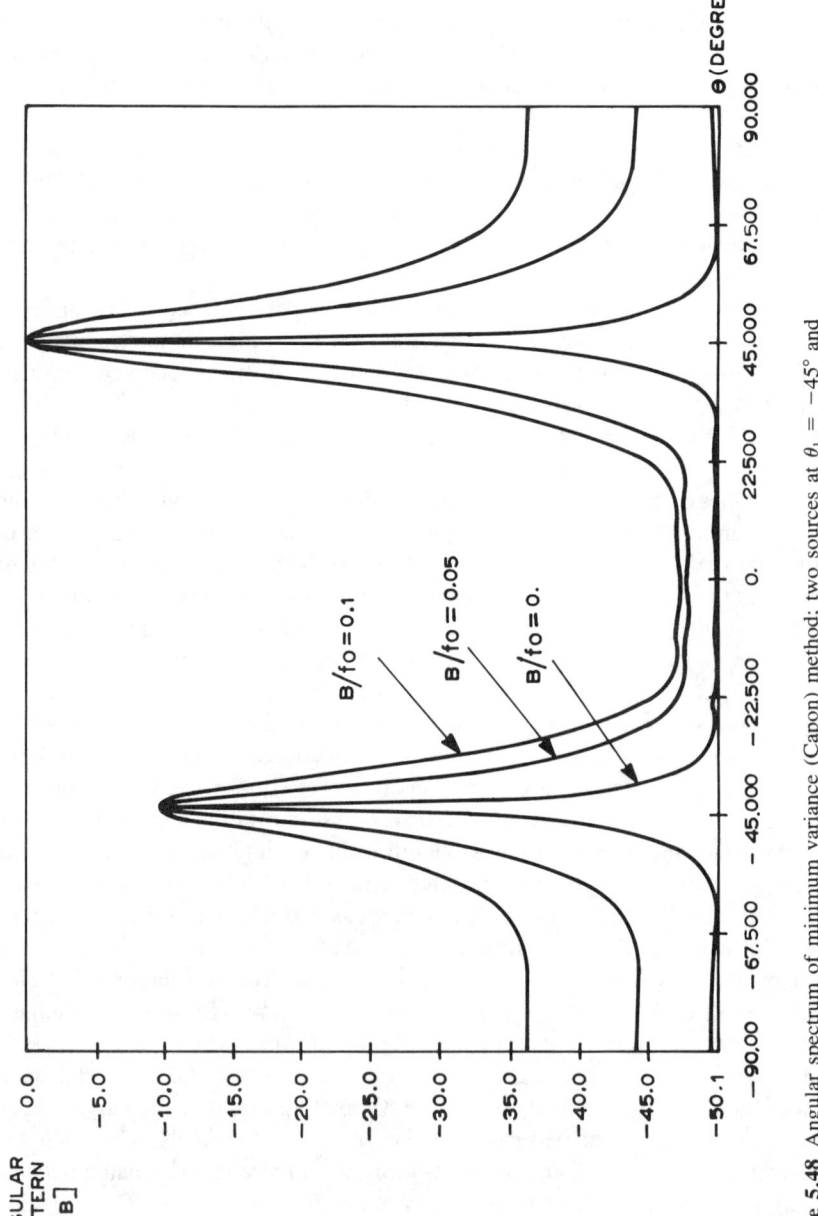

Figure 5.48 Angular spectrum of minimum variance (Capon) method; two sources at $\theta_1 = -45°$ and $\theta_2 = +45°$ with $JNR_1 = 30$ dB and $JNR_2 = 40$ dB; B/f_0 is the fractional bandwidth of the receivers downstream from the array elements, number of array elements, $N = 10$.

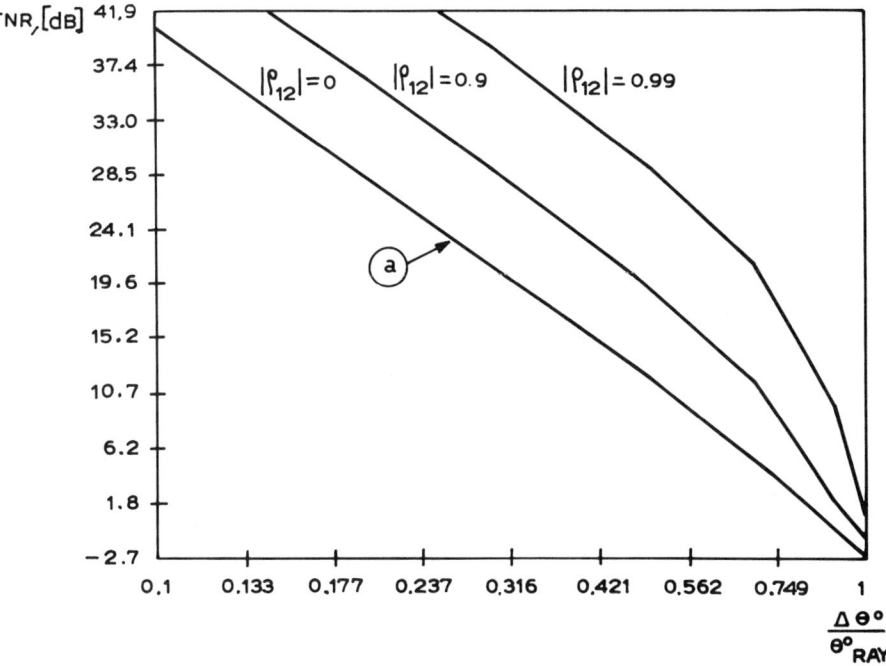

Figure 5.49 Angular resolution limit of the minimum variance (Capon) method against two monochromatic correlated sources.

limits the angular resolution capabilities of the method. This behavior is similar to that experienced in the linear prediction method of Section 5.7.2.

5.7.4 DOA Estimate Based on Eigenanalysis

The rationale for this class of algorithms is related to the division of information in the autocorrelation matrix of the data received by the array elements into two vector subspaces, one a signal subspace, the other a noise subspace. This can be easily visualized by remembering the eigenvalue spectrum, depicted in Figure 5.11, for instance, which shows a clear separation between the source and noise eigenvalues. These algorithms assume that the signal of interest lies in a lower dimensional space than the full N-dimensional space spanned by the vectors of data samples received by the array elements. Physically, this means that I narrowband sources impinge on the array with $I < N$, where N is the number of array elements.

We describe the signal subspace with a set of I orthonormal vectors $(\mathbf{x}_1, \mathbf{x}_2, \ldots, \mathbf{x}_I) \triangleq \mathbf{X}$. If $\mathbf{S}^*(\theta)$ is directed along a source DOA, $\mathbf{S}^H(\theta) \mathbf{X}\mathbf{X}^H \mathbf{S}(\theta)$ will have a maximum equal to $\mathbf{S}^H(\theta) \mathbf{S}(\theta)$ and $\mathbf{S}^H(\theta) (\mathbf{I} - \mathbf{X}\mathbf{X}^H)\mathbf{S}(\theta)$ will be zero. The matrix $\mathbf{P} =$

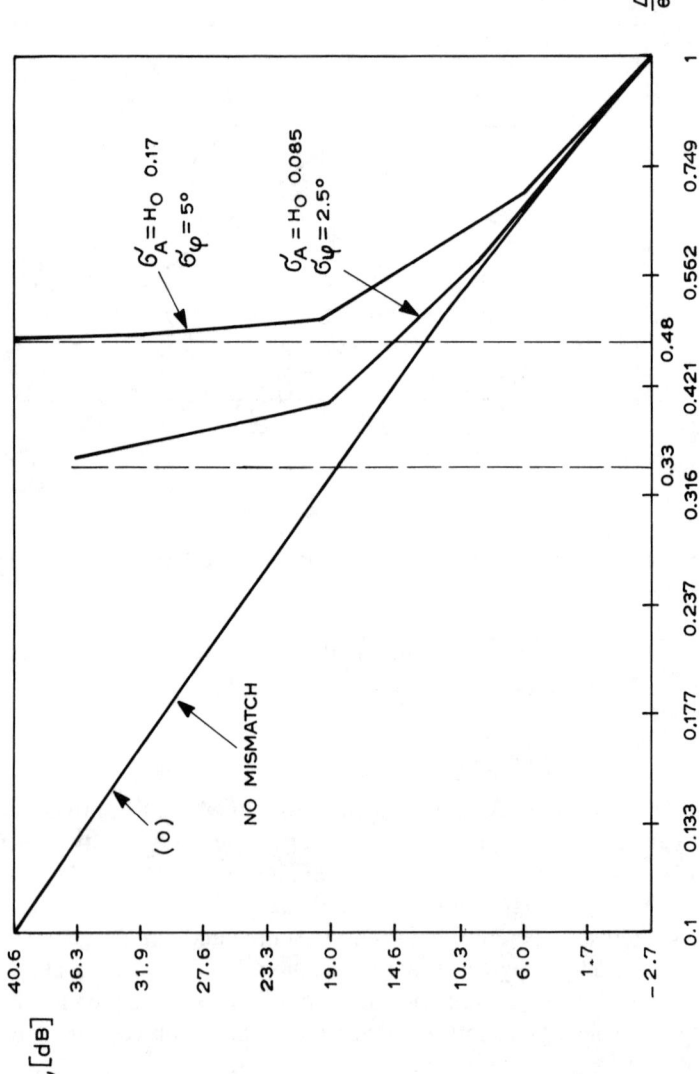

Figure 5.50 Angular resolution limit of the minimum variance (Capon) method; two equal-power nonmonochromatic ($B/f_0 = 0.05$) sources impinging on an array of $N = 10$ receiving elements; amplitude mismatching, σ_A/H_0, and phase mismatching, σ_ϕ, of the receiving channels as parameters of the curves.

XX^H is a projection operator onto the space spanned by the columns of X, called the *signal subspace*. $P^\perp = (I - XX^H)$ is a projection onto the *noise subspace* orthogonal to the signal subspace. In general, the DOA estimate is obtained by viewing the peaks of the following function of the scanning angle, θ:

$$P(\theta) = \frac{1}{S^H(\theta) P^\perp S(\theta)} \tag{5.86}$$

If the columns of X are the eigenvectors corresponding to the largest eigenvalues of M, the previous equation gives the mathematic expression of the famous MUSIC algorithm. Among others of the same class, we mention those due to Hung and Turner (1982; 1983), Kumaresan and Tufts (1983), Reddi (1979), ESPRIT (*estimation of signal parameters by rotationally invariance techniques*) (Paulray et al., 1986). See also (Brandwood, 1987; Yeh, 1986, 1987). In the following, only the MUSIC algorithm will be described in detail.

The *multiple signal classification* (MUSIC) method was proposed by Schmidt (1986) to provide asymptotically unbiased estimates of the following parameters: (1) number of sources, (2) DOAs, (3) source strengths, and (4) crosscorrelation among the sources. To derive the MUSIC algorithm, we need to remember the concepts of eigenvalue decomposition for the covariance matrix M of the received data by the array in principal eigenvalues and noise eigenvalues (refer to Section 5.3.2). Assume I monochromatic, uncorrelated sources impinging on the array (5.10). Under ideal conditions, i.e., if the covariance matrix, M, has been estimated by using an infinite number of data snapshots, all noise eigenvalues will be equal to σ^2 (refer to Section 5.3.2). Consequently, the matrix, M, can be defined as follows:

$$M = \sum_{i=1}^{q} (\lambda_i - \sigma^2) q_i q_i^H + \sigma^2 I \tag{5.87}$$

where q is the number of principal eigenvectors, ideally $q = I$. Comparing (5.87) with (5.10) for the snapshot data model, we derive the following equalities:

$$VPV^H = \sum_{i=1}^{I} |p_i|^2 v_i v_i^H = \sum_{i=1}^{q} (\lambda_i - \sigma^2) q_i q_i^H \tag{5.88}$$

According to this equation, the principal eigenvectors q_i ($i = 1, \ldots, q$) are linear combinations of the source directional vectors v_i ($i = 1, \ldots, I$) and *vice versa*. Equivalently, the vectors v_i and q_i span the same vector subspace (referred to as the *signal subspace*). Furthermore, as the noise eigenvectors are always orthogonal to the principal eigenvectors, the noise eigenvectors thus must occupy a subspace (the *noise subspace*) that is orthogonal to the source vector space.

Figure 5.51 shows the relationship among the source directions $\mathbf{v}_1$ and $\mathbf{v}_2$ (in the example, two sources are considered impinging on an array of three elements) and the two principal eigenvectors $\mathbf{q}_1$ and $\mathbf{q}_2$, and the noise eigenvector $\mathbf{q}_3$ (Huizing and DeValk, 1987). In Section 5.3.2, if the noise eigenvectors are used as antenna array element weights, they will produce pattern nulls at source DOAs because of their orthogonality to source eigenvectors (see Figure 5.17). With this background, the MUSIC algorithm may be derived in the two following ways: (1) substituting for $\mathbf{P}^\perp$ of (5.86) the expression:

$$\mathbf{P}^\perp = \sum_{i=q+1}^{N} \mathbf{q}_i \mathbf{q}_i^H \tag{5.89}$$

where $\mathbf{q}_i$ ($i = q + 1, \ldots, N$) are the noise eigenvectors (remember that $\sum_{i=1}^{N} \mathbf{q}_i \mathbf{q}_i^H = \mathbf{I}$); (2) by approximating $\mathbf{M}^{-1}$ in (5.80) for the minimum variance algorithm with

$$\mathbf{M}^{-1} \approx \sum_{i=q+1}^{N} \frac{1}{\lambda_i} \mathbf{q}_i \mathbf{q}_i^H \tag{5.90}$$

In conclusion, the expression of the MUSIC algorithm is

$$P_{\text{MUSIC}}(\theta) = \frac{1}{\mathbf{S}^T(\theta) \sum_{i=q+1}^{N} \mathbf{q}_i \mathbf{q}_i^H \mathbf{S}^*(\theta)} \tag{5.91}$$

where the weighting coefficient λ_i^{-1} has been omitted because it is approximately constant. Again, $P_{\text{MUSIC}}(\theta)$ is not a true psd, but only a very good indicator of the DOAs.

The performance as estimator of DOAs is far superior to the linear prediction and minimum variance methods. A relationship has been found between Capon and MUSIC methods (Nickel, 1988a). As has been demonstrated, MUSIC can be interpreted as a minimum variance method, but one which uses a covariance matrix that corresponds to an infinite SNR. This explains the improved resolution of MUSIC over the Capon method.

We should stress that MUSIC provides a pattern with peak values, which are not related to the power values of the sources impinging on the array. Once the DOA values have been estimated, we may calculate the strength of sources and their cross-correlation values as follows. By using a known noise eigenvalue, the noise-subtracted covariance matrix is

$$\mathbf{M} - \sigma^2 \mathbf{I} = \mathbf{VPV}^H \tag{5.92}$$

which is instrumental in calculating the matrix, $\mathbf{P}$:

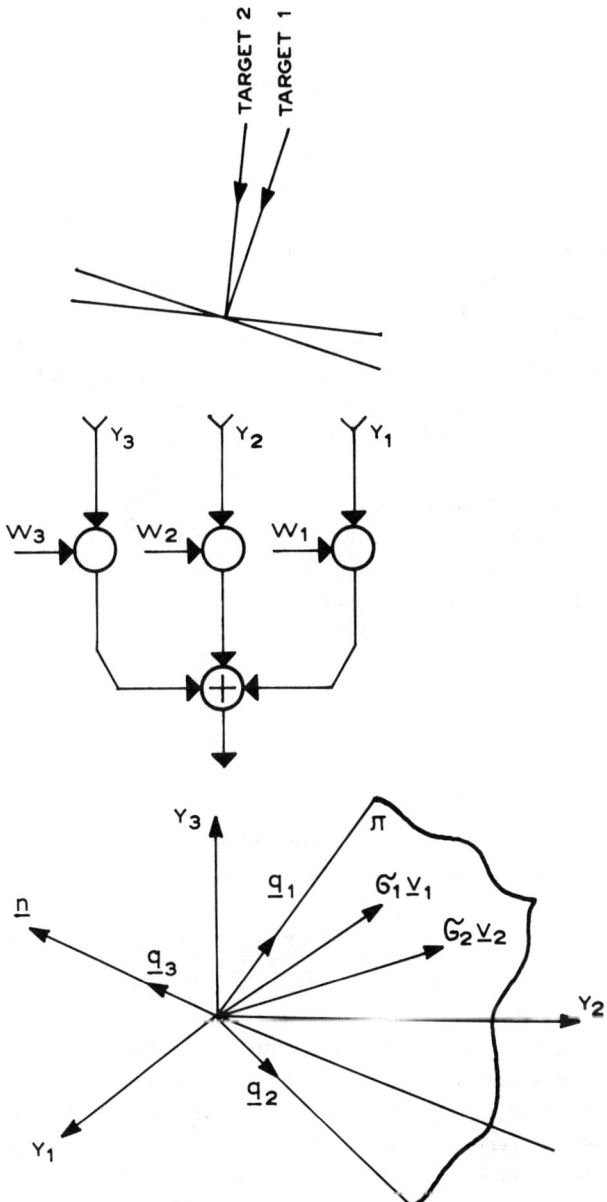

Figure 5.51 Relationships between source directions and eigenvector decomposition of the covariance matrix **M**; $N = 3$, $I = 2$ (**n** is normal to the plane π). (From Huizing and DeValk, 1987, p. 69).

$$\mathbf{P} = \mathbf{V}^{-1}(\mathbf{M} - \sigma^2\mathbf{I})\mathbf{V}^{-H} \tag{5.93}$$

providing us with the desired information.

When working with a finite number of data snapshots, the covariance matrix must be estimated as well as the corresponding eigenvalues and eigenvectors. The simple idea of separating the eigenvectors into signal and noise subspaces, based on examination of the ordered eigenvalues of the covariance matrix does not work well in practice, especially for short data records. Therefore, we need to resort to statistical criteria to determine the number of principal eigenvalues. This is a problem similar to the order selection with AR-based psd estimators. Consequently, we may use the AIC and *minimum description length* (MDL) criteria. See, for mathematical details, (Haykin, 1986, p. 376; 1989, pp. 335–336; Marple, 1987, p. 374). In addition, the noisy estimates of the principal eigenvectors and noise eigenvectors are only approximately orthogonal and only approximately the noise eigenvectors are orthogonal to the source directional vectors. Nevertheless, the MUSIC algorithm works well, as shown by the ensuing analysis.

A theoretical performance prediction of the MUSIC algorithm has been done in (Forrier *et al.*, 1988), where the expected value of the MUSIC psd has been calculated by using a second-order perturbation analysis. In more detail, the covariance matrix $\mathbf{M}$ is estimated by averaging L data snapshots, $\mathbf{Z}(n)$, $n = 1, 2, \ldots, L$, as follows[8]:

$$\hat{\mathbf{M}} = \frac{1}{L}\sum_{n=1}^{L} \mathbf{Z}(n)\mathbf{Z}^H(n) \tag{5.94}$$

and the aforementioned reference shows the average of $P_{\text{MUSIC}}(\theta)$, which is accurate to order $1/L^2$. The following result is found:

$$E\left\{\frac{1}{P_{\text{MUSIC},L}(\theta)}\right\} = \frac{1}{P_{\text{MUSIC},\infty}(\theta)} + \mathbf{S}^H(\theta)\left|\sum_{i=1}^{I}\frac{\lambda_i(N-I)}{L(\lambda_i - \sigma^2)^2}\left(1 + \frac{I}{L}\right)\mathbf{q}_i\mathbf{q}_i^H\right|\mathbf{S}(\theta) \tag{5.95}$$

where $P_{\text{MUSIC},L}(\theta)$ and $P_{\text{MUSIC},\infty}(\theta)$ are the expressions for the MUSIC algorithm, calculated by averaging L data snapshots and an infinite number of data snapshots (which corresponds to the ideal value of the MUSIC function), respectively. In addition λ_i and $\mathbf{q}_i$ are the ith eigenvalue and ith eigenvector of the ideal covariance matrix $\mathbf{M}$. Note that the effect of the limited number L of data snapshots introduces the principal eigenvectors in the correction term (the second on the right-hand side of (5.95)). The presence of the principal eigenvectors reduces the spiky nature of the ideal MUSIC pattern, which uses only the noise eigenvectors. In deriving (5.95), we assumed to

---

[8] The covariance matrix is defined as $\mathbf{M} = E\{\mathbf{ZZ}^H\}$.

know *a priori* the number of sources. This is a strong hypothesis; nevertheless, the analysis provides useful insight into the number of data snapshots needed to achieve a certain degree of angular resolution.

Figure 5.52 shows an example of angular pattern provided by the MUSIC algorithm. Two equal-power, uncorrelated sources ($JNR_1 = JNR_2 = 40$ dB) impinge on an array of $N = 10$ elements. Two curves are shown, one referring to the case of monochromatic sources (i.e., $B/f_0 = 0$), and the other concerning wideband sources with flat spectrum in the channel receiving bandwidth, $B = 0.05 f_0$. The two curves were derived by assuming *a priori* knowledge of the covariance matrix **M** and the number of sources.

Subsequently, the effect of the limited number of data snapshots to estimate **M** is considered by using (5.95). Specifically, Figure 5.53 shows a kind of angular resolution limit of the MUSIC algorithm for two equal-power, uncorrelated, monochromatic sources. The x-axis indicates the minimum angular separation of the two sources divided by the Rayleigh limit, θ_{Ray}. The y-axis gives the source-to-noise power ratio (JNR) necessary for the two peaks to be correctly positioned on the sources' DOAs (i.e., the power needed to have an unbiased estimate of DOA values). Three curves are reported for three different values of data snapshots L, namely; $L = 50, L = 100, L = 200$. Note the progressive increase of MUSIC's performance with the increase of L.

The next two figures illustrate the performance degradation of the MUSIC method against correlated sources and mismatching of receiving channels. Figure 5.54 shows the angular resolution limiting curves for two monochromatic but correlated sources. The parameter of the curves is the correlation coefficient, ρ_{12} (the phase of the correlation is taken as equal to zero). The mismatching limitation effect is analyzed in Figure 5.55. Again, we assume the mismatching model of Figure 5.26 and (5.37). The number of data snapshots is $L = 100$, while the parameters of the curves are the standard deviations σ_A and σ_ϕ of the mismatching errors between the receiving channels. The deleterious effects of source correlation and channel mismatching are obvious. Methods to compensate for these phenomena should be seriously considered (see, for instance, Subsection 4.3.2.4 for mismatching, and Section 5.7.6 for the correlated sources). Finally, we note that many theoretical analyses have been performed on the MUSIC algorithm. See, for example, Friedlander (1990), Li and Vaccaro (1990), and Stoica and Nehorai (1990).

5.7.5 Application of Singular Value Decomposition

Part of our previous analysis has assumed that the covariance matrix **M** is known *a priori*. One way to approach the practical problem of a finite-length data record was briefly indicated in Section 5.3.1, where the data were used to estimate the covariance matrix (5.20), which, in turn, can be used in lieu of the ideal covariance matrix

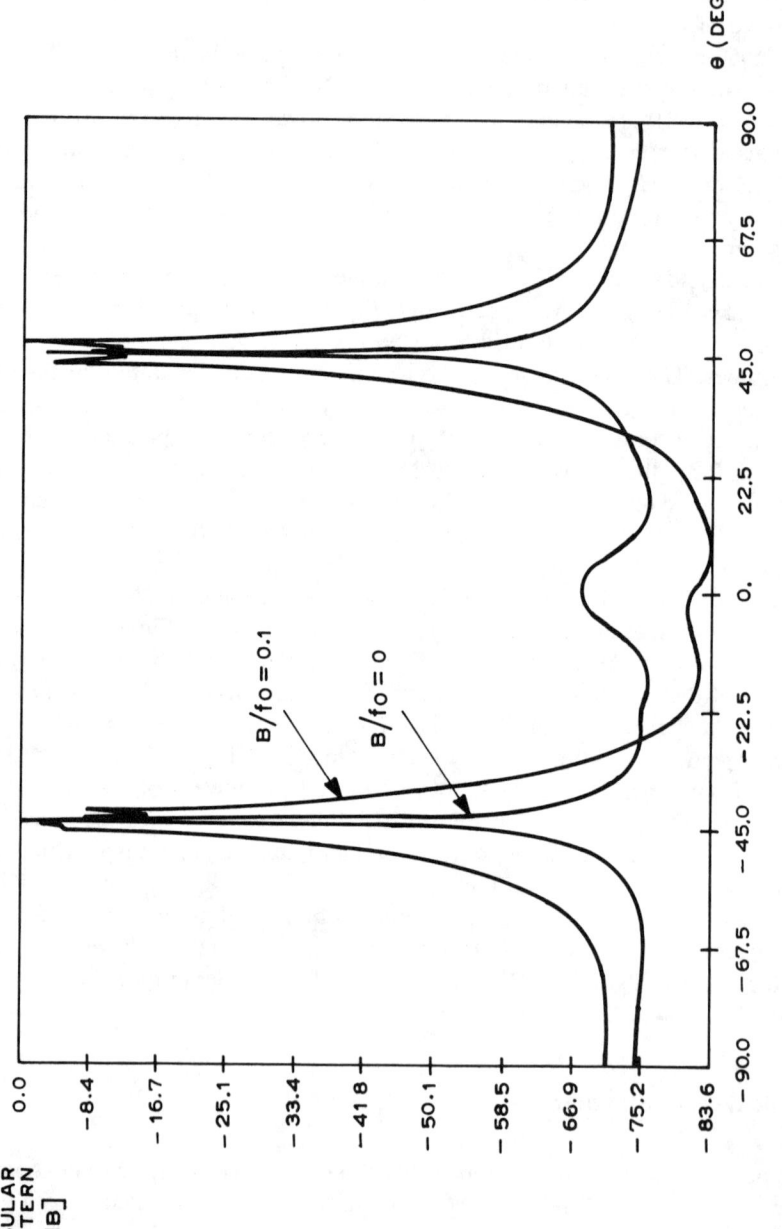

Figure 5.52 Angular pattern for the MUSIC algorithm; two uncorrelated equal-power sources (JNR$_1$ = JNR$_2$ = 40 dB) with DOAs $\theta_1 = -45°$ and $\theta_2 = +45°$, number of array elements, N = 10; B/f_0 as parameter.

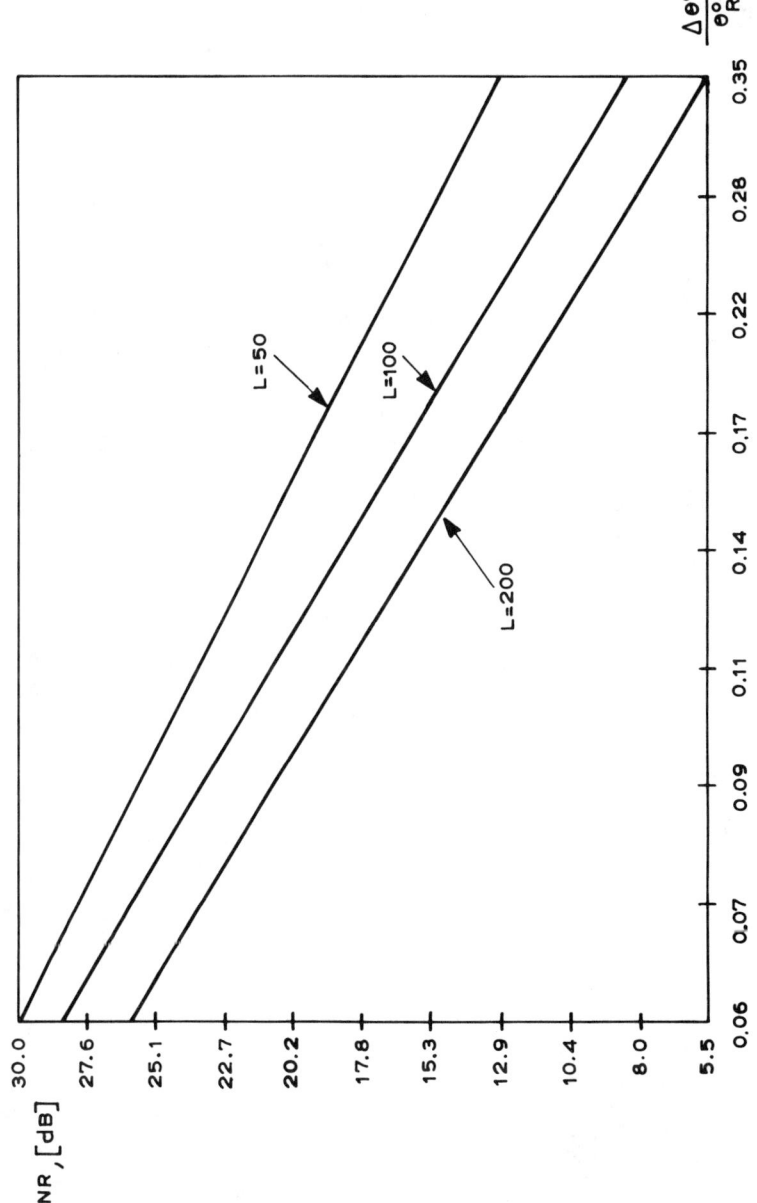

Figure 5.53 Angular resolution of MUSIC as a function of the number L of data snapshots; two equal-power uncorrelated monochromatic sources.

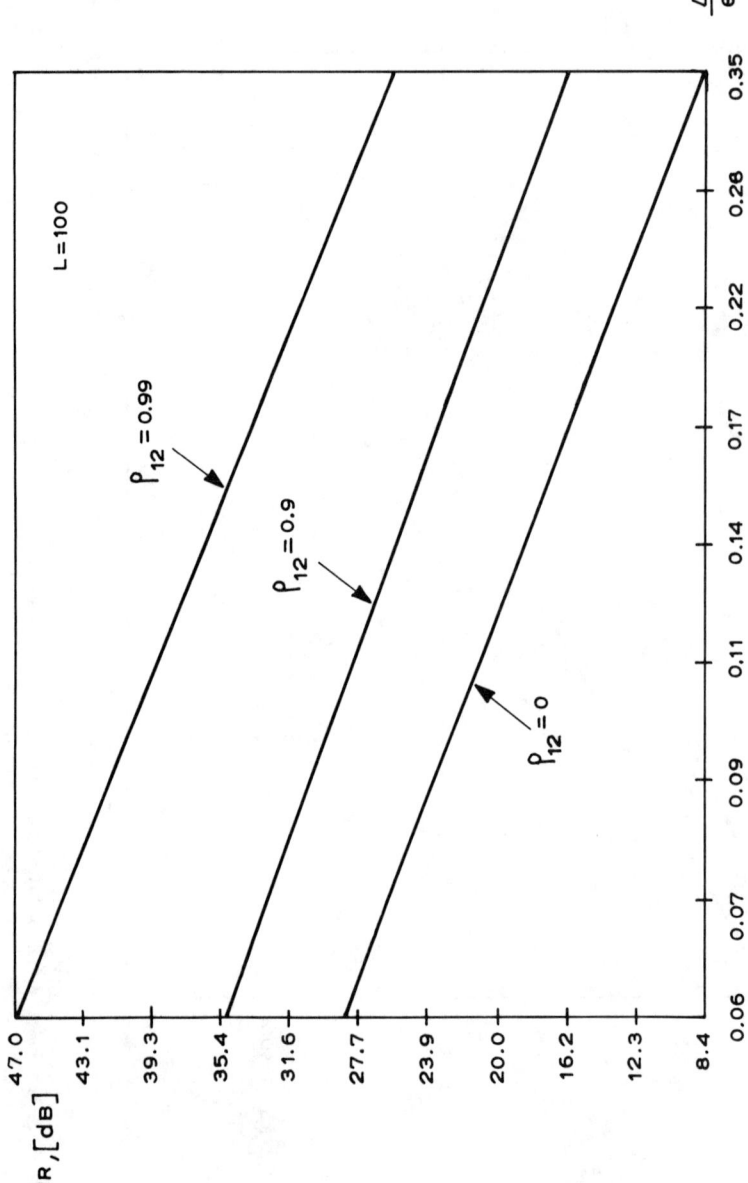

Figure 5.54 Angular resolution capability of MUSIC for two correlated (correlation coefficient, ρ_{12}) monochromatic equal-power sources; number of data snapshots, $L = 100$.

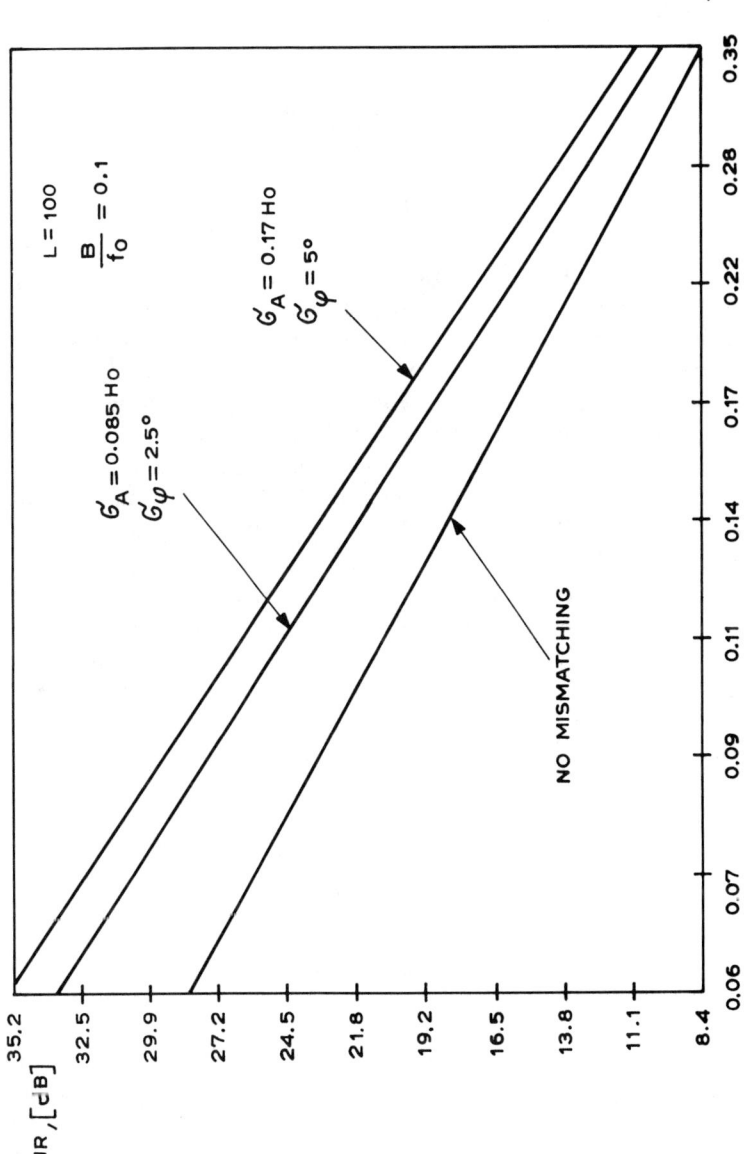

Figure 5.55 Angular resolution capability of MUSIC; two equal-power wideband ($B/f_0 = 0.1$) uncorrelated sources; effects of receiving channel mismatching, σ_A and σ_ϕ as parameters.

in the various algorithms already presented. An alternative, highly efficient method that works directly on the L data snapshots without calculating the covariance matrix is based on the *singular value decomposition* (SVD) of the rectangular matrix ($L \times N$) containing the data snapshots:

$$\mathbf{Z} = \begin{bmatrix} \mathbf{Z}^T(1) \\ \mathbf{Z}^T(2) \\ \vdots \\ \mathbf{Z}^T(L) \end{bmatrix} \qquad (5.96)$$

This method should be considered as a complement to the data-domain methods, discussed in Section 4.2.3. The calculation of the inverse of the estimated covariance matrix (a mathematical operation found in the angular superresolution algorithms as well as the jammer cancellation algorithms) by using the mathematical expression $(\mathbf{Z}^H \mathbf{Z})^{-1}$ is not recommended because this calculation has potential numerical instabilities, as it introduces more sensitivity to perturbations than is already inherent in the problem. Before showing how to use SVD in the angular resolution algorithm with a finite data set, let us recall the theorem of SVD. See, for details, (Marple, 1987, pp. 76–78; Haykin, 1989, Chapter 12), the cited references, and the additional references: (Biglieri and Yao, 1989; Klema and Laub, 1980).

The SVD theorem can be seen as a generalization to a rectangular matrix of the well known spectral decomposition of a squared definite-positive Hermitian matrix (Section 5.3.2). Consider an arbitrary ($L \times N$) complex-valued matrix $\mathbf{Z}$ of rank k. The singular value decomposition theorem states that there exist positive-real numbers, $\sigma_1 \geq \sigma_2 \geq \ldots \geq \sigma_k > 0$ (the so-called singular values of $\mathbf{Z}$), an $L \times L$ unitary matrix, $\mathbf{U} = [\mathbf{u}_1 \ldots \mathbf{u}_L]$, and an $N \times N$ unitary matrix, $\mathbf{F} = [\mathbf{f}_1 \ldots \mathbf{f}_N]$ such that the matrix $\mathbf{Z}$ can be expressed as

$$\mathbf{Z} = \mathbf{U}\mathbf{\Sigma}\mathbf{F}^H = \sum_{i=1}^{k} \sigma_i \mathbf{u}_i \mathbf{f}_i^H \qquad (5.97)$$

where the $L \times N$ matrix $\mathbf{\Sigma}$ has the structure:

$$\mathbf{\Sigma} = \begin{vmatrix} \mathbf{D} & \mathbf{0} \\ \mathbf{0} & \mathbf{0} \end{vmatrix} \qquad (5.98)$$

and $\mathbf{D} = \text{diag}(\sigma_1, \ldots, \sigma_k)$ is a $k \times k$ diagonal matrix. A consequence of the unitary nature of the matrices $\mathbf{U}$ and $\mathbf{F}$, i.e., $\mathbf{U}^H \mathbf{U} = \mathbf{I}$ and $\mathbf{F}^H \mathbf{F} = \mathbf{I}$, is that the two following equations hold:

$$\mathbf{Z}^H\mathbf{Z}\mathbf{f}_i = \sigma_i^2 \mathbf{f}_i$$
$$\mathbf{Z}\mathbf{Z}^H\mathbf{u}_i = \sigma_i^2 \mathbf{u}_i, \quad 1 \leq i \leq k \qquad (5.99)$$

The matrix products $\mathbf{Z}^H\mathbf{Z}$ and $\mathbf{Z}\mathbf{Z}^H$ are Hermitian matrices of size $N \times N$ and $L \times L$, respectively. The columns of $\mathbf{U}$ are therefore the orthonormal eigenvectors of $\mathbf{Z}\mathbf{Z}^H$, and the columns of $\mathbf{F}$ are the orthonormal eigenvectors of $\mathbf{Z}^H\mathbf{Z}$. Both share the same eigenvalues σ_i^2 for $1 \leq i \leq k$. Thus, the singular values are simply the positive square roots of the nonzero eigenvalues of $\mathbf{Z}^H\mathbf{Z}$ and $\mathbf{Z}\mathbf{Z}^H$. At this point, it is relevant to note that, instead of calculating the eigenvalues of $\mathbf{Z}^H\mathbf{Z}$ (the covariance matrix of our data matrix), in view of an improved numerical stability and implementation of the algorithm on systolic array computers, a more convenient approach is to calculate the singular values of $\mathbf{Z}$ by using *ad hoc* numerical algorithms, such as the Jacobi, Householder, Givens rotations, Gram-Schmidt, and QR decomposition (Golub and Van Loan, 1983). These numerical algorithms and their mapping onto parallel computers are at the forefront of research, and practical applications will be considered in Section 5.8. Proper numerical algorithms are able to diagonalize the data matrix $\mathbf{Z}$, i.e., to calculate the diagonal matrix $\mathbf{D}$ formed of the singular values and the so-called left singular vectors of $\mathbf{Z}$, columns of $\mathbf{U}$, and right singular vectors of $\mathbf{Z}$, columns of $\mathbf{F}$.

Let us now discuss how to use the SVD theorem in the MUSIC algorithm (Haykin, 1989, pp. 330–345). From the singular values σ_i, $1 \leq i \leq k$, of the data matrix $\mathbf{Z}$, we calculate the eigenvalues of the estimated covariance matrix $\mathbf{Z}^H\mathbf{Z}$ by squaring them and dividing by the number of data snapshots. The eigenvectors of the estimated covariance matrix are coincident with the right singular vectors $\mathbf{f}_i$, $1 \leq i \leq k$, of the data matrix $\mathbf{Z}$. Assuming that the number I of sources is known *a priori*, we partition the space spanned by the N columns of the $L \times N$ data matrix $\mathbf{Z}$ into two subspaces:

1. The signal subspace, which is spanned by the right singular vector, $\mathbf{f}_1, \mathbf{f}_2, \ldots$, $\mathbf{f}_I$, associated with the largest singular values of $\mathbf{Z}$;
2. The noise subspace, which is spanned by the right singular vectors, $\mathbf{f}_{I+1}, \ldots$, $\mathbf{f}_N$, associated with the smallest singular values of $\mathbf{Z}$. The MUSIC algorithm, based on the SVD of the data matrix $\mathbf{Z}$, now has the following expression:

$$P_{\text{MUSIC,SVD}}(\theta) = \frac{1}{\mathbf{S}^T(\theta) \sum_{i=I+1}^{N} \mathbf{f}_i \mathbf{f}_i^H \mathbf{S}^*(\theta)} \qquad (5.100)$$

When the number of sources are not known *a priori*, the usual AIC and MDL methods can be applied by using the eigenvalues of $\mathbf{Z}^H\mathbf{Z}$, calculated by means of the singular values of $\mathbf{Z}$. See, for mathematical details, (Haykin, 1989, pp. 335–336).

In the following, a computer exercise is presented, dealing with the application of SVD and MUSIC algorithms to an adaptive array system for (DiBartolo, 1988):
1. Estimating the jammer DOAs;
2. Jammer suppression;
3. Detection of a useful target;
4. Estimation of target DOA.

The system considered (see Figure 5.56) is a linear array with $N = 8$ isotropic antennas spaced 0.5 λ_0. Each antenna is equipped with a receiving channel (RCVR), having a fractional bandwidth B/f_0, and an analog-to-digital converter for sampling the incoming waveforms, and thus providing the data snapshots. A digital memory collects L data snapshots; only one of the snapshots may contain a useful target, while the whole set of L snapshots is corrupted by jamming signals. The target is assumed to be received at the array boresight. We consider the interesting case of main-beam jammer in addition to the more conventional problem of a sidelobe jammer. The data matrix **Z** of dimension $L \times N$ is first processed for estimation of the number and DOAs of the jammers. This task, accomplished by the block "SVD analysis," results in the computation of the ranked singular values (matrix **D**) and the right singular vectors (matrix **F**) of the data matrix **Z**. The eigenvalues allow us to estimate the number, I, of jammers and to separate the eigenvectors into the noise subspace and jammer subspace. From the noise-subspace eigenvectors, the jammers' DOAs are evaluated by the MUSIC algorithm. The same noise eigenvectors, their number is $(N - I)$, are used to form $(N - I)$ nondirective beams having nulls along the jammers' DOAs. The generation of one beam of this type is accomplished by considering the elements of the generic noise eigenvector to be the weights of the adaptive array and making a linear combination of the weights with the data **Z**. The complementary set of I jammer eigenvectors, when applied to the data, provide I beams in the direction of the jammers. The next step is the detection and angle estimation of the useful target. This can be done by proper combination of the jammer-free nondirective beams to form the directive sum (Σ) and difference (Δ) beams. Target detection is done by thresholding the sum beam, while the angle estimation is done using the monopulse technique.

The performance of the scheme described so far has been evaluated by Monte Carlo simulation in terms of jammer strobe estimation, target detection capability, and accuracy of target angular location. Figure 5.57 shows the patterns of the adapted array when the weights used for linearly combining the data snapshots are respectively two jammer eigenvectors and a noise eigenvector. The jammer eigenvectors provide two patterns aligned with the jammer DOAs, whereas the noise eigenvector gives a uniform pattern with nulls along the jammer DOAs. Remember that noise eigenvectors and jammer eigenvectors are orthogonal. Subsequently, we consider the case of one jammer against a target to be detected. Figure 5.58 shows the detection probability P_D, in percent, *versus* the jammer DOA, the target being at boresight.

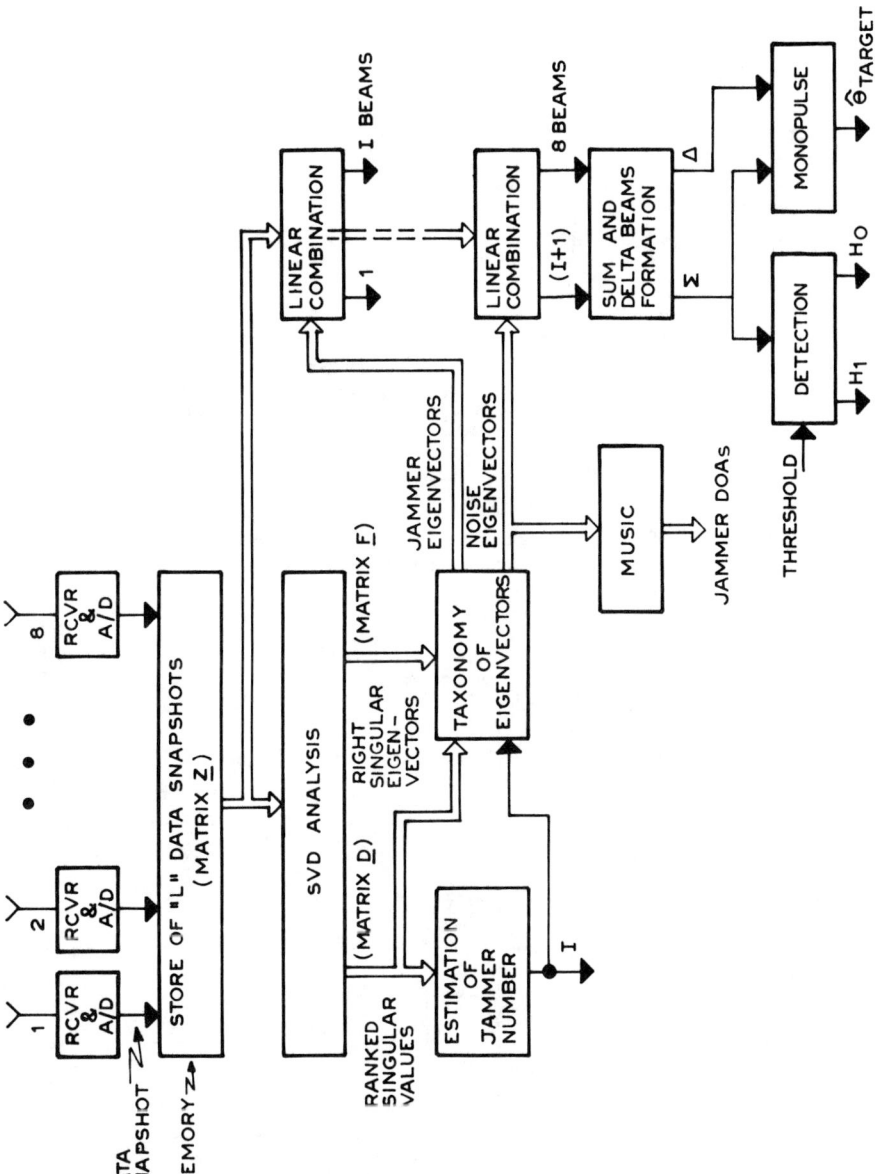

Figure 5.56 An example of processing scheme for the application of SVD and MUSIC algorithms.

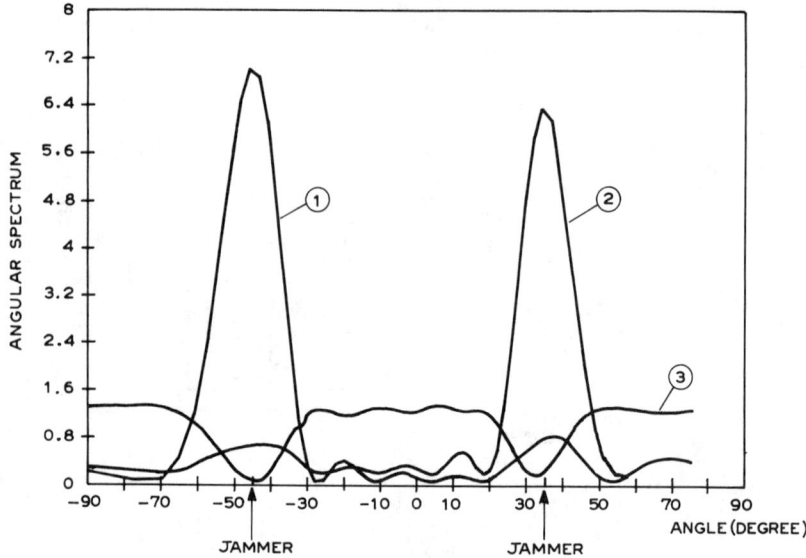

Figure 5.57 Three patterns of an adaptive array which uses as array weights the components of two jammer eigenvectors (patterns 1 and 2) and one noise eigenvector (pattern 3).

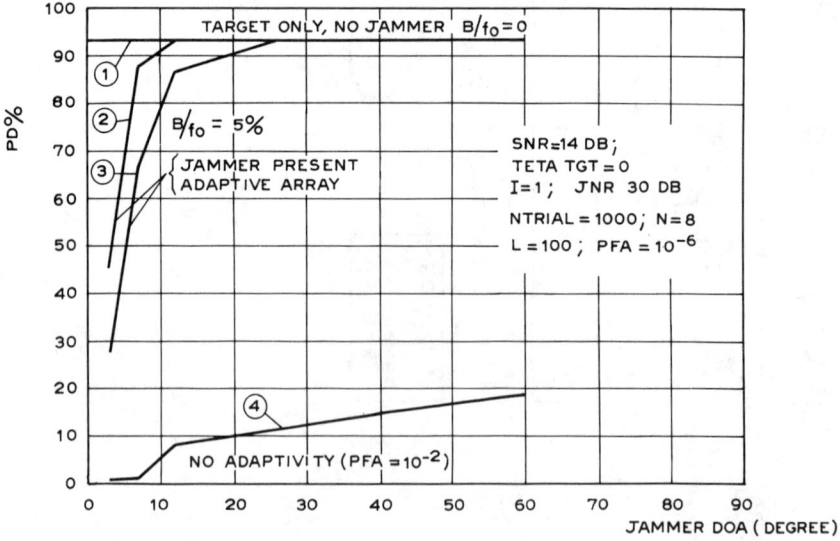

Figure 5.58 Detection probability *versus* jammer DOA.

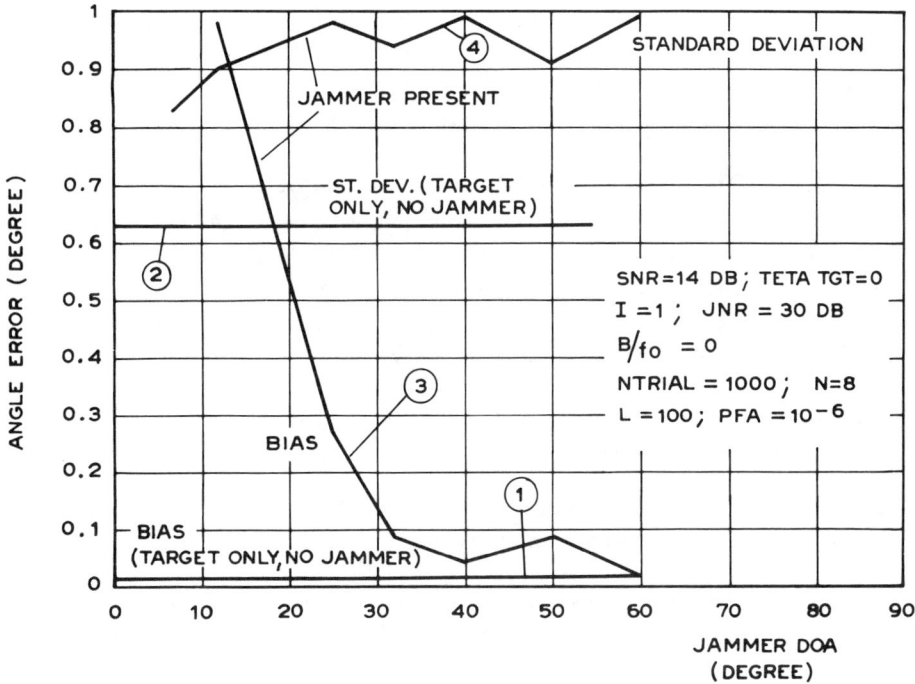

Figure 5.59 Bias and standard deviation of the estimation of target DOA *versus* the jammer DOA; $B/f_0 = 0$.

Relevant parameters of the experiment are the following. The target SNR is 14 dB, while the JNR is 30 dB. The number of snapshots is $L = 100$, and the number of Monte Carlo trials to estimate P_D is NTRIAL = 1000. Because the target model is assumed to be Swerling 1, and the jammer and thermal-noise pdf values are Gaussian, the detection threshold on the sum channel can be evaluated analytically to maintain a value of $P_{FA} = 10^{-6}$. Four curves are shown in the figure:

1. The case in which only the target is present, while the jammer is absent;
2. The case of target and jammer simultaneously present with $B/f_0 = 0$;
3. The same as curve (2), except for the receiver channel bandwidth $B/f_0 = 5\%$;
4. The case of no adaptivity in the array.

Note that in curve (4) the P_{FA} has been increased to 10^{-2} to obtain modest P_D values. Because the jammer DOA varies in the [3°, 60°] range, and the half-power main-beamwidth of the array is 12.75°, the important case of main-beam jamming is also considered in addition to the case of sidelobe jamming.

Figure 5.59 complements the previous figure because it shows the accuracy of the monopulse technique in locating the useful target when a jammer is present. The parameters of the problem are the same as in Figure 5.58. Four curves are shown:
- 1 and 2—The bias and standard deviation errors, respectively, when only the target is present, while the jammer is absent;
- 3 and 4—The bias and standard deviation errors, respectively, when the jammer is present in addition to the useful target.

Figure 5.60 differs from the previous figure for the B/f_0 value: now 5%, before 0%. The detection probability *versus* the JNR value, when the jammer is constantly in the array main beam, is shown in Figure 5.61. Figure 5.62 differs in SNR value from the previous one: now we have SNR = 8 dB; before SNR was 14 dB.

5.7.6 Superresolution of Coherent Sources by Adaptive Array Techniques: Problems and Remedies

The resolution of coherent sources is a problem for the previously described adaptive array techniques. Two signals are said to be coherent if one is a scaled and delayed

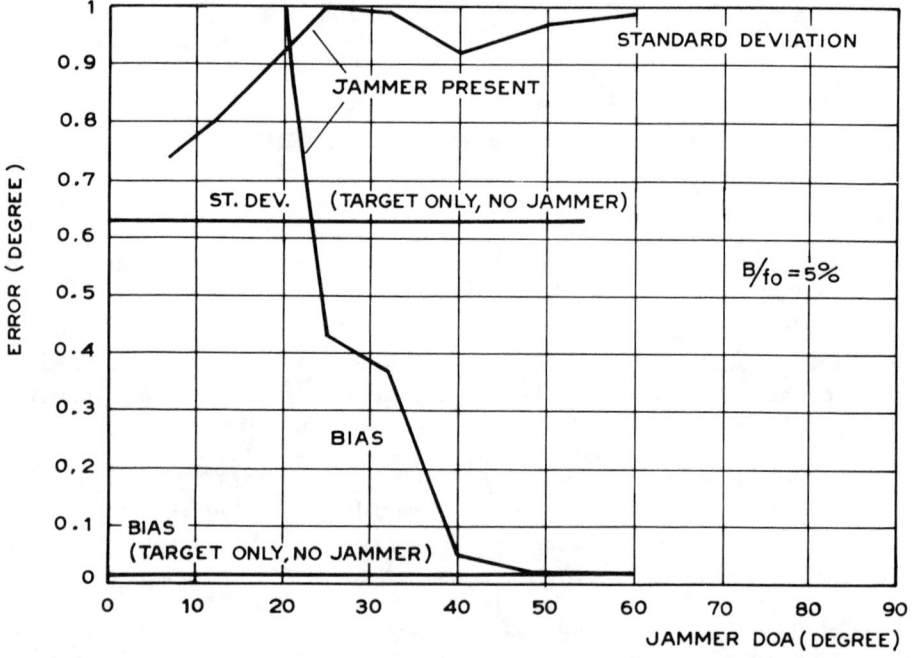

Figure 5.60 Bias and standard deviation of the estimation of target DOA *versus* the jammer DOA; $B/f_0 = 5\%$.

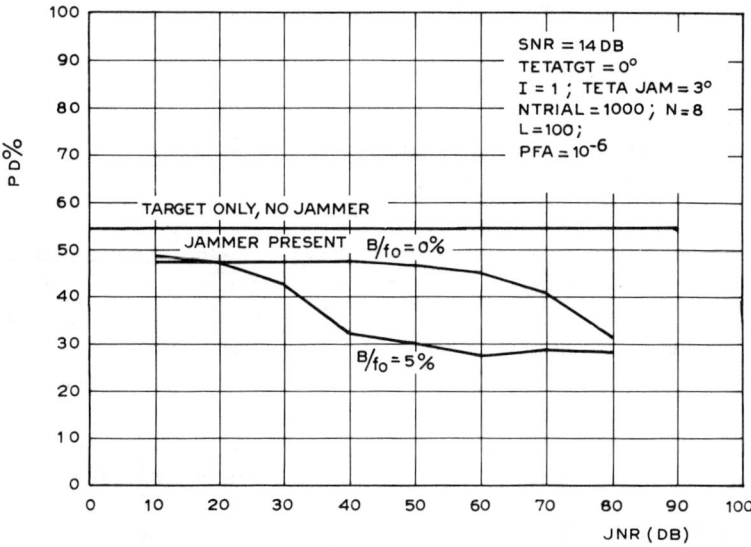

Figure 5.61 Detection probability *versus* the JNR value; the jammer is in the main beam; SNR = 14 dB.

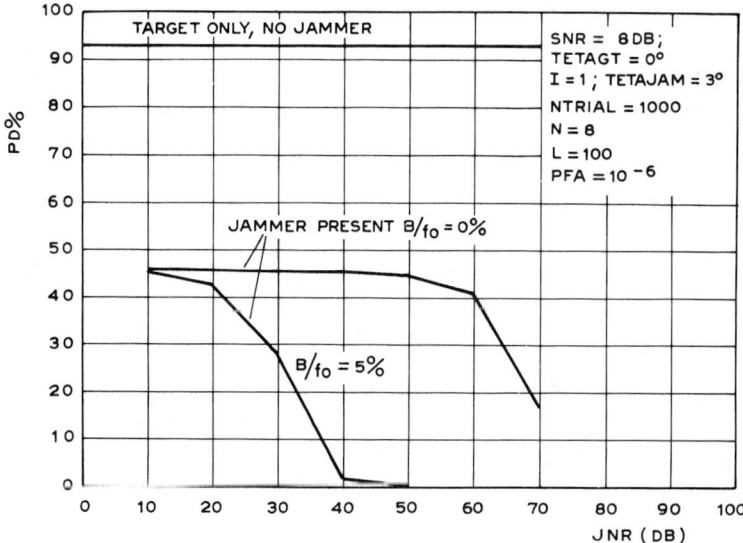

Figure 5.62 Detection probability *versus* the JNR value; the jammer is in the main beam; SNR = 8 dB.

replica of the other. In particular, the phase difference between the two coherent signals is fixed. Noncoherent signals do not have fixed phase differences. This problem, initially pointed out by Gabriel (1980a), primarily occurs in the following situations:

1. Multipath, where the reflected signal is correlated with the direct path signal;
2. In the presence of smart jammers;
3. When superresolution is attempted with only one data snapshot to process.

In the third case, although the sources are noncoherent, their relative phase is fixed, i.e., they are coherent simply because only one snapshot is available.

The loss of angular resolution due to correlation between the sources is reported in many references. See, for instance, (Gabriel, 1980b). Also remember the computer exercises in Sections 5.7.2–5.7.4. The mathematical motivations of this resolution degradation are related to the non-Toeplitz nature of the array covariance matrix **M** (refer to Example 3 of Section 5.2), and the fact that two correlated sources blend into a single eigenvalue instead of two, as is demonstrated by the examples in Figures 5.13 and 5.28. The first shows the collapsing of the two eigenvalues into one eigenvalue as the correlation coefficient moves from zero to unity. The second shows instead how the nulls of the adaptive array along the jammer DOAs are progressively reduced in depth as the correlation coefficient goes to unity. Therefore, the accuracy in locating the jammer DOAs also reduces as the correlation between the sources increases.

To alleviate the problem, several spatial averaging techniques have been proposed to reduce the modulation phase in the covariance matrix (see Example 3 of Section 5.2) to restore the Toeplitz nature of **M**. Spatial averaging refers to a class of preprocessing schemes for modifying the estimated covariance matrix of the array prior to the use of superresolution algorithms. Spatial averaging techniques reduce the correlation between the sources. The greater is the degree of this reduction, the greater will be the restoration of the resolution to levels achieved when no correlation is present. Spatial averaging can be implemented by using the synthetic motion of a smaller subaperture along the single-snapshot data sample. As noted by Gabriel (1980a, p. 664), such synthetic motion of a subaperture is very similar to that which occurs in the MEM of Burg, where a k-point linear prediction filter is moved across a larger data sequence of N samples with $k < N$. Recommended references on spatial averaging techniques are Gabriel, 1980b; Linebarger and Johnson, 1990; Reddy *et al.*, 1987; Evans, 1979a, 1979b; Pei *et al.*, 1988; White 1979a, 1979b; Zoltowski, 1988; Shan *et al.*, 1985; Rao and Hari, 1990; Williams *et al.*, 1988; Haber and Jaggard, 1989. The ensuing material follows the theory developed by Haber and Jaggard.

There are four spatial averaging decorrelation techniques:

1. Forward-backward averaging (FBAV);
2. Spatial smoothing;

3. Doppler cycle averaging (DCA);
4. Element pattern diversity.

We briefly describe the rationale for each method and the expected performance improvement.

The FBAV requires symmetry in the deployment of the array elements. The method involves averaging two array output covariance matrices, both obtained by the same array and input, but with reverse indexing of the array elements. The modified covariance matrix is

$$\mathbf{M} = \mathbf{V}\text{Re}(\mathbf{P})\mathbf{V}^H + \sigma^2\mathbf{I} \qquad (5.101)$$

where the assumed model of the original covariance matrix is given by (5.7). Because the $(i\text{-}j)$th element of $\text{Re}(\mathbf{P})$ is $(\mathbf{P})_{ij} \cos\phi_{ij}$, this off-diagonal term may drop significantly, or not, depending on the phase ϕ_{ij} of the correlation coefficient between the two sources i and j.

The spatial smoothing consists of averaging a number of array covariance matrices, computed from overlapping subarrays to form a smoothed (hence, better) covariance matrix. The smoothed matrix will be of full rank, I, if the number of subarrays is $L \geq I$, where I is the number of sources received by the array. Limitations of this method are related to the need for operating with uniformly spaced arrays and reduction of the effective aperture size of the array to that of the subarray. The latter limitation is related to the always existing trade-off between resolution and accuracy of the angular estimate (see Section 5.7.1). The spatially smoothed covariance matrix is as follows:

$$\mathbf{R}_m = \frac{1}{L}\sum_{l=1}^{L} \mathbf{R}_l = \mathbf{V}_m \mathbf{P}_m \mathbf{V}_m^H + \sigma^2\mathbf{I} \qquad (5.102)$$

where $\mathbf{R}_l$ is the covariance matrix of the lth subarray. The magnitude of the off-diagonal element (ij) of $\mathbf{P}_m$ is

$$|(\mathbf{P}_m)_{ij}| = |(\mathbf{P})_{ij}|\left|\frac{\sin(2\pi L f_{ij})}{L \sin(2\pi f_{ij})}\right| \qquad (5.103)$$

with $f_{ij} = d/\lambda \, (\cos\theta_i - \cos\theta_j)$, where θ_i and θ_j are the DOAs of the two generic sources i and j $(i, j = 1, 2, \ldots, I)$ impinging on the array, and f_{ij} is a kind of spatial frequency modulation introduced by the correlation between the sources (see also (5.14)). The equation translates the above-mentioned trade-off into mathematical terms. In fact, a relevant reduction of the correlation magnitude implies a large value of L, which, in turn, corresponds to a reduction of the effective aperture of the array. Also note that the spatial averaging technique is not very effective for small separation of DOAs (i.e., f_{ij} small).

The DCA technique applies to situations where the array elements are mounted on a moving platform, such as an aircraft. DCA involves averaging the doppler cycle frequencies incurred during the motion of the array. If the integration time is large, a considerable decorrelation is achieved. In contrast to the previous technique, DCA is not restricted to some special array deployment. Examine how much integration time, T, is needed for adequately decorrelating the coherent signals. To this end, the array is assumed to move in the direction of its length with constant speed, v. In addition, we assume that the DOAs are invariant during the integration time. The magnitude of the crosscorrelation element ij in the matrix $\mathbf{P}$ is found to be

$$|\rho_{ij}| = \left| \frac{\sin\left[\pi(\cos\theta_i - \cos\theta_j)\dfrac{vT}{\lambda}\right]}{\pi(\cos\theta_i - \cos\theta_j)\dfrac{vT}{\lambda}} \right| \qquad (5.104)$$

which reflects the decorrelation effect of the motion. To have $|\rho_{ij}| = 0$, we need

$$\frac{vT}{\lambda} = \frac{n}{|\cos\theta_i - \cos\theta_j|}, \quad n = 1, 2, \ldots \qquad (5.105)$$

or $vT/\lambda \to \infty$. Figure 5.63 shows the quantity $|\rho_{ij}|$ for two generic equal-power sources i, j ($i, j = 1, 2, \ldots, I$) impinging on an array of $N = 10$ elements spaced 0.5λ; $|\rho_{ij}|$ is depicted as a function of vT/λ and having the difference $\Delta\theta$ between the two DOAs as a parameter. By using this figure in conjunction with Figures 5.45, 5.49, and 5.54 giving the angular resolution limits for correlated sources, we may derive the amount of integration time needed to achieve the required angular resolution.

The element pattern diversity involves averaging the covariance matrices of the outputs of several, independent, array element patterns. This technique does not require any constraint on the deployment of array elements; neither does it need an array in motion. The only requirement is the existence of different choices of element patterns during the acquisition of the data snapshots. In practice, this is obtained by grouping the array elements in subarrays and using relative phase shifts to generate different element patterns.

5.8 EXPERIMENTAL SYSTEMS AND REALIZATIONS WITH ADAPTIVE ARRAYS

This section briefly describes several experimental systems with adaptive arrays, as they appear in the open literature.

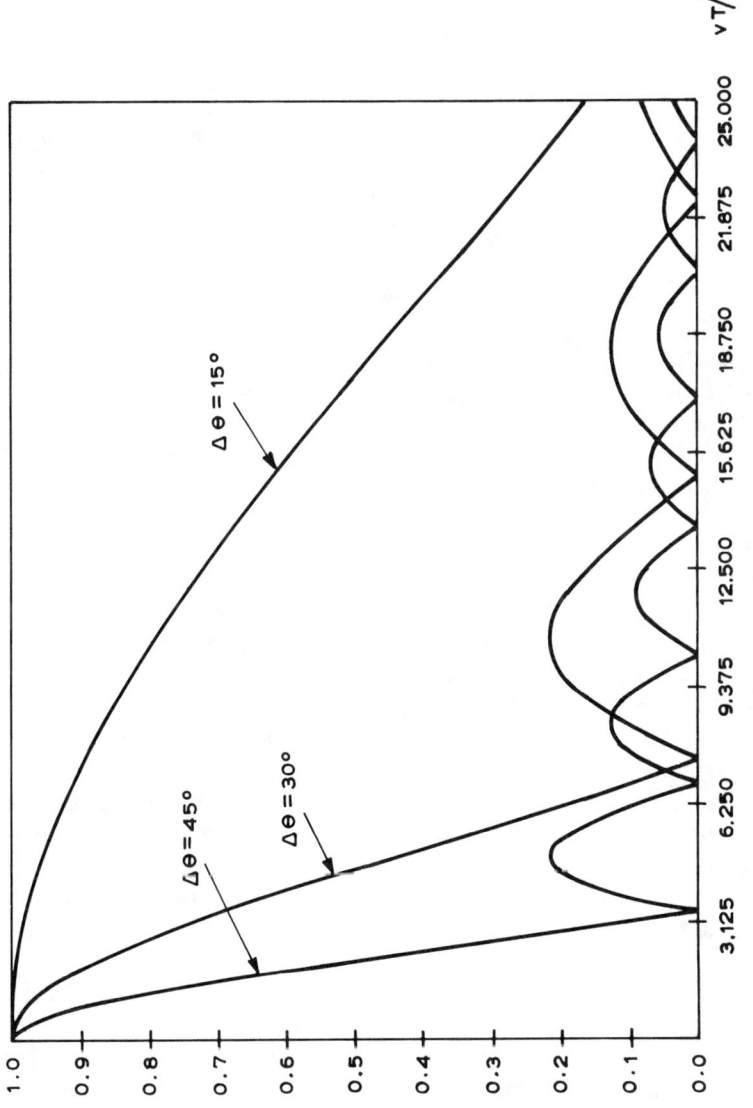

Figure 5.63 The reduction of correlation between sources for the DCA technique.

The first experimental system dates back to the late 1970s. Baldwin *et al.* (1979) describes a linear arrangement of eight dipoles operating at X-band. The signal processor is an analog implementation of a simple gradient descent algorithm. Experimental results show an SNR gain on the order of 35 dB against one CW interference source.

A four-element array, where the adaptive weights are calculated by the SMI algorithm, is described by Horowitz *et al.* in (1979a,b). SNR improvements of 19 dB are reported against one and two jammers.

One of the earliest adaptive arrays for superresolution is described by Evans in (1979a). The system is a vertical linear array with 11 elements operating at L-band. The scope of the experiment was to collect data on the angular distribution of the scattered power in a low-angle target tracking situation. To this purpose, the minimum variance (Capon method) and the MEM were used. An interesting technical discussion appeared in the *IEEE Transactions on AES* (Evans, 1979a,b; White, 1979a,b), related to the correlated nature of the direct and reflected echoes in the multipath and on the ability of superresolution to operate in this situation.

An experimental adaptive array for superresolution has been developed at the FGAN-FFM (Germany) to test an algorithm called *parametric target model fitting* (PTMF) (Nickel, 1988b, 1989a, 1989c). The method hypothesizes an M-point target model for the radar echoes. The amplitude and DOA values of the M-point model are efficiently estimated by resorting to stochastic approximation methods. The PTMF is able to resolve even completely correlated targets, and thus is of interest to counter multipath. Detection of the number of sources is automatic. The experimental system is an eight-element planar array at S-band. The processing of the digitized snapshots is by a microprocessor. The targets are simulated by three antennas in front of the array. The achieved degree of superresolution is better than 0.5 beamwidth for two equal-power sources. Targets with unequal power values require a longer convergence time for the estimation of the direction of the weak target; more than two targets also require longer convergence time. The limiting factor for superresolution is mainly due to array inaccuracy factors, while inadequate source power-to-noise ratio may simply produce slower convergence. Among the array inaccuracies, we mention the channel offset, amplification and orthogonality errors in the I and Q channels, interchannel amplitude and phase errors. These errors can be properly compensated by software. To this end, different points should be available in the receiving chains of each channel for inserting a test signal. An additional hardware requirement is represented by the number of bits to have a sufficient dynamic range for each channel.

The paper by Haykin *et al.* (1985) describes the application of a linear prediction method (specifically, the modified forward-backward linear prediction) for angle-of-arrival estimation in a low-angle tracking radar environment. The method is applied to real data pertaining to an angular separation of less than one standard beamwidth between the direct and image components impinging on the radar antenna. The receiving antenna is formed by eight elements operating at 9.6 GHz (the system is from STL Ltd. in the United Kingdom). The algorithm was tested on real flying aircraft approaching the radar at an altitude of 45 m, yielding elevation angles of less than 0.5 beamwidths under conditions where specular multipath predominated. The results of the analysis indicated that the linear prediction algorithm was capable of resolving target and image signals separated by a fraction of a beamwidth at SNR values typical of the low-angle tracking environment.

Another experimental adaptive array for solving multipath problems is that described by Theil (1989). The system is a vertical array of eight elements operating at 10.5 GHz. The eight outputs are linked to a common receiver channel through multiplexing. The snapshots are digitized with 12 bit words. Several modern super-resolution algorithms have processed the recorded live data. The multipath experiments were performed at an inland lake and at a coastal site. Noticeable angular resolution performances have been reported.

A further experimental system for high-resolution direction-finding (DF) is described by Starkey (1986). It is a 24-element, S-band, phased-array receiver linked to a digital beamformer. The in-phase and quadrature components of each received signals are coded with 12 bits. One hundred data snapshots are processed to estimate the covariance matrix. The algorithms used for DOA estimates are the Capon, autoregressive, and MUSIC methods. One major result of the analysis is the comparison of simulated with experimental data relevant to similar operational conditions. Also shown is that sources can be distinguished at angular separations of less than 0.5 beamwidth.

An interesting practical application of the MUSIC algorithm is due to Schmidt (the inventor of MUSIC) and Franks (1986). The experimental system is made of eight circularly polarized antennas, located in a circle as a sparse array. The system operates at 1.8 GHz; the received signals are sampled at 16 kHz with 8 bits. The computations are carried out by a minicomputer with direct memory access. The system is able to calculate, using the MUSIC algorithms, the DOAs, strength, polarization, and correlation of the sources. In addition, the system is able to isolate one signal among others and reconstruct its waveform. Experiments have been conducted with two, three, and four signals incident on the array. The experimental results reported that the DOA and all parameters of a signal source can be accurately measured, and the signal waveform can be recovered in the presence of other sources within a beamwidth. In particular, two equally polarized sources can be resolved to one-third beamwidth. The resolution is further improved to one-sixth beamwidth if the two sources have orthogonal polarization.

An advanced adaptive array system has been implemented in the United Kingdom by BNR Europe Ltd. and RSRE Electronics Division of Defense Research Agency (Ward et al., 1986; Davie et al., 1986; Hargrave et al., 1988). This is a test bed to demonstrate the capability of advanced processing algorithms mapped onto systolic array computers to solve the adaptive array functions of jammer nulling and estimating source DOAs. The implemented algorithms pertain to the realm of data-domain techniques, while the parallel processor is a triangular systolic array (see Section 4.2.3). The STL-RSRE test bed is formed of a six-element L-band antenna from which digital samples are fed to a wavefront array processor. The 33 processing nodes are mounted on 17 double Eurocards, housed in a 4-foot rack. The data rate of the processor is 10^4 samples per second. The processing cell is programmable, giving a high flexibility to the whole array processor. The experimental tests show the capability of the system to cancel up to five jammers, the achieved null depth has been 40 dB. Also, superresolution algorithms, such as the MEM, have been programmed onto the array processor, and resolution of one-half beamwidth has been obtained in the field. A recent improvement of the test bed foresees the use of a VLSI node chip processor, designed as part of the very high performance integrated circuit (VHPIC) application demonstrator program (Ward and Hargrave, 1990).

A relevant system has been implemented by MIT Lincoln Laboratory (Teitelbaum, 1991). A UHF planar array is made of a vertical stack of 14 corporate-fed rows. An adaptive beamformer allows shaping the elevation pattern to produce null depths in the 60 to 70 dB range. Significant features of the system that provide such formidable performance are (1) digital baseband quadrature sampling by direct conversion of the IF signal with only one ADC (see also Section 2.7.3) to overcome the limited accuracy of conventional analog quadrature video detectors; (2) a set of adaptive FIR equalizing filters, employed in all channels to match them to the one selected as a reference; and (3) an adaptive beamformer, based on the Givens rotation algorithm, mapped onto a linear systolic array. Each systolic processing node is a floating-point digital processor chip of 16 million floating-point operations per second (MFLOPS).

Another parallel architecture specialized for real-time adaptive nulling is called MUSE (matrix update systolic experiment) also developed by MIT Lincoln Laboratory (Rader *et al.*, 1990). A specific realization of MUSE can update 64-element nulling weights, based on 300 snapshots, in only 6.7 ms. This is equivalent to 2.88 giga-operations for a conventional processor. The computations are accurate enough to support a 50 dB SNR improvement. The architecture is a linear systolic array, each processing node using the CORDIC algorithm to realize the Givens transformations. This system is quite interesting from a technological point of view because it can be realized on a single large wafer, using restructurable VLSI.

The last experimental system we wish to mention is described in (Dilsavor and Gupta, 1991), a three-element sample matrix inversion adaptive array antenna, which operates in a signal scenario consisting of a desired signal and a weak interfering signal (i.e., below thermal noise). This is not a common scenario for radar, but rather one common to communication applications. Nevertheless, this realization is technically interesting. Conventional adaptive arrays are designed to operate against strong interference sources, which are reduced to the thermal-noise level. Thus, weak interfering signals are not suppressed. However, they continue to disturb the desired signal. In a modified SMI algorithm, the covariance matrix is redefined to reduce the effect of thermal noise on the weights of the adaptive array. The experimental system described by Dilsavor and Gupta (1991) operates at 69 MHz with 6 MHz bandwidth. The experimental results show the need for several thousand snapshots to cancel the weak interference.

5.9 CONCLUDING REMARKS

Several techniques and algorithms for jammer nulling and angular superresolution with adaptive arrays have been described in this chapter. These techniques are moving from theory to practical application, as demonstrated by the experimental test beds described in Section 5.8.

For the near future, we expect additional efforts toward practical systems being able to operate in real-time. In this respect, the trend is toward efficient numerical algorithms operating directly on the data matrix **Z** without forming and inverting the covariance matrix. Efficient, numerically stable algorithms are the SVD and the QR decomposition among others. Additional algorithms instrumental to SVD and QR decomposition are the Jacobi, Householder, Givens rotation, and Gram-Schmidt (Bucciarelli *et al.*, 1982; Farina and Studer, 1984; Gerloch and Kretschmer, 1990, 1991; Yuen, 1989, 1991). For an extensive mathematical description of these topics, the reader is encouraged to read the following references: Haykin, 1986, 1989; Golub and VanLoan, 1983; Steinhardt, 1988; Ward *et al.*, 1986. An important point in favor of this type of algorithm is that it can be efficiently mapped onto parallel computers as a systolic array (Kung *et al.*, 1981; Kung, 1985; Tang *et al.*, 1991) and, for large arrays, onto massive parallel machines, such as the Connection Machine (Adams, 1991). Also very important to achieve good performance for both jammer cancellation and angular superresolution is the capability of accurate compensation of the many error sources inevitably present in practical systems. Calibration techniques and channel equalization means are needed, see, for instance, (Teitelbaum, 1991), for a notable example of a high-performance experimental system.

We now highlight the most recent research trends in adaptive array theory. Some topics find application in both jammer cancellation and angular superresolution, while others, are more related to one specific application.

Despite the many studies done (Ahmed and Evans, 1983; Cantoni and Godara, 1982; Griffiths *et al.*, 1985; Ko, 1987, 1989a, 1989b, 1990; Ko *et al.*, 1989; Rodgers and Compton, 1979; Mayhan *et al.*, 1980; Takao and Komiyama, 1980; Tseng, 1980), a research area which needs further effort is that of broadband arrays. The many curves presented here have demonstrated the performance reduction of adaptive arrays in jammer nulling and angular superresolution. When a source having a flat receiver spectrum impinges on the array, the resulting received samples are correlated in time and space (it is a two-dimensional problem) and cannot be handled with an adaptive system operating only in the spatial domain. Gabriel (1989) emphasizes interaction between the time (frequency) and spatial domains by considering a linear array with uniform aperture illumination and with a beam pointed in the spatial direction, θ_0, at center frequency, f_0. When the angle, θ_0, is different from zero, the beam will point in a different direction, θ, when the frequency, f, differs from f_0:

$$(f/f_0) \sin\theta = \sin\theta_0 \qquad (5.106)$$

The shift *versus* frequency, which occurs throughout the sidelobe regions of the antenna pattern, affects the null depths and the structure of lobes. As a consequence, a spatial point source appears to the array as a cluster of monochromatic, spatially distributed sources. From the nulling point of view, a broadband jammer can be suppressed by using more spatial degrees of freedom (i.e., more array elements). The same technique does not provide comfortable performance when we wish to estimate the jammer DOA. This example shows the difference between source estimation and source cancellation systems. We need to separate the two domains and to operate adaptively on each if superresolution is required. This result can be achieved by appending to each receiving element an adaptive tapped delay line with proper weights. See Monzingo and Miller (1980), and Subsections 4.3.1.2 and 4.3.2.4. In this way, the linear combination of spatial samples is accomplished by frequency-dependent coefficients in contrast to the case in which frequency-independent weights are applied in the absence of a tapped delay line. The benefit of using tapped delay lines is demonstrated in Gabriel (1989) to estimate the DOA of jammers that do not have spatial overlap within their operating frequency bandwidths. When a significant spatial overlap occurs, the tapped delay-line approach is no longer able to resolve the sources in angle. The technique suggested in this case is an array utilizing time-delay steering, i.e., weights varying with frequency. Other technical approaches to the problem of broadband adaptive arrays are, for instance, those described by Indukumar and Reddy (1990), Wang and Kaveh (1984), Yang and Kaveh (1990), which are somewhat related to Gabriel's approach.

Let us show the previous concepts with an application example. Consider the processing scheme of Figure 5.64. The number of antennas is $M = 6$, and the interspace between antenna elements is $d = 0.5\lambda_0$, where λ_0 is the wavelength of the carrier. Each antenna has k taps so that the overall available data are Mk. The tap duration is the inverse of twice the receiver bandwidth. The minimum number of taps needed may be found in Gabriel (1989, p. 18). The MUSIC method is used to estimate the jammers' DOAs. Two wideband jammers impinge on the array. Their JNR value is 0 dB and their bandwidth is $0.4f_0$. Two cases are considered: (1) the jammer DOAs are $\theta_1 = 0°$ and $\theta_2 = 10°$; (2) the jammer DOAs are $\theta_1 = 30°$ and θ_2

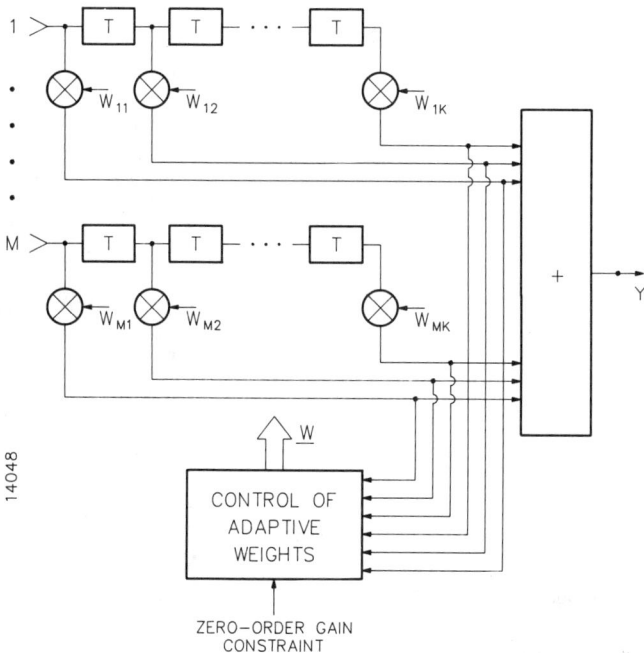

Figure 5.64 A linear broadband array with M sensors, k taps and the adaptive control of weights by the MUSIC method.

$= 40°$. In the first case, there is no spatial overlap between the two jammers; consequently, the processor of Figure 5.64 improves the jammers' resolution, as shown in Figure 5.65. In fact, the width of the peaks and sidelobe levels are reduced. In the second case, the spatial overlap of the two jammers is significant and the processor of Figure 5.64 is not helpful. The angular spectrum for this case is shown in Figure 5.66. The occupation on the $\sin\theta$ axis of the jammer with $\theta_1 = 30°$ is [0.416, 0.625], while the occupation on the same axis of the jammer with $\theta_2 = 40°$ is [0.535, 0.803]; indeed the two intervals overlap.

Additional areas of investigation in the disturbance cancellation field with adaptive arrays are:

1. The cancellation of specular and diffuse jammer multipath (Fante, 1991);
2. The simultaneous cancellation of clutter and jammer when the array is fixed or on-board a moving platform (Farina, 1987a, pp. 150–160; D'Addio et al., 1989; Barbarossa and Farina, 1991).

Again, tapped delay line networks downstream from each antenna element are needed for these problems. The immediate result is the increased number of degrees

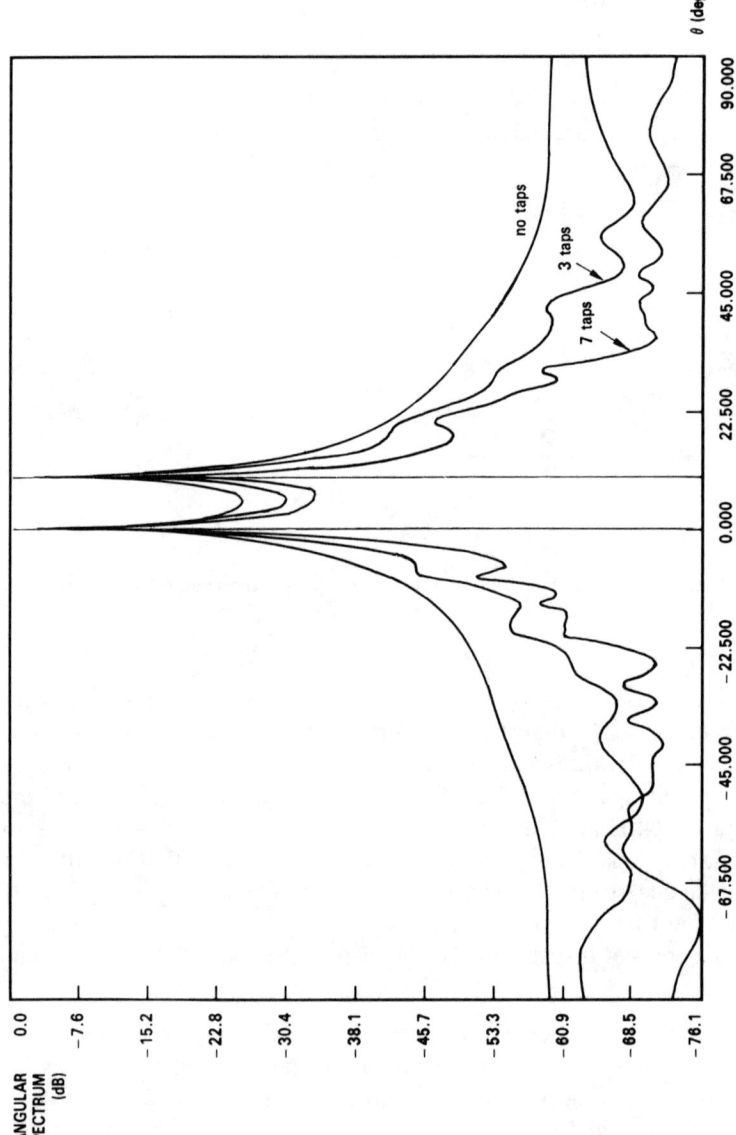

Figure 5.65 Angular spectrum of two broadband sources: $\theta_1 = 0°$, $\theta_2 = 10°$, $JNR_1 = JNR_2 = 0$ dB, $B/f_0 = 40\%$.

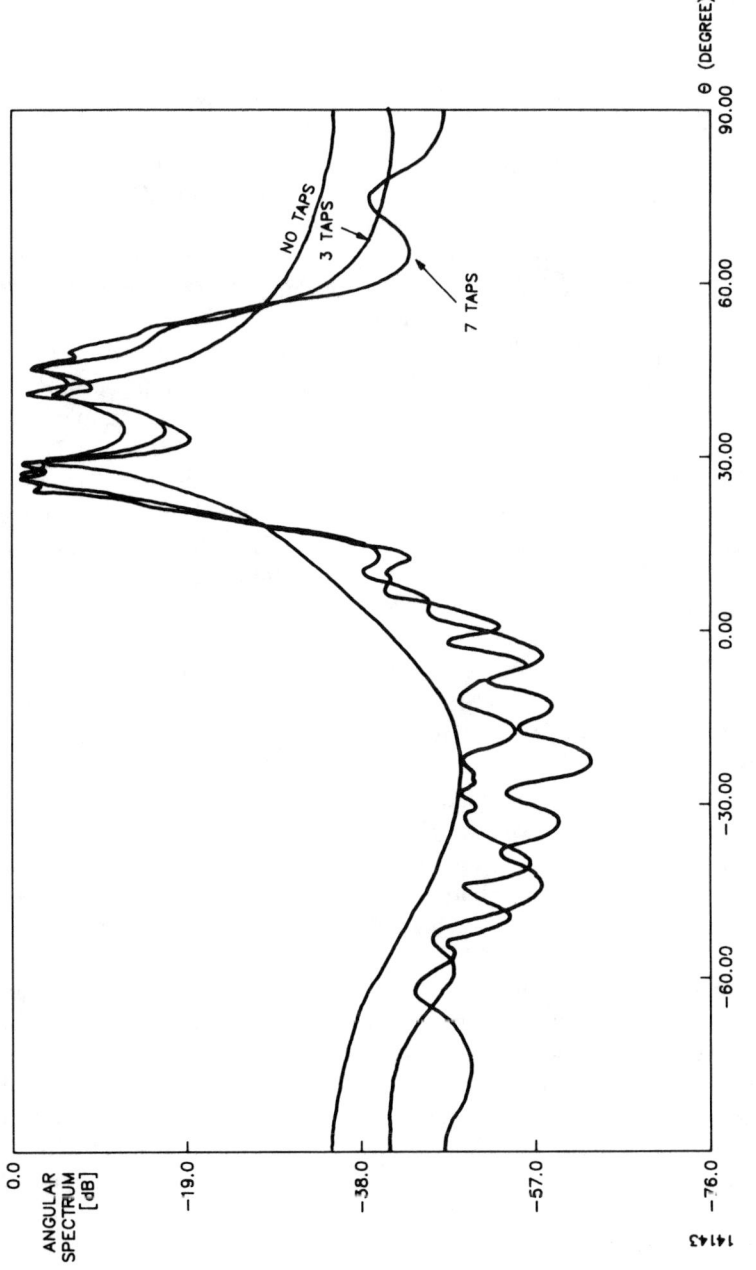

Figure 5.66 Angular spectrum of two broadband sources: $\theta_1 = 30°$, $\theta_2 = 40°$, $JNR_1 = JNR_2 = 0$ dB, $B/f_0 = 40\%$.

of freedom with the consequent increase of computational burden and a slower reaction time to reach the steady-state. An effective solution does not yet seem to be available.

An interesting area of investigation is described in (Haimovich and Bar-Nes, 1991), which is somewhat related to (Lin, 1982, p. 216). With common adaptive arrays, the cancellation is effected by global optimization procedures that include the interference and thermal noise. However, the proposed technique focuses on the interference only, resulting in superior cancellation performance. A kind of "supercancellation" is gained, i.e., while the conventional array has some finite response to interference, the proposed approach theoretically has zero response. Furthermore, the method achieves full effectiveness, almost immediately after the number of snapshots exceeds the number of interference sources.

The angular superresolution problem can be viewed within the basic framework of the parameter estimation of superimposed exponential signals in noise. The accuracy of the maximum-likelihood estimate (MLE), which is the optimum estimation method, approaches the unbiased Cramer-Rao lower bounds (Rife and Boorstyn, 1976; 1978; Marple, 1987, p. 113) with SNR values and length of data records that are significantly lower than the respective values of the methods described in Section 5.7. Nevertheless, the suboptimal methods of Section 5.7, all of which can be derived from the MLE (Stoica and Sharman, 1990) are preferred over the exact MLE because of computational problems. Direct maximization of the likelihood function, which is highly nonlinear in the unknown signal parameters (amplitude, DOA, *et cetera*), requires a computationally expensive, multidimensional search. Recently, Bresler and Macovski (1986a, 1986b) have proposed an efficient algorithm to calculate the MLE. In particular, an iterative algorithm, termed the *iterative quadratic maximum likelihood* (IQML), allows us to solve the optimization problem associated with MLE by solving only linear equations. The algorithm also is applicable in the case of coherent interference sources. Although the convergence properties of the algorithm require further theoretical and experimental investigation, the simulation results show superior accuracy for short data records. This observation suggests that the algorithm may be used for the estimation of DOAs from as little data as a single snapshot.

Superresolution is helpful for two applications: (1) angular superresolution on the target when multiple phase centers of reflection are present in the same range cell; and (2) superresolution for accurate angular estimation of jammers. In the first situation, there may be very few snapshots available, perhaps only one, whereas many are generally available for the jammers. These two functions usually have different solutions. For example, if the jammer angle estimation is desired, it can be economically derived as a byproduct of the jammer cancellation calculations. The choice of algorithm may be a function of the number of snapshots available and the presence or absence of correlated multipath. Another key issue that involves the choice of algorithm is whether two-dimensional angular superresolution is being considered because many of the one-dimensional algorithms are illposed for two dimensions. See Marple (1987, Chapter 16).

Another distinction can be made concerning algorithms that make use of the full aperture, such as MLE, MUSIC, and Capon, and those that operate on reduced aperture dimensions (and suffer the immediate resolution loss), such as the aperture smoothing and nearly all of the single-snapshot Prony-motivated methods (Marple, 1987, Chapter 11), including ARMA, Pisarenko, reduced-rank subspace methods of Kumaresan and Tufts (1983), ESPRIT method (Roy *et al.*, 1985; 1986), matrix pencil method (Ouibrahim, 1989), and others. Yet another classification refers to the algorithms designed to cope with signal bandwidth, such as the many variations of ESPRIT (Hung and Kaveh, 1990; Ottersten and Kailath, 1990), and focusing matrices (Hung and Kaveh, 1988), and the very general approach suggested by Cadzow *et al.* (1989).

Despite the great wealth of available algorithms, there is still a problem with this technique. Superresolution is fundamentally quite sensitive (for strong signals) to modeling errors. Practical arrays will suffer in this regard because of manufacturing errors, near-field effects, edge effects, and nonuniform mutual coupling (Friedlander, 1990b; Friedlander and Weiss, 1991).

New investigative areas in superresolution are related to the decentralized approach to DOA estimate by using non-collocated subarrays, each providing local DOA estimates that are combined in a fusion center (Lee *et al.*, 1990) and the use of neural network for maximum-likelihood estimation of DOAs (Jelonek and Reilly, 1990). Neural networks also have been proposed for jammer cancellation (Sung *et al.*, 1990).

A new application of superresolution techniques recently has been suggested by W.F. Gabriel (1989; 1990). Specifically, high resolution techniques have been applied to the radar imaging of rotating objects, i.e., to *inverse synthetic aperture radar* (ISAR) systems, in both the doppler and range domains. As is well known, the rotation of an object relative to the radar generates a doppler frequency gradient and permits the extraction of cross-range resolution, which is better than that obtainable by the radar's beamwidth. By application of superresolution to the doppler domain, instead of the conventional windowed Fourier transformation, the result is either higher resolution images from the same data samples, or equal-quality images from significantly fewer data samples (Gabriel, 1989). The resolution in range is currently obtained by transmitting a "chirped" signal and processing the echoes by using a pulse compression filter. The application of superresolution in the range domain utilizes adaptive weighting, rather than conventional deterministic matched filter weighting (Gabriel, 1990). Simulation examples illustrate the possibility of separating two or more point targets within the same range cell without increasing the radar bandwidth.

Bibliography

Adam, J.A. (1988), "Pinning Defense Hopes on Aegis," *IEEE Spectrum*, June 1988, pp. 24–27.
Adams, G., Finn, A.M., and Griffin, M.F. (1991), "A Fast Implementation of the Complex Singular Value Decomposition on the Connection Machine," *Proc. of IEEE 1991 Intl. Conf. on Acoustic, Speech, and Signal Processing*, May 14–17, 1991, Toronto, Canada.
Ahmed, K.M., and Evans, R.J. (1983), "Broadband Adaptive Array Processing," *IEE Proc.*, Vol. 130, Pt.F, No. 5, August 1983, pp. 433–440.
Andrews, G.A., and Gerlach, K. (1989), "SBR Clutter and Interference," Ch. 11 of *Space-Based Radar Handbook*, L.J. Cantafio, Editor, Artech House, 1989, pp. 373–479.
Applebaum, S.P. (1976), "Adaptive Arrays," Syracuse University Research Corporation, Rep. SPL TR 66-1, 1966. This report is reproduced in *IEEE Trans. on Antennas and Propagation, Special Issue on Adaptive Antennas*, Vol. AP-24, 1976, pp. 585–598.
Applebaum, S.P., and Chapman, D.J. (1976), "Adaptive Arrays with Main Beam Constraints," *IEEE Trans. on Antennas and Propagation*, Vol. AP-24, 1976, pp. 650–662.
Applebaum, S.P., Howells, P.W., and Kovarik, C. (1977), "Multiple Intermediate Frequency Side-Lobe Canceler," U.S. Patent No. 4,044,359, August 23, 1977.
Arancibia, P.O. (1978), "A Sidelobe Blanking System Design and Demonstration," *Microwave Journal*, Vol. 21, No. 3, March 1978, pp. 69–73.
Baird, C.A. (1974), "Kalman-Type Processing Adaptive Antenna Arrays," *IEEE Intl. Conf. on Communications*, Minneapolis, 1974, pp. 10G-1–10G-4.
Baird, C.A., and Rassweiler, G.G. (1978), "Adaptive Nulling Using Digitally Controlled Phase-Shifters," *IEEE Trans. on Antennas and Propagation*, Vol. AP-24, 1978, pp. 638–649.
Baldwin, P.J., Denison, E., and O'Connor, S.F. (1979), "An Experimental Adaptive Array for Radar Applications," *Proc. of 9th European Microwave Conf.*, Brighton (UK), September 17–20, 1979, pp. 667–671.
Baozheng *et al.* (1983), "Improvement in the Adaptive Sidelobe-Canceler Performance: Theoretical Analysis and Experimental Results," *Proc. of Intl. Radar Symposium*, India, Bangalore, October 9–12, 1983, pp. 115–120.
Barbarossa, S., and Farina, A. (1991), "Detection and Imaging of Moving Objects with SAR by a Joint Space-Time-Frequency Processing," 1991 CIE Intl. Conf. on Radar (CICR'91), October 22–24, 1991, Beijing, China.
Barton, D.K. (1988), *Modern Radar System Analysis*, Artech House, Norwood, MA, 1988.
Barton, P. (1980), "Digital Beam Forming for Radar," *IEE Proc.*, Vol. 127, Pt. F, No 4, August 1980, pp. 266–277.
Barton, P., and Blair, P.K. (1982), "Adaptive Beam Forming for Radar," *Electrical Communication*, Vol. 57, No. 1, 1982, pp. 62–69.

Biglieri, E., and Yao, K. (1989), "Some Properties of Singular Value Decomposition and their Applications to Digital Signal Processing," *Signal Processing*, Vol. 18, 1989, pp. 277–289.

Billetter, D.R. (1989), *Multifunction Array Radar*, Artech House, 1989.

Bock, H.W. (1978), "Digital Signal Processing in Adaptive Sidelobe Canceler," *Intl. Radar Conf.*, Paris, December 1978.

Bodnar, D.G. (1987), "MMW Antennas," Ch. 11 in *Principles and Applications of Millimeter-Wave Radar*, Edited by N.C. Currie and C.E. Brown, Artech House, Norwood, MA, 1987.

Borgiotti, G.V. (1978), "Superresolution of Uncorrelated Interference Sources by Using Adaptive Array Techniques," *Proc. of National Radio Science Meeting* (URSI), Boulder, Colorado, November 5–10, 1978, pp. 1–11.

Boroson, D.M. (1980), "Sample Size Consideration for Adaptive Array," *IEEE Trans. on Aerospace and Electronic Systems*, Vol. AES-16, No. 4, July 1980, pp. 446–451.

Bowers, L., and Perry, E. (1980), "Analysis of Tapped Delay Line Processing for Adaptive Sidelobe Cancellation," *1980 Symposium Digest Antennas and Propagation*, Quebec, Canada, 1980, pp. 114–117.

Brandwood, D.H., and Tarran, C. (1982), "Adaptive Arrays for Communication," *IEE Proc.*, Pt. F, Vol. 129, No. 3, June 1982, pp. 223–232.

Brandwood, D. H. (1983), "Convergence Time of Sidelobe Cancellation Systems," *IEE Proc.*, Part F, Vol. 130, No. 1, February 1983, pp. 46–56.

Brandwood, D.H. (1987), "Noise-Space Projection: MUSIC without Eigenvectors," *IEE Proc.*, Vol. 134, Pt. H, No. 3, June 1987, pp. 303–309.

Brennan, L.E., Pugh, E.L., and Reed, I.S. (1971), "Control-Loop Noise Adaptive Array Antennas," *IEEE Trans. on Aerospace and Electronic Systems*, Vol. AES-7, No. 2, March 1971, pp. 254–262.

Brennan, L.E., and Reed, I.S. (1971), "Effect of Envelope Limiting in Adaptive Array Control Loops," *IEEE Trans. on Aerospace and Electronic Systems*, Vol. AES-7, July 1971, pp. 698–700.

Brennan, L.E., and Reed, I.S. (1973), "Theory of Adaptive Radar," *IEEE Trans. on Aerospace and Electronic Systems*, Vol. AES-9, No. 2, March 1973, pp. 237–252.

Brennan, L.E., Reed, I.S., and Swerling, P. (1979), "Method and Apparatus for Sidelobe Reduction in Radar," U.S. Patent No. 4,146,889, Mar. 28, 1979.

Bresler, Y., and Macovski, A. (1986a), "On the Number of Signals Resolvable by a Uniform Linear Array," *IEEE Trans. on Acoustics, Speech, and Signal Processing*, Vol. ASSP-34, No. 6, December 1986, pp. 1381–1375.

Bresler, Y., and Macovski, A. (1986b), "Exact Maximum Likelihood Parameter Estimation of Superimposed Exponential Signals in Noise," *IEEE Trans. on Acoustics, Speech, and Signal Processing*, Vol. ASSP-34, No. 5, October 1986, pp. 1081–1089.

Bresler, Y., Reddy, V.U., and Kailath, T. (1988), "Optimum Beamforming for Coherent Signal and Interferences," *IEEE Trans. on Acoustics, Speech, and Signal Processing*, Vol. ASSP-36, No. 6, June 1988, pp. 833–843.

Bristow, T.A. (1983), "Experimental Open-Loop Canceler for Radar," *IEE Proc.*, Part. F, Vol. 130, No. 1, February 1983, pp. 109–113.

Brookner, E., and Howell, J.M. (1986), "Adaptive-Adaptive Array Processing," *IEEE Proc.*, Vol. 74, No. 4, April 1986, pp. 602–604.

Brookner, E., and Howell, J.M. (1987), "Adaptive-Adaptive Array Processing," *Proc. of Intl. Radar Conference, Radar'87*, IEE London, October 19–21, 1987, pp. 257–263.

Brown, A.R. (1981), "Effects of Element Cross-Polarization in Adaptive Antennas," *Antennas and Propagation*, IEE Conf. Publication, No. 195, April 13–16, 1981, Univ. of York (UK), pp. 1-144–1-148.

Bucciarelli, T., Esposito, M., Farina A., and Losquadro, G. (1982), "The Gram-Schmidt Sidelobe Canceler," *Proc. of Intl. Radar Conf., Radar'82*, IEE, London, October 18–20, 1982, pp. 486–490.

Buckley, K.M., and Griffiths, L.D. (1986), "An Adaptive Generalized Sidelobe Canceler with Derivative Constraints," *IEEE Trans. on Antennas and Propagation,* Vol. AP-34, No. 3, March 1986, pp. 311–319.

Buhring, W. (1978a), "Adaptive Orthogonal Projection for Rapid Converging Interference Suppression," *Electronic Letters,* 14, August 1978, pp. 515–516.

Buhring, W. (1978b), "Adaptive Antenna with Rapid Convergence," *IEE Conf. on Antennas and Propagation Record,* No. 169, November 1978, pp. 51–54.

Buhring, W. (1980), "Coherent Signal Processing in a Frequency Agile Jittered Pulse Radar," *IEEE Intl. Radar Conf.,* Washington, D.C., 1980, pp. 188–193.

Burg, J.P. (1968), "A New Analysis Technique for Time Series Data," NATO Advanced Study Institute on Signal Processing, Enschede, Netherlands, 1968.

Burgess, L.W., and Berkowitz, R.S. (1981), "Design Limitations on Adaptive Array Control Loop Nulling Time," IEEE 1981 *Intl. Symposium Digest Antennas and Propagation,* Los Angeles CA, June 16–19, 1981, Vol. 1, pp. 294–297.

Cadzow, J.A., Kim, Y.S., and Shiue, D.C. (1989), "General Direction-of-Arrival Estimation: A Signal Subspace Approach," *IEEE Trans. on Aerospace and Electronic Systems,* Vol. 25, No. 1, January 1989, pp. 31–47.

Cantoni, A., and Godara, L.C. (1982), "Fast Algorithms for Time Domain Broadband Adaptive Array Processing," *IEEE Trans. on Aerospace and Electronic Systems,* Vol. AES-18, No. 5, September 1982, pp. 682–699.

Capon, J. (1969), "High-Resolution Frequency-Wavenumber Spectrum Analysis," *Proc. IEEE,* Vol. 57, 1969, pp. 1408–1418.

Carhoun, D.O. (1991), "Adaptive Nulling and Spatial Spectral Estimation Using an Iterated Principal Components Decomposition," *IEE Intl. Conf. on Acoustic, Speech, and Signal Processing,* IEASSP 91, May 14–17, 1991, Toronto (CA).

Carlier, R. (1979), "Random Excitation Errors in Array Antennas and their Influence on the Choice of Array Distribution," *The Marconi Review,* Second Quarter, 1979, pp. 119–134.

Carlson, B.D. (1988), "Covariance Matrix Estimation Errors and Diagonal Loading in Adaptive Arrays," *IEEE Trans. on Aerospace and Electronic Systems,* Vol. AES-24, No. 4, July 1988, pp. 397–401.

Carlson, B.D., Goodman, L.M., Austin, J., Ganz, M.W., and Upton, L.O. (1990), "An Ultralow-Sidelobe Adaptive Array Antenna," *The Lincoln Laboratory Journal,* Vol. 3, No. 2, 1990, pp. 291–310.

Chan, K.K., Faubert, D., Martin, R., and Turner, R. (1989), "Space-fed Lens Phased Array for Advanced Radar Systems," *Intl. Conf. on Radar,* April 24–28, 1989, Paris, pp. 70*–75*.

Chapman, D.J. (1976), "Partial Adaptivity for the Large Array," *IEEE Trans. on Antennas and Propagation,* Vol. AP-24, No. 5, September 1976, pp. 685–695.

Cheston, T.C. (1970), "Array Antennas," Ch. 11 in *Radar Handbook,* M.I. Skolnik, Editor, McGraw-Hill Book Company, 1970. See also Cheston, T.C., and Frank, J., "Phased Array Radar Antennas," Ch. 7 in *Radar Handbook,* 2nd Ed., M.I. Skolnik, Editor, McGraw-Hill Book Company, 1990.

Childers, D.G., Editor (1978), *Modern Spectrum Analysis,* IEEE Press, New York, 1978.

Chin, J.E., Liebman, P.M., and Fleming, J.E. (1989), "Reducing the Interference Between Sidelobe Cancelers and Sidelobe Blankers in Electronic Scanning Array Radars," *Proc. of 1989 IEEE National Radar Conf.,* March 29–30, 1989, Dallas, pp. 141–146.

Chrzanowski, E.J. (1990), *Active Radar Electronic Countermeasures,* Artech House, Norwood, MA, 1990.

Clark, C.R. (1989), "Main Beam Jammer Cancellation and Target Angle Estimation with a Polarimetric Agile Monopulse Antenna," *Proc. of 1989 IEEE National Radar Conf.,* March 29–30, 1989, Dallas, pp. 95–100.

Clergeot, H., Ouamri, A., and Tressens, S. (1985), "High Resolution Spectral Methods for Spatial Discrimination of Closely Spaced Correlated Sources," *Proc. of Intl. Conf. on Acoustics, Speech and Signal Processing*, ICASSP 1985, pp. 15.3.1–15.3.4.

Cofer, J.W., Martin, G.P., and Ralph, S.E. (1981), "Adaptive Null Steering by Reflector Antennas," IEE Conf. Publ. No. 195, *Antennas and Propagation*, Univ. of York (UK), April 13–16, 1981, pp. 169–173.

Cohen, M., Magnan, J.C., and Combes, S.P. (1985), "Adaptive Array Antenna Performances," *Fourth Intl. Conf. on Antennas and Propagation* (ICAP 85), April 16–19, 1985, Coventry (UK), pp. 241–245.

Compton, R.T. (1979), "The Power-Inversion Adaptive Arrays: Concept and Performance," *IEEE Trans. on Aerospace and Electronic Systems*, Vol. AES-15, No. 6, November 1979, pp. 803–814.

Compton, R.T. (1981), "The Effect of Integrator Pole Position on the Performance of an Adaptive Array," *IEEE Trans. on Aerospace and Electronic Systems*, Vol. AES-17, No. 4, July 1981, pp. 598–602.

Compton, R.T. (1982), "The Effect of a Pulsed Interference Signal on an Adaptive Array," *IEEE Trans. on Aerospace and Electronic Systems*, Vol. AES-18, No. 3, May 1982, pp. 297–309.

Compton, R.T. (1988), *Adaptive Antennas: Concepts and Performance*, Prentice Hall, Englewood Cliffs, NJ, 1988.

Curt, C., and Medynski, D. (1989), "Comparison of Iterative and Semi-Direct CCM Methods Applied to Sparse Arrays," *Intl. Conf. on Radar*, April 24–28, 1989, Paris, pp. 261–266.

D'Addio, E., Farina, A., and Morabito, C. (1989), "The Applications of Multidimensional Processing to Radar Systems," *Intl. Conf. on Radar*, April 24–28, 1989, Paris, pp. 62–78.

Davie, E.B., Higgins, D.G., Cawthorne, C.D. (1986), "An Advanced Adaptive Antenna Test-Bed Based on a Wavefront Array Processor System," STC Technology Ltd, UK, 1986.

Davis, R.C., Brennan, L.E., and Redd, I.S. (1976), "Angle Estimation with Adaptive Arrays in External Noise Field," *IEEE Trans. on Aerospace and Electronic Systems*, Vol. AES-12, No. 2, March 1976, pp. 179–186.

Di Bartolo, S. (1988), "Application of Angular Superresolution Algorithms to Search Radar in UHF-VHF Bands," Doctoral Thesis, University of Rome, "La Sapienza," 1988 (in Italian).

Di Carlo, D.M., and Compton, R.T. (1978), "Reference Loop Phase Shift in Adaptive Arrays," *IEEE Trans. on Aerospace and Electronic Systems*, Vol. AES-14, No. 4, July 1978, pp. 599–607.

DiVito, A., and Iovino, D. (1986), *Analysis of Errors in the Receiving Chain of a DBF*, Selenia Technical Memorandum, (in Italian), July 28, 1986.

DiVito, A., and Iovino, D. (1987), *Digital Synthesis Techniques of Beams for Phased-Array Antennas*, Selenia Technical Memorandum, February 18, 1987 (in Italian).

Dicken, L.W. (1977), "The Use of Null Steering in Suppressing Main Beam Interference," *Proc. of IEE Intl. Conf. on Radar*, Radar'77, London, October 25–28, 1977, pp. 226–231.

Dilsavor, R.S., and Gupta, I.J. (1991), "An Experimental SMI Adaptive Antenna Array Simulator for Weak Interfering Signals," *IEEE Trans. on Antennas and Propagation*, Vol. 39, No. 2, February 1991, pp. 236–243.

Docter, R.A., Masenten, W.R., and Kinkel, J.F. (1979), "Trends in Adaptive Antenna Circuit Design," 1979 Electro Professional Conf., April 24–26, 1979, p. 11/2.

Dollinger, K. (1977), "Sidelobe Cancellation System," U.S. Patent No. 4,057,802; November 8, 1977.

Dolph, C.L. (1946), "A Current Distribution for Broadside Arrays which Optimizes the Relationship Between Beamwidth and Sidelobe Levels," *Proc. of the IRE*, Vol. 34, No. 6, June 1946, pp. 335–348.

Downie, J.W. (1986a), "Wide Band Multiple Side-Lobe Canceler," U.S. Patent No. 4,586,045, April 29, 1986.

Downie, J.W. (1986b), "Side-Lobe Canceler," U.S. Patent No. 4,586,048, April 29, 1986.

Drabowitch, S., Aubry, C., and Roger, J. (1981), "Application des Proprietes Spatiales des Antennes aux Techniques d'Opposition des Lobes Secondaires," (in French), Nice, June 1981, pp. 371–776.

Drane, C., and McIlvenna, J. (1970), "Gain Maximization and Controlled Null Placement Simultaneously Achieved in Aerial Array Patterns," The Radio and Electronic Engineer, Vol. 39, No. 1, January 1970, pp. 49–57.

Dufort, E.C. (1989), "Pattern Synthesis Based on Adaptive Array Theory," *IEEE Trans. on Antennas and Propagation*, Vol. 37, No. 8, August 1989, pp. 1011–1018.

Durrani, T.S., and Sharman, R.C. (1983), "Eigenfilter Approaches to Adaptive Array Processing," *IEE Proc.*, Vol. 130, Pts. F and H, No. 1, February 1983, pp. 22–28.

Eggestad, M., and Heier, R. (1980), "A Combined Programmed and Adaptive Null Steering Technique," *IEEE Trans. on Aerospace and Electronic Systems*, Vol. AES-16, No. 5, September 1980, pp. 640–647.

Ekholdt, R. (1975), "Adaptive Interference Cancellation in Nyquist Rate Scanned Arrays," *IEEE Trans. on Aerospace and Electronic Systems*, Vol. AES-11, No. 3, May 1975, pp. 359–362.

Esposito, R., and Wilson, L.R. (1973), "Statistical Properties of Two Sine Waves in Gaussian Noise," *IEEE Trans. on Information Theory*, Vol. IT-19, March 1973, pp. 176–183.

Evans, J.E. (1979a), "Aperture Sampling Techniques for Precision Direction Finding," *IEEE Trans. on Aerospace and Electronic Systems*, Vol. AES-15, No. 6, November 1979, pp. 892–895.

Evans, J.E. (1979b), "Comments on Angular Spectra in Radar Applications," *IEEE Trans. on Aerospace and Electronic Systems*, Vol. AES-15, No. 6, November 1979, pp. 899–903.

Fante, R. (1980), "Sidelobe and Nulling Properties of Overlapped, Subarrayed Scanning Antennas," *1980 Intl. Symposium Digest Antennas and Propagation*, Quebec, Canada, 1980, pp. 138–140.

Fante, R. L. (1991), "Cancellation of Specular and Diffuse Jammer Multipath Using a Hybrid Adaptive Array," *Proc. of the 1991 National Radar Conf.*, Los Angeles, March 12–13, 1991, pp. 54–57.

Farina, A. (1977), "Single Sidelobe Canceler: Theory and Evaluation," *IEEE Trans. on Aerospace and Electronic Systems*, Vol. AES-13, No. 6, November 1977, pp. 690–699.

Farina, A., Giusto, R., and Losquadro, G. (1980), "Analysis of a Device for the Generation of a Wide Null in an *a priori* Known Direction in the Receiving Pattern of a Phased-Array," *Selenia Technical Report*, 11 December 1980 (in Italian).

Farina, A., and Giusto, R. (1981), "Performance of Sidelobe Cancellation Techniques," *Intl. Conf. on Digital Signal Processing*, Florence, September 2–5, 1981, pp. 922–929.

Farina, A., and Studer, F.A. (1982), "Evaluation of Sidelobe-Canceler Performance," *IEE Proc.*, Vol. 129, Pt. F, No. 1, February 1982, pp. 52–58.

Farina, A., and Studer, F.A. (1984), "Application of Gram-Schmidt Algorithm to Optimum Radar Signal Processing," *IEE Proc.*, Vol. 131, Pt. F, No. 2, April 1984, pp. 139–145.

Farina, A. (1985), "Interference Suppressor for an Electronically or Mechanically Scanning Monopulse Radar Generating Sum and Difference Signals from Received Microwave Energy," US Patent Number 4,549,183, October 22, 1985.

Farina, A., and Galati, G. (1985), "Surveillance Radars: State of Art. Research and Perspectives," *Alta Frequenza*, Vol. 54, No. 4, July–August 1985, pp. 243–260.

Farina, A. (1986), "Adaptive System for Suppressing Interference from Directional Jammers in Electronically Scanning Radar," US Patent Number 9,573,051, February 25, 1986.

Farina, A. (1987a), *Optimised Radar Processors*, Peter Peregrinus, Ltd., London, 1987.

Farina, A., and Studer, F.A. (1988), "Adaptive Radar Signal Processor for the Detection of Useful Echo and the Cancellation of Clutter," US Patent, Number 4,719,466, January 12, 1988.

Farina, A. (1990), "Electronic Counter-Counter Measures," Ch. 9 of *Radar Handbook*, 2nd Ed. 1990, M.I. Skolnik, Editor, McGraw-Hill Book Company.

Farina, D.J. (1987b), "Adaptive Array Performance Analysis," *Proc. of IEE Intl. Radar Conf.*, Radar'87, London, October 19–21, 1987, pp. 264–268.
Farina, D.J., and Flam, R.P. (1990), "Simplified Covariance Matrix Measurement in Adaptive Arrays," *IEEE Trans. on Aerospace and Electronic Systems*, Vol. 26, No. 4, July 1990, pp. 669–672.
Farrier, D.R., Jeffries, D.J., and Mardani, R. (1988), "Theoretical Performance Prediction of the MUSIC Algorithm," *IEE Proc.*, Vol. 135, Pt. F, No. 3, June 1988, pp. 218–224.
Fenn, A.J., and Tsandoulas, G.N. (1989), "Near Field Adaptive Nulling," *Intl. Conf. on Radar*, April 24–28, 1989, Paris, pp. 59*–64*.
Ficker, A.J., and Bresler, Y. (1989), "Sensor-Efficient Wideband Source Location," *Proc. of 32nd Midwest Symposium on Circuits and Systems*, Champaign, IL, August 14–16, 1989, pp. 586–589.
Fielding, J.G. (1976), "Signal Processing in Electronic Warfare," *Intl. Specialist Seminar on the Impact of New Technologies in Signal Processing*, Aviemore, Scotland, September 20–24, 1976, pp. 112–120.
Fielding, J.G., et al. (1977), "Adaptive Interference Cancellation in Radar Systems," *Proc. of IEE Intl. Conf. on Radar*, Radar'77, London, October 25–28, 1977, pp. 212–217.
Flam, R.P. (1984), "Design Rules for Adaptive Arrays," *IEEE Trans. on Aerospace and Electronic Systems*, Vol. AES-20, No. 2, March 1984, pp. 199–202.
Floyd, F.W., and Mayhan, F.T. (1980), "Some Effects of Hard Limiting in Adaptive Antenna Systems," *IEEE Trans. on Aerospace and Electronic Systems*, Vol. AES-16, No. 6, November 1980, pp. 839–850.
Friedlander, B. (1988), "A Signal Subspace Method for Adaptive Interference Cancellation," *IEEE Trans. on Acoustics, Speech, and Signal Processing*, Vol. ASSP-36, No. 12, December 1988, pp. 1835–1845.
Friedlander, B., and Porat, B. (1989), "Performance Analysis of a Null-Steering Algorithm Based on Direction of Arrival Estimation," *IEEE Trans. on Acoustics, Speech, and Signal Processing*, Vol. ASSP-37, No. 4, April 1989, pp. 461–466.
Friedlander, B. (1990a), "A Sensitivity Analysis of the MUSIC Algorithm," *IEEE Trans. on Acoustic, Speech, and Signal Processing*, Vol. ASSP-38, No. 1, October 1990, pp. 1740–1751.
Friedlander, B. (1990b), "Sensitivity Analysis of the Maximum Likelihood Direction-Finding Algorithm," *IEEE Trans. on Aerospace and Electronic Systems*, Vol. AES-25, No. 6, November 1990, pp. 953–968.
Friedlander, B., and Weiss, A.J. (1991), "Direction Finding in the Presence of Mutual Coupling," *IEEE Trans. on Antennas and Propagation*, Vol. AP-39, No. 3, March 1991, pp. 273–284.
Frost, O.L. (1982), "An Algorithm for Linearly Constrained Adaptive Array Processing," *Proc. IEEE*, Vol. 60, No. 8, August 1982, pp. 926–935.
Gabriel, W.F. (1976), "Adaptive Arrays: An Introduction," *Proc. IEEE*, Vol. 64, 1976, pp. 239–272.
Gabriel, W.F. (1979), "Maximum Entropy Spacial Resolution Adaptive Array," *1979 Int. Symp. Digest on Antennas and Propagation*, University of Washington, Seattle, pp. 203–206.
Gabriel, W.F. (1980a), "Spectral Analysis and Adaptive Array Superresolution Techniques," *Proc. IEEE*, Vol. 68, 1980, pp. 654–666; and NRL Report 8345, July 24, 1979.
Gabriel, W.F. (1980b), "Superresolution of Coherent Sources by Adaptive Array Techniques," *Proc. of IEEE Intl. Radar Conf.*, Washington, DC, 1980, pp. 182–187.
Gabriel, W.F. (1984), "A High Resolution Target-Tracking Concept Using Spectral Estimation Techniques," NRL Report 8797, May 31, 1984.
Gabriel, W.F. (1986), "Using Spectral Estimation Techniques in Adaptive Processing Antenna Systems," *IEEE Trans. on Antennas and Propagation*, Vol. AP-34, No. 3, March 1986, pp. 291–300; and NRL Report 8920.
Gabriel, W.F. (1989a), "Large-Aperture Sparse Array Antenna Systems of Moderate Bandwidth for Multiple Emitter Location," *IEEE Trans. on Antennas and Propagation*, Vol. 37, No. 1, January 1989, pp. 16–29.

Gabriel, W.F. (1989b), "Superresolution Techniques and ISAR Imaging," *Proc. of the 1989 IEEE National Radar Conf.,* March 29–30, 1989, Dallas, pp. 48–55.

Gabriel, W.F. (1990), "Superresolution Techniques in the Range Domain," *Proc. of the IEEE 1990 Intl. Radar Conf.,* May 7–10, 1990, Washington, DC, pp. 263–267.

Ganz, M.W., Moses, R.L., and Wilson, S.L. (1990), "Convergence of the SMI and the Diagonally Loaded SMI Algorithms with Weak Interference," *IEEE Trans. on Antennas and Propagation,* Vol. 38, No. 3, March 1990, pp. 394–399.

Gaston, F.M.F., Irwing, G.W., and McWhirter, J.G. (1990), "Systolic Square Root Covariance Kalman Filtering," *Journal of VLSI Signal Processing,* Vol. 2, 1990, pp. 37–49.

Gentleman, W.M. (1973), "Least Squares Computations by Givens Transformations without Square Roots," *J. Inst. Math and Appl.,* Vol. 12, 1973, pp. 329–336.

Gerlach, K., and Lang, R.H. (1986), "Exact Solution for Control Loop Adaptive Antenna Weights in Band-Limited Noise," *IEEE Trans. on Antennas and Propagation,* Vol. AP-34, No. 3, March 1986, pp. 395–403.

Gerlach, R. (1986), "Fast Orthogonalization Networks," *IEEE Trans. on Antennas and Propagation,* Vol. AP-34, No. 3, March 1986, pp. 458–462.

Gerlach, K., and Kretschmer, F.F. (1990), "Convergence Properties of Gram-Schmidt and SMI Adaptive Algorithms," *IEEE Trans. on Aerospace and Electronic Systems,* Vol. 26, No. 1, January 1990, pp. 44–56.

Gerlach, K., and Kretschmer, F.F. (1991), "Convergence Properties of Gram-Schmidt and SMI Adaptive Algorithms: Part II," IEEE Trans. on Aerospace and Electronic Systems, Vol. 27, No. 1, January 1991, pp. 83–91.

Gerlach, K. (1990a), "The Effects of IF Bandpass Mismatch Errors on Adaptive Cancellation," *IEEE Trans. on Aerospace and Electronic Systems,* Vol. 26, No. 3, May 1990, pp. 455–468.

Gerlach, K. (1990b), "Adaptive Array Transient Sidelobe Levels and Remedies," *IEEE Trans. on Aerospace and Electronic Systems,* Vol. 26, No. 3, May 1990, pp. 560–568.

Gerlach, K. (1990c), "Equivalent Transformation for a Partially Adaptive Array," *IEEE Trans. on Aerospace and Electronic Systems,* Vol. AES-26, No. 6, November 1990, pp. 1031–1034.

Gerlack, K. (1991), "Adaptive Canceler and Pulse Compressor Interactions," IEEE Trans. on Aerospace and Electronic Systems, Vol. 27, No. 2, March 1991, pp. 331–342.

Giaccari, E., and Penazzi, C.A. (1989), "A Family of Radars for Advanced Systems," *Alta Frequenza,* Vol. LVIII, No. 2, March–April 1989, p. 97–114.

Giraudon, C. (1975), "Adaptive Antenna Processing," US Patent No. 3,876,947, April 8, 1975.

Giusto, R., and DeVincenti, P. (1983), "Phase-only Optimization for the Generation of Wide Deterministic Nulls in the Radiation Pattern of Phased-Arrays," *IEEE Trans. on Antennas and Propagation,* Vol. AP-31, 1983, pp. 814–817.

Goggins, W.B. (1976), "Adaptive Clutter Cancellation and Interference Rejection System for AMTI Radar," US Patent No. 3,995,271, November 30, 1976.

Golub, G.H., and VanLoan, C.F. (1983), *Matrix Computations,* Johns Hopkins University Press, Baltimore, 1983.

Grant, P.M., and Cowan, C.F.N. (1981), "Adaptive Antennas Find Military and Civilian Applications," *Microwave Systems News,* September 1981, pp. 97–107.

Griffiths, H.D., et al. (1985), "Broadband Nulls from a Circular Array," *Fourth Intl. Conf. on Antennas and Propagation* (ICAP '85), Proc. No. 248, Coventry (UK), April 16–19, 1985, pp. 304–308.

Groger, I. (1989), "Antenna Pattern Shaping," *Intl. Conf. on Radar,* April 24–28, 1989, Paris, pp. 283–287.

Groger, I., Sander, W., and Wirth, W.D. (1989), *Contributions to Problems of Active Phased Array Antenna,* FFM Report No. 392, June 1989 (in German).

Gupta, I.J. (1984), "Performance of a Modified Applebaum Adaptive Array," *IEEE Trans. on Aerospace and Electronic Systems,* Vol. AES-20, No. 5, September 1984, pp. 583–593.

Gupta, I.J. (1986a), "SMI Adaptive Antenna Arrays for Weak Interfering Signals," *IEEE Trans. on Antennas and Propagation*, Vol. AP-34, No. 10, October 1986, pp. 1237–1242.

Gupta, I.J. (1986b), "Two-State Feedback Loop Adaptive Arrays for Pulsed Interference Signals," *IEEE Trans. on Aerospace and Electronic Systems*, Vol. AES-22, No. 6, November 1986, pp. 716–724.

Haber, F., Bar-Ness, Y., and Yeh, C.C. (1983), "An Adaptive Interference Canceling Array Utilizing Hybrid Techniques," *IEEE Trans. on Aerospace and Electronic Systems*, Vol. AES-19, No. 5, September 1983, pp. 795–804.

Haber, F., and Zoltowski, M. (1986), "Spatial Spectrum Estimation in a Coherent Signal Environment Using an Array in Motion," *IEEE Trans. on Antennas and Propagation*, Vol. AP-34, No. 3., March 1986, pp. 301–310.

Haber, F., and Jaggard, D.L. (1989), "On the Decorrelation Effects of Spatial Averaging for Direction of Arrival Estimation in a Correlated Signal Scene," Quarterly Progress Report No. 57, Jan-Feb-March 1989, Valley Forge Research Centre, University of Pennsylvania, pp. 49–61.

Haimovich, A.M., and Bar-Ness, Y. (1991), "An Eigenanalysis Interference Canceler," *IEEE Trans. on Signal Processing*, Vol. SP-39, No. 1, January 1991, pp. 66–84.

Hendon, E.J., and Reed, I.S. (1990), "A New CFAR Sidelobe Canceler Algorithm for Radar," *IEEE Trans. on Aerospace and Electronic Systems*, Vol. AES-26, No. 5, September 1990, pp. 792–803.

Hara, Y., Fujii, K., Ishikawa, H., and Kasamaki, K. (1989), "Fast Response Digital Processing Antenna," *Intl. Conf. on Radar*, April 24–28, 1989, Paris, pp. 65*–69*.

Hargrave, P.J., and Ward, C.R. (1985), "New Approach to Adaptive Beamforming," *Electrical Communication*, Vol. 59, No. 3, 1985, pp. 300–305.

Hargrave, P.J., Massey, D.R., and Cawthorne, C.D. (1988), "The Experimental Verification of the Performance of a Systolic Array Adaptive Processor," *Proc. Military Microwave Conf.*, 1988.

Hargrave, P. (1990), "Adaptive Antenna Signal Processing," Tutorial at the Intl. Conf. on Radar, Washington, DC, May 1990.

Harris, F.J. (1978), "On the Use of Windows for Harmonic Analysis with the Discrete Fourier Transform," *Proc. of the IEEE*, Vol. 66, No. 1, January 1978, pp 51–83.

Harvey, D.H., and Wood, T.L. (1980), "Designs for Sidelobe Blanking Systems," *IEEE Intl. Radar Conf.*, April 1980, Washington, DC, pp. 410–416.

Haupt, R.L., O'Brien, M.J., and Shore, R.A. (1984), "Using the Phase Shifters in an Experimental Array for Adaptive Nulling," *Proc. of the 1984 Intl. Symposium on Noise and Clutter Rejection in Radars and Imaging Sensors*, Tokyo, Japan, 1984, pp. 579–584.

Hauptmann, R. (1983), "Circuit Arrangement for Sidelobe Suppression in Radar Apparatuses," U.S. Patent No. 4,367,472, January 4, 1983.

Haykin, S. (1980) *Array Processing: Applications to Radar*, Dowden, Hutchinson and Ross, Inc., 1980.

Haykin, S., Greenlay, T., and Litva, J. (1985). "Performance Evaluation of the Modified FBLP Method for Angle of Arrival Estimation Using Real Radar Multipath Data," *IEE Proc.*, Vol. 132, Pt. F, No. 3, June 1985, pp. 159–174.

Haykin, S. (1986) *Adaptive Filter Theory*, Prentice Hall, Englewood Cliffs, NJ, 1986.

Haykin, S. (1989) *Modern Filters*, Macmillan Publishing Company New York, 1989.

Hicks, D., and Raymond, G. (1978) "Adaptive Arrays and Sidelobe Cancelers for Communications and Radar," *Military Microwave Conf.*, London, October 1978, pp. 366–378.

Horowitz, L.L., et al. (1979a) "Implementation of the Sample Matrix Inversion Algorithm for Controlling Adaptive Antenna Arrays," *1979 Electro Professional Conf.*, New York, April 24–26, 1979, p. 11/5.

Horowitz, L.L., et al. (1979b) "Controlling Adaptive Antenna Arrays with the Sample Matrix Inversion Algorithm," *IEEE Trans. on Aerospace and Electronic Systems*, Vol. AES-15, No. 6, November 1979, pp. 840–847.

Horowitz, L.L., et al. (1980) "Convergence Rate of Extended SMI Algorithm for Narrowband Adaptive Arrays," *IEEE Trans. on Aerospace and Electronic Systems*, Vol. AES-16, No. 5, September 1980, pp. 738–740.

Howard, D.D. (1969), "Sidelobe Cancelling System for Array Type Target Detection," U.S. Patent Office, Patent No. 3,435,453, March 25, 1969.

Howard, R.L., III, Belcher, M.L., and Corey, L.E. (1989), "The Relationship Between Dispersion Loss and Sidelobe Levels in Wideband Phased-Array Radars," *Intl. Conf. on Radar*, April 24–28, 1989, Paris, pp. 93*–98*.

Howells, P.W. (1965), "Intermediate Frequency Sidelobe Canceler," US Patent No. 3,202,990, August 24, 1965.

Howells, P.W. (1976), "Explorations in Fixed and Adaptive Resolution at GE and SURC," *IEEE Trans. Antennas and Propagation*, Special Issue on Adaptive Antennas, Vol. AP-24, 1976, pp. 575–584.

Hsiao, J.K. (1984a), "Normalized Relationship Among Errors and Sidelobe Levels," *Radio Science*, Vol. 19, No. 1, January–February 1984, pp. 292–302.

Hsiao, J.K. (1984b), "Phased Array Sidelobe Level, Gain, Beamwidth and Error Tolerance," *Intl. Conf. on Radar*, May 21–24, 1984, Paris, pp. 304–308.

Hudson, J.E. (1979), "A Kalman-type Algorithm for Adaptive Radar Arrays and Modeling of Non-Stationary Weights," in "Case Studies in Advanced Signal Processing," IEE Conf. Publ. No. 180, 1979.

Hudson, J.E. (1981), *Adaptive Array Principles*, Peter Peregrinus Ltd, London, 1981.

Hudson, J.E. (1985a), "Fast Near-Optimum Resolution of Coherent Sources by Line Arrays," *Fourth Intl. Conf. on Antennas and Propagation* (ICAP '85), Coventry (UK), April 16–19, 1985, pp. 237–240.

Hudson, J.E. (1985b), "Antenna Adaptivity by Direction Finding and Null Steering," *IEE Proc.*, Vol. 132, Pt. H, No. 5, August 1985, pp. 307–311.

Huizing, A.G., and DeValk, G.C. (1987) "High Resolution Parameter Estimation," Report No. FEL 1987-82, Physics and Electronics Laboratory, TNO, The Netherlands.

Hung, E.K.L., and Turner, R.M. (1982) "A Fast Beamforming Algorithm for Large Arrays," *Proc. of Intl. Conf. on Radar*, Radar '82, London, October 18–20, 1982, pp. 92–96.

Hung, E.K.L., and Turner, R.M. (1983) "A Fast Beamforming Algorithm for Large Arrays," *IEEE Trans. on Aerospace and Electronic Systems*, Vol. AES-19, No. 4, July 1983, pp. 598–607.

Hung, E.K.L. (1985), "Adaptive Sidelobe Cancellation by Linear Prediction," *Proc. of Intl. Radar Conf.*, Washington, DC, May 6–9, 1985, pp. 286–291.

Hung, H., and Kaveh, M. (1988), "Focussing Matrices for Coherent Signal-Subspace Processing," *IEEE Trans. on Acoustics, Speech, and Signal Processing*, Vol. ASSP-36, No. 8, August 1988, pp. 1272–1281.

Hung, H., and Kaveh, M. (1990), "Coherent Wide-Band ESPRIT Method for Direction of Arrival Estimation of Multiple Wide-Band Source," *IEEE Trans. on Acoustics, Speech, and Signal Processing*, Vol. 38, No. 2, February 1990, pp. 354–356.

Inatsune, S., Fujisaka, T., Oh-Hashi, Y., Kondo, M., Ito, R., and Takeuchi, N. (1989), "Digital Beamforming Using a Planar Array," *Proc. of the 1989 Intl. Symp. on Noise and Clutter Rejection in Radars and Imaging Sensors*, November 14–16, 1989, Kyoto, Japan, pp. 377–382.

Indukumar, K.C., and Reddy, V.U. (1990), "Broad-Band DOA Estimation and Beamforming in Multipath Environment," *Proc. of IEEE Intl. Radar Conf.*, May 1990, Washington D.C., pp. 532–537.

Iovino, D. (1987), "Performance of a Sidelobe Blanking System," Selenia Technical Report, TR-E87143 In, December 17, 1987.

Jablon, N.K. (1986), "Steady State Analysis of the Generalized Sidelobe Canceler by Adaptive Noise Cancelling Techniques," *IEEE Trans. Antennas and Propagation*, Vol. AP-34, No. 3, March 1986, pp. 330–337.

Jackson, M.C., and Matthewson, P. (1986), "Digital Processing of Bandpass Signals," *GEC Journal of Research*, Vol. 4, No. 1, 1986, pp. 32–41.

Jelonek, T.M., and Reilly, J.P. (1990), "Maximum Likelihood Estimation for Direction of Arrival Using A Nonlinear Optimising Neural Network," *Proc. of Intl. Joint Conference on Neural Networks*, San Diego, 1990, pp. I-253–I-257.

Johnson, M.A., and Stoner, D.C. (1978), "ECCM from the Radar Designers View Point," *Microwave Journal*, Vol. 21, No. 3, March 1978, pp. 59–63.

Johnson, J.R., Fenn, A.J., Aumann, H.M., and Willwerth, F.G. (1991), "An Experimental Adaptive Nulling Receiver Utilizing the Sample Matrix Inversion Algorithm with Channel Equalization," *IEEE Trans. on Microwave Theory and Techniques*, Vol. 39, No. 5, May 1991, pp. 798–808.

Johnson, R.C., and Jasik, H., Eds. (1984), *Antenna Engineering Handbook*, 2nd Ed., McGraw-Hill Book Company, New York, 1984.

Johnston, S.L., Editor (1979), *"Radar Electronic Counter-Countermeasures,"* Artech House, Norwood, MA, 1979.

Kaplan, P.D. (1986), "Predicting Antenna Sidelobe Performance," *Microwave Journal*, September 1986, pp. 201–206.

Kay, S.M., and Marple, S.L. (1981), "Spectrum Analysis. A Modern Perspective," *Proc. of the IEEE*, Vol. 69, No. 11, November 1981, pp. 1380–1419.

Kay, S.M. (1988), *Modern Spectral Estimation: Theory and Application*, Prentice Hall, Englewood Cliffs, NJ, 1988.

Kelly, E.J. (1986), "An Adaptive Detection Algorithm," *IEEE Trans. on Aerospace and Electronic Systems*, Vol. AES-22, No. 1, March 1986, pp. 115–127.

Kelly, E.J. (1989), "Performance of an Adaptive Detection Algorithm; Rejection of Unwanted Signals," *IEEE Trans. on Aerospace and Electronic Systems*, Vol. AES-25, No. 2, March 1989, pp. 122–133.

Kennedy, P.D. (1978), "Main Lobe Signal Canceler in a Null Steering Array Antenna," US Patent No. 4,129,873, December 12, 1978.

Khanna, R., and Madan, B.B. (1983), "Adaptive Beam Forming Using a Cascade Configuration," *IEEE Trans. on Acoustics, Speech, and Signal Processing*, Vol. ASSP-31, No. 4, August 1983, pp. 940–945.

Kikuma, N., and Takao, K. (1986), "Effect of Initial Values of Adaptive Arrays," *IEEE Trans. on Aerospace and Electronic Systems*, Vol. AES-22, No. 6, November 1986, pp. 688–694.

Klema, V.C., and Laub, H.J. (1980), "The Singular Value Decomposition: Its Computation and Some Applications," *IEEE Trans. on Automatic Control*, Vol. AC-25, No. 2, April 1980, pp. 164–178.

Klemm, R. (1975), "Suppression of Jammers by Multiple Beam Signal Processing," *Proc. of IEEE Intl. Radar Conf.*, Washington, DC, 1975, pp. 176–180.

Klemm, R. (1977), "Design Considerations for a Digital Adaptive Least Squares Filter for Real-Time Processing," *Proc. of the Intl. Radar Conf.*, Radar '77, London, 1977, pp. 271–274.

Knetsch, H.D. (1977), "Adaptive Spatial Filtering for Cancelling Jamming Signals," *Siemens Forsch*, Vol. 6, No. 55, 1977, pp. 300–307.

Ko, C.C. (1980), "Power Inversion Array in a Rotating Source Environment," *IEEE Trans. on Aerospace and Electronic Systems*, Vol. AES-16, No. 6, November 1980, pp. 755–762.

Ko, C.C. (1986), "Adaptive Arrays Processing Using the Davies Beamformer," *IEE Proc.*, Vol. 133, Pt. H, No. 6, December 1986, pp. 467–483.

Ko, C.C. (1987), "Tracking Performance of a Broadband Tapped Delay Line Adaptive Array Using LMS Algorithm," *IEE Proc.*, Vol. 134, Pt. F, No. 3, June 1987, pp. 295–304.

Ko, C.C. (1989a), "Broadband Power Inversion Array with Maximally Flat Response at Null Directions," *IEE Proc.*, Vol. 136, Pt. F, No. 4, August 1989, pp. 161–167.

Ko, C.C. (1989b), "Nulling Performance of a Broadband Linearly Constrained Array," *Signal Processing*, Vol. 17, No. 4, August 1989, pp. 313–320.

Ko, C.C., Quek, T.S., and Ser, W. (1989), "A Signal Fast Algorithm for Null Steering Adaptive Arrays," *Signal Processing,* Vol. 18, No. 1, September 1989, pp. 55–62.

Ko, C.C. (1990), "Fast Null-Steering Algorithm for Broadband Power Inversion Array," *IEE Proc.,* Vol. 137, Pt. F, No. 5, October 1990, pp. 377–383.

Kretschmer, F.F. (1975), "Effects of Cascading Sidelobe Canceler Loops," *Proc. of IEEE Intl. Radar Conf.,* Washington, DC, 1975, pp. 181–185.

Kretschmer, F.F., and Lewis, B.L. (1978a), "A Digital Open-Loop Adaptive Processor," *IEEE Trans. on Aerospace and Electronic Systems,* Vol. AES-14, No. 1, January 1978, pp. 165–171.

Kretschmer, F.F., and Lewis, B.L. (1978b), "An Improved Algorithm for Adaptive Processing," *IEEE Trans. on Aerospace and Electronic Systems,* Vol. AES-14, No. 1, January 1978, pp. 172–177.

Krucker, K. (1981), "On an Application of Kalman Filtering in an Adaptive Spatial Filter with Rapid Convergence," *Intl. Conf. on Digital Signal Processing,* Florence, September 2–5, 1981, pp. 878–885.

Krucker, K. (1983), "Rapid Interference Suppression Using a Kalman Filter Technique," *IEE Proc.,* Vol. 130, Pts. F and H, No. 1, February 1983, pp. 36–40.

Krucker, K. (1986), "On a Specific Performance Limitation of Partially Adaptive Antennas," *IEEE Trans. on Aerospace and Electronic Systems,* Vol. AES-22, No. 4, July 1986, pp. 466–468.

Kumaresan, R., and Tufts, D.W. (1983), "Estimating the Angles of Arrival of Multiple Plane Waves," *IEEE Trans. on Aerospace and Electronic Systems,* Vol. AES-19, No. 1, January 1983, pp. 134–139.

Kumaresan, R., and Shaw, A.K. (1985), "High Resolution Bearing Estimation without Eigen Decomposition," *Proc. of Intl. Conf. on Acoustic, Speech, and Signal Processing, ICASSP-85,* pp. 576–579.

Kung, H.T., and Gentleman, W.M. (1981), "Matrix Triangulation by Systolic Arrays," *Proc. SPIE,* No. 298, *Real Time Signal Processing IV,* 1981.

Kung, S.Y. (1985), "VLSI Array Processors," *IEEE ASSP Magazine,* July 1985, pp. 4–22.

Kurth, R.R. (1974), "Optimization of Array Performance Subject to Multiple Power Pattern Constraints," *IEEE Trans. on Antennas and Propagation,* January 1974, pp. 103–105.

Kwok, P.C.K., and Brandon, P.S. (1979a), "Eigenvalues of Noise Covariance Matrix of Linear Array in the Presence of two Directional Interferences," *Electronic Letters,* January 18, 1979, Vol. 15, No. 2, pp. 50–51.

Kwok, P.C.K., and Brandon, P.S. (1979b), "Optimal Radiation Pattern of an Array in the Presence of Two Directional Interferences," *Electronics Letters,* April 26, 1979, Vol. 15, No. 9, pp. 251–252.

Lacomme, P. (1984), "Sidelobe Narrowband Interference Cancellation," *Proc. of the 1984 Intl. Symposium of Noise and Clutter Rejection in Radars and Imaging Sensors,* Tokyo, Japan, 1984, pp. 563–568.

Lank, G.W., and Brennan, L.E. (1972), "Effect of Single-bit Digitization in Adaptive Array Control Loop," *IEEE Trans. on Aerospace and Electronic Systems,* Vol. AES-8, July 1972, pp. 547–549.

Lee, D.D., Kashyap, R.L., and Madan, R.N. (1990), "Robust Decentralized Direction of Arrival Estimation in Contaminated Noise," *IEEE Trans. on Acoustics, Speech, and Signal Processing,* Vol. 38, No. 3, March 1990, pp. 496–505.

Lee, J.H., and Wu, J.F. (1989), "Adaptive Beam Forming Without Signal Cancellation in the Presence of Coherent Jammers," *IEE Proc.,* Vol. 136, Pt. F, No. 4, August 1989, pp. 169–173.

Lewis, B.L., and Olin, D. (1976), "Sidelobe Canceler System," US Patent No. 3,933,153, February 10, 1976.

Lewis, B.L., and Evans, J.B. (1983), "A New Technique for Reducing Radar Response to Signals Entering Antenna Sidelobes," *IEEE Trans. on Antennas and Propagation,* Vol. AP-31, No. 6, November 1983, pp. 993–996.

Lewis, B.L., and Kretschmer, F. (1984), "Cascaded Adaptive Loops," U.S. Patent No. 4,439,770, March 27, 1984.
Lewis, B.L., Kretschmer, F.F., and Shelton, W.W. (1986), *Aspects of Radar Signal Processing,* Artech House, Norwood MA 1986.
Li, Fu, and Vaccaro, R.J. (1990), "Analysis of Min-Norm and MUSIC with Arbitrary Array Geometry," *IEEE Trans. on Aerospace and Electronic Systems,* Vol. AES-25, No. 6, November 1990, pp. 976–985.
Lin, F.L.C., and Kretschmer, F.F. (1990), "Angle Measurement in the Presence of Mainbeam Interference," *Proc. of Intl. Radar Conf.,* Washington D.C., May 1990, pp. 444–450.
Lin, Heng-Cheng (1982), "Spatial Correlations in Adaptive Arrays," *IEEE Trans. on Antennas and Propagation,* Vol. AP-30 No. 2, March 1982, pp. 212–223.
Linebarger, D.A., and Johnson, D.H. (1990), "The Effect of Spatial Averaging on Spatial Correlation Matrices in the Presence of Coherent Signals," *IEEE Trans. on Acoustics, Speech, and Signal Processing,* Vol. 38, No. 5, May 1990, pp. 880–884.
Ling, F., Manolakis, D., and Proakis, J.G. (1986), "A Recursive Modified Gram-Schmidt Algorithm for Least Squares Estimation," *IEEE Trans. on Acoustic, Speech, and Signal Processing,* Vol. ASSP-34, August 1986, pp. 829–836.
Loomis, J. (1990), "Digital Beamforming," Tutorial, IEEE 1990 *Intl. Radar Conf.,* Washington, DC, May 7–10,1990.
Lu Zhongliang, et al. (1986), "Applications of Digital Open-Loop Sidelobe Canceler to High-Rotation-Rate Radar," *Intl. Radar Conf.,* Nanjing, China, November 4–7, 1986, pp. 624–629.
Madden, J.N. (1987), "The Adaptive Suppression of Interference in HF Ground Wave Radar," *Proc. of IEE Intl. Radar Conf.,* Radar '87, London, October 19–21, 1987, pp. 98–102.
Maisel, L. (1968), "Performance of Sidelobe Blanking Systems," *IEEE Trans. on Aerospace and Electronic Systems,* Vol. AES-4, No. 2, March 1968, pp. 174–180.
Marple, S.L. (1982), "Frequency Resolution of Fourier and Maximum Entropy Spectral Estimates," *Geophysics,* Vol. 47, No. 9, September 1982, pp. 1303–1307.
Marple, S.L. (1985), "Spectral Estimation with Applications," in *Modern Signal Processing,* T. Kailath, Ed. Hemisphere Publishing, 1985, pp. 129–168.
Marple, S.L. (1987), *Digital Spectral Analysis with Applications,* Prentice Hall, Signal Processing Series, Englewood Cliffs, NJ, 1987.
Marr, J.D. (1986), "A Selected Bibliography on Adaptive Antenna Arrays," *IEEE Trans. on Aerospace and Electronic Systems,* Vol. AES-22, No. 6, November 1986, pp. 781–798.
Martin, G.P., and Mikenas, V.A. (1979), "Comparison of Analog and Digital Control Techniques for Adaptive Array Systems," *1979 Electro Professional Conf.,* New York, April 24–26, 1979, p. 11/3.
Masak, R.J. (1976), "Interference Rejection System for Multi-Beam Antenna," U.S. Patent No. 3,981,014, September 14, 1976.
Masak, R.J. (1979), "Multiplexing of Multiple Loop Sidelobe Cancelers," U.S. Patent No. 4,177,464, December 4, 1979.
Masak, R.J. (1983), "Combined Sidelobe Canceler and Frequency Selective Limiter," U.S. Patent No. 4,370,655, January 25, 1983.
Masenten, W.K. (1979), "Adaptive Signal Processing," IEE Seminar "Case Studies in Advanced Signal Processing," Peebles (Scotland), September 1979.
Mayhan, J.T. (1978), "Adaptive Nulling with Multiple-Beam Antennas," *IEEE Trans. on Antennas and Propagation,* Vol. AP-26, No. 2, March 1978, pp. 267–273.
Mayhan, J.T., Simmons, A.J., and Cummings, W.C. (1980), "Wideband Nulling with Adaptive Arrays Using Tapped Delay Lines," *1980 Intl. Symposium Digest Antenna and Propagation,* Quebec, Canada, 1980, pp. 110–113.
McQueen, J.G. (1977), "Adaptive Cancellation Arrangement," Patent Specification No. 1599035, UK Patent Office, London, March 31, 1977.

McWhirter, J.G. (1983a), "Recursive Least-Squares Minimization Using a Systolic Array," *Proc. SPIE*, Vol. 431 on Real-Time Signal Processing VI, 1983, pp. 105–112.

McWhirter, J.G. (1983b), "Systolic Array for Recursive Least-Squares Minimization," *Electronic Letters*, September 1, 1983, Vol. 19, No. 18, pp. 729–730.

McWhirter, J.G. (1989), "Parallel Signal Processing," *Intl. Conf. on Radar*, April 24–28, 1989, Paris, pp. 1*–10*.

Mendelovicz, E., and Oestreich, E.T. (1979), "Phase-only Adaptive Nulling with Discrete Values," 1979 *Intl. Symp. Digest Antennas and Propagation*, Vol. I, Univ. of Washington, Seattle, pp. 193–197.

Mesiwala, H., and Widrow, B. (1979), "Compression of Eigenvalue Range by Using a Surplus of Adaptive Antenna Elements," *1979 Electro Professional Conf.*, New York, April 24–26, 1979, p. 11/1.

Monzingo, R.A., and Miller, T.W. (1980), *Introduction to Adaptive Arrays*, John Wiley and Sons, New York, 1980.

Morgan, D.R. (1978), "Partially Adaptive Array Techniques," *IEEE Trans. on Antennas and Propagation*, Vol. AP-26, No. 6, November 1978, pp. 823–833.

Morgan, D.R. (1980), "Analysis of Multiple Correlation Cancellation Loops with a Filter in the Auxiliary Path," *IEEE Trans. on Acoustics, Speech, and Signal Processing*, Vol. ASSP-28, No. 4, August 1980, pp. 454–467.

Morooka, T., and Kawabata, K. (1980), "Frequency Characteristics of an Adaptive MSLC," *1980 Intl. Symp. Digest Antennas and Propagation*, Quebec, Canada, 1980, pp. 122–125.

Nickel, U.R.O. (1982), "Superresolution Using an Active Antenna Array," *Proc. IEE Intl. Radar Conf.*, London, 1982, pp. 87–91.

Nickel, U.R.O. (1987a), "Angle Estimation with Adaptive Arrays and its Relation to Superresolution," *IEE Proc.*, Vol. 134, Pt. H, No. 1, February 1987, pp. 77–82.

Nickel, U.R.O. (1987b), "Angular Superresolution with Phased Array Radar: A Review of Algorithms and Operational Constraints," *IEE Proc.*, Vol. 134, Pt. F, No. 1, February 1987, pp. 53–59.

Nickel, U.R.O. (1988a), "Algebraic Formulation of Kumaresan-Tufts Superresolution Method, Showing Relation to ME and MUSIC Methods," *IEE Proc.*, Vol. 135, Pt. F, No. 1, February 1988, pp. 7–10.

Nickel, U.R.O. (1988b), "Application of Array Signal Processing to Phased Array Radar," Signal Processing IV: Theories and Applications, J.L. Lacoume *et al.* (Editors), Elsevier Science B.V. (North Holland), 1988, pp. 467–474.

Nickel, U.R.O. (1989a), "Angular Superresolution by Antenna Array Processing," *Intl. Conf. on Radar*, April 24–28, 1989, Paris, pp. 48*–58*.

Nickel, U.R.O. (1989b), "Subarray Configurations for Interference Suppression with Phased Array Radar," *Intl. Conf. on Radar*, April 24–28, 1989, Paris, pp. 82*–86*.

Nickel, U.R.O. (1989c), "Application of Superresolution Methods for Airborne Radar," Agard Conf. Proc. No. 459, *High Resolution Air and Spaceborne Radar*, The Hague, May 8–12, 1989, pp. 7-1–7-6.

Nickel, U.R.O. (1990), "Comparison of a Non-Adaptive Low-Sidelobe Array Antenna with an Adaptive Phased Array Antenna," *IEEE 1990 Intl. Radar Conf.*, May 7–10, 1990, Washington, DC, pp. 487–490.

Nicolau, E., and Zaharia, D. (1989), *Adaptive Arrays*, Elsevier, Studies in Electrical and Electronic Engineering, Vol. 35, 1989.

Nitzberg, R. (1973), "Effects of Errors in Adaptive Weights," *IEEE Trans. on Aerospace and Electronic Systems*, Vol. AES-12, May 1976, pp. 369–373.

Nitzberg, R. (1981), "Canceler Performance Degradation Due to Estimation Noise," *IEEE Trans. on Aerospace and Electronic Systems*, Vol. AES-17, No. 5, September 1981, pp. 684–692.

Nitzberg, R. (1984), "Detection Loss of the Sample Matrix Inversion Technique," *IEEE Trans. on Aerospace and Electronic Systems*, Vol. AES-20, No. 6, November 1984, pp. 824–827.

Nitzberg, R. (1985), "Application of the Normalized LMS Algorithm to MSLC," *IEEE Trans. on Aerospace and Electronic Systems,* Vol. AES-21, No. 1, January 1985, pp. 79–91.

Nitzberg, R. (1986), "Normalized LMS Algorithm Degradation Due to Estimation Noise," *IEEE Trans. on Aerospace and Electronic Systems,* Vol. AES-22, No. 6, November 1986, pp. 740–758.

Nitzberg, R. (1989), "Effect of Interchannel Mismatches Upon Adaptive Array Cancellation," *Proc. of 1989 IEEE National Radar Conf.,* March 29–30, 1989, Dallas, pp. 119–124.

Oestrich, E.T., and Mendelovicz, E. (1979), "Phase Only Adaptive Nulling with Discrete Values," *European Microwave Conf.,* Brighton (UK), September 17–20, 1979, pp. 164–168.

Ogawa, Y., Ohmiya, M., and Itoh, R. (1983), "An Analog Open-Loop Adaptive-Array Antenna System," *IEEE Trans. on Aerospace and Electronic Systems,* Vol. AES-19, No. 1, January 1983, pp. 89–101.

Old, J.C. (1984a), "Multiple Open-Loop Interference Canceler for a Rotating Search Radar," *IEE Proc.,* Vol. 132, Pt. F, No. 2, April 1984, pp. 203–207.

Old, J.C. (1984b), "Sidelobe Canceler for Radar Systems," U.S. Patent No. 4,434,424, February 28, 1984.

Omuro, T., Tachibana, Y., and Kondo, M. (1984), "Sidelobe Canceler Response for Pulse Interference," *IEEE Intl. Conf. on Acoustic, Speech, and Signal Processing,* ICASSP-1984, San Diego, March 19–21, 1984, pp. 47.7.1–47.7.4.

O'Sullivan, M.R. (1987), "A Comparison of Sidelobe Blanking Systems," *Proc. of IEE Intl. Radar Conf.,* Radar '87, London, 19–21 October 1987, pp. 345–349.

O'Sullivan, M.R., Harrington, K.M., and Nevin, R.L. (1989), "Sidelobe Blanking Systems Performance for Swerling Target," *Intl. Conf. on Radar,* April 24–28, 1989, Paris, pp. 569*–573*.

Ott, R.H. (1990), "The Application of Howells-Applebaum Adaptive Control Theory To Continuously Variable Time-Domain Reflector Antenna Response," *IEEE Trans. on Antennas and Propagation,* Vol. 38, No. 3, March 1990, pp. 359–368.

Ottersten, B., and Kailath, T. (1990), "Direction of Arrival Estimation for Wide-Band Signals Using the ESPRIT Algorithm," *IEEE Trans. on Acoustics, Speech, and Signal Processing,* Vol. ASSP-38, No. 2, February 1990, pp. 317–327.

Ouibrahim, H. (1989), "Prony, Pisarenko, and the Matrix Pencil: A Unified Approach," *IEEE Trans. on Acoustics, Speech, and Signal Processing,* Vol. ASSP-37, No. 1, January 1989, pp. 133–134.

Palumbo, B. (1989), "Some New Ideas in Radar Antenna Technology," *Microwave Journal,* Vol. 32, 1989, pp. 95–104.

Patton, W.T. (1980), "Low Sidelobe Antennas for Tactical Radars," *Record IEEE Intl. Radar Conf.,* April 1980, Washington, DC, pp. 243–247.

Paulray, A., and Kailath, T. (1985), "On Beamforming in Presence of Multipath," *Proc. of Intl. Conf. on Acoustic, Speech, and Signal Processing, ICASSP,* 1985, pp. 564–567.

Paulray, A., Roy, R., and Kailath, T. (1986), "A Subspace Rotation Approach to Signal Parameter Estimation," *Proc. IEEE,* Vol. 74, No. 7, July 1986, pp. 1044–1045.

Pearson, A., Barton, P., Waddoup, W.D., and Sherwell, R.J. (1982), "An X-Band Array Signal Processing Radar for Tracking Targets at Low Elevation Angles," *Proc. of IEE Intl. Conf. on Radar,* Radar '82, London 1982, pp. 439–443.

Pei, S.C., Yeh, C.C., and Chiu, S.C. (1988), "Modified Spatial Smoothing for Coherent Jammer Suppression without Signal Cancellation," *IEEE Trans. on Acoustics, Speech, and Signal Processing,* Vol. ASSP-36, No. 3, March 1988, pp. 412–414.

Picardi, G., and Spina, G. (1985), "The Superresolution Applications in Tracking Radar," *Proc. of Intl. Radar Conf.,* Washington, DC, May 1985, pp. 95–100.

Picardi, G. (1988), *"Elaborazione del segnale Radar Metodologie ed Applicazioni"* (in Italian), Franco Angeli Editore, 1988.

Pourally, J.L., DeReffye, J., and Legendre, C. (1982), "Spatial Digital Processing Application to Radar Antennas," *Intl. Conf. on Acoustics, Speech, and Signal Processing,* ICASSP, 1982, pp. 361–370.

Powell, N.F., Lee, E., and Guillen, F.J. (1982), "Hybrid Adaptive Sidelobe Canceling System," U.S. Patent No. 4,313,116, January 26, 1982.

Prasad, S., and Charan, R. (1984), "On the Constrained Synthesis of Array Patterns with Applications to Circular and Arc Arrays," *IEEE Trans. on Antennas and Propagation,* Vol. AP-32, No. 7, July 1984, pp. 725–730.

Rader, C.M., and Steinhardt, A.O. (1986), "Hyperbolic Householder Transformations," *IEEE Trans. on Acoustics, Speech, and Signal Processing,* Vol. ASSP-34, No. 6, December 1986, pp. 1589–1602.

Rader, C.M., Allen, D.L., Glasco, D.B. and Woodward, C.E. (1990), "MUSE—A Systolic Array for Adaptive Nulling with Sixty-four Degrees of Freedom, Using Givens Transformations and Wafer-Scale Integration," MIT Lincoln Laboratory, Technical Report 886, May 18, 1990.

Rafik, T.A., and Griffiths, J.W.R. (1988), "Feasibility Study of Implementing High Resolution DF Techniques in Real Time Using Transputers," in Signal Processing IV: Theories and Applications, J.L. Lacoume *et al.,* Editors, Elsevier Science B.V. (North-Holland), 1988, pp. 1377–1380.

Rao, B.D., and Hari, K.V.S. (1990), "Effect of Spatial Smoothing on the Performance of MUSIC and the Minimum Norm Method," *IEE Proc.,* Vol. 137, Pt. F, No. 6, December 1990, pp. 449–458.

Reddi, S.S. (1979), "Multiple Source Location. A Digital Approach," *IEEE Trans. on Aerospace and Electronic Systems,* Vol. AES-15, 1979, pp. 95–105.

Reddy, V.U., Paulray A., and Kailath, T. (1987), "Performance Analysis of the Optimum Beamformer in the Presence of Correlated Sources and its Behaviour Under Spatial Smoothing," *IEEE Trans. on Acoustics, Speech, and Signal Processing,* Vol. ASSP-35, No. 7, July 1987, pp. 927–936.

Reed, I.S., Mallet, J.D., and Brennan, L.W. (1974), "Rapid Convergence Rate in Adaptive Arrays," *IEEE Trans. on Aerospace and Electronic Systems,* Vol. AES-10, November 1974, pp. 853–863.

Rickman, J.D. (1981), "Results from a Four-Element Adaptive Antenna Experiment," *IEEE Trans. on Aerospace and Electronic Systems,* Vol. AES-17, No. 1, January 1981, pp. 35–47.

Ries, G., and Krucker, K. (1976), "Problems of Adaptive Sidelobe Suppression," AGARD Conf., June 1976, The Hague, Conf. Proc. No. 197 on "New Devices Techniques and Systems in Radar," pp. 23-1–23-12.

Rife, D.L., and Boorstyn, R.R. (1976), "Multiple Tone Parameter Estimation from Discrete-Time Observations," *The Bell System Technical Journal,* Vol. 55, No. 9, November 1976, pp. 1389–1409.

Rife, D.C. and Boorstyn, R.R. (1978), "Single Tone Parameter Estimation from Discrete-Time Observations," *IEEE Trans. on Information Theory,* Vol. IT-20, September 1978, pp. 591–598.

Rodgers, W.E., and Compton, R.T. (1979), "Adaptive Array Bandwidth with Tapped Delay-line Processing," *IEEE Trans. on Aerospace and Electronic Systems,* Vol. AES-15, No. 1, January 1979, pp. 21–28.

Roy, R., Paulray, A., and Kailath, T. (1985), "Estimation of Signal Parameters via Rotational Invariance Techniques—ESPRIT," *Proc. of 19th Asilomar Conf. on Circuits, Systems, and Computing,* Asilomar, CA, November 1985.

Roy, R., Paulray, A., and Kailath, T. (1986), "Comparative Performance of ESPRIT and MUSIC for Direction of Arrival Estimation," *Proc. of 20th Asilomar Conf. on Circuits, Systems, and Computing,* Asilomar, CA, November, 1986.

Ruze, J. (1966), "Antenna Tolerance Theory—A Review," *Proc. IEEE,* Vol. 54, No. 4, April 1966, pp. 633–640.

Schleher, D.C. (1986), *Introduction to Electronic Warfare,* Artech House, Norwood, MA, 1986.
Schmidt, R.O. (1986), "Multiple Emitter Location and Signal Parameter Estimation," *IEEE Trans. on Antennas and Propagation,* Vol. AP-34, N. 3, March 1986, pp. 276–280.
Schmidt, R.O., and Franks, R.E. (1986), "Multiple Source DF Signal Processing: An Experimental System," *IEEE Trans. on Antennas and Propagation,* Vol. AP-34, N. 3, March 1986, pp. 281–290.
Schrank, H.E. (1988), "Low-Sidelobe Phased Array and Reflector Antennas," Ch. 6 of *Aspects of Modern Radar,* edited by E. Brookner, Artech House, Norwood, MA, 1988.
Schreiber, R.J., and Kuekes P.J. (1985), "Systolic Linear Algebra Machines in Digital Signal Processing," in *VLSI and Modern Signal Processing,* edited by S.Y. Kung, H.J. Withehouse, and T. Kailath, Prentice Hall, Englewood Cliffs, NJ, 1985, pp. 389–405.
Shan, T.J., and Kailath, T. (1985), "Adaptive Beamforming for Coherent Signals and Interference," *IEEE Trans. on Acoustics, Speech, and Signal Processing,* Vol. ASSP-33, No. 3, 1985, pp. 527–536.
Shan, T., Wax, M., and Kailath, T. (1985), "On Signal Smoothing for Direction of Arrival Estimation of Coherent Signals," *IEEE Trans. on Acoustics, Speech, and Signal Processing,* Vol. ASSP-33, No. 4, August 1985, pp. 806–811.
Shan, T.J., Paulray, A., and Kailath, T. (1987), "On Smoothed Rank Profile Tests in Eigenstructure Methods for Direction of Arrival Estimation," *IEEE Trans. on Acoustics, Speech, and Signal Processing,* Vol. ASSP-35, No. 10, October 1987, pp. 1377–1385.
Sharman, K.C., and Durrani, T.S. (1983), "Spatial Lattice Filter for High-Resolution Spectral Analysis of Array Data," *IEE Proc.,* Vol. 130, Pt. F, N. 3, April 1983, pp. 279–287.
Shenoy, R.P. (1989), "Active Aperture Phased Arrays," *Alta Frequenza,* Vol. LVIII, No. 2, March–April 1989, pp. 137–149.
Sherman, J.W. (1970), "Aperture-Antenna Analysis," Ch. 9 in *Radar Handbook,* 1st Ed., edited M.I. Skolnik, McGraw-Hill Book Company, New York, 1970.
Sielman, P.F. (1979), "Digital Processing Adaptive Arrays," *1979 Electro Professional Conf.,* New York, April 24–26, 1979, p. 11/4.
Silverstein, S.D., and Pimbley, J.M. (1988), "Robust Spectral Estimation: Autocorrelation Based Minimum Free Energy Method," *Proc. of 22nd Asilomar Conference on Signals, Systems and Computer,* Asilomar, CA, 1988.
Skolnik, M.I. (1990), *Radar Handbook,* 2nd Ed., McGraw-Hill, New York, 1990.
Soule, H.H., and Jureller, J.F. (1976), "Sidelobe Canceler with Programmable Correlation Signal Weighting," U.S. Patent No 488,395, January 27, 1976.
"Special Issue on Adaptive Systems" (IEEE, 1976a), *Proc. IEEE,* Vol. 64, No. 8, 1976, pp. 1123–1240.
"Special Issue on Adaptive Antennas" (IEEE, 1976b), *IEEE Trans. on Antennas and Propagation,* Vol. AP-24, No. 5, September 1976, pp. 573–764.
"Special Issue on Phased Arrays" (IEE, 1980), *IEE Proceedings,* Vol. 127, Pt. F, No. 4, August 1980, pp. 241–336.
"Special Issue on Spectral Estimation" (IEEE, 1982), *Proc. IEEE,* Vol. 70, No. 9, 1982, pp. 883–1125.
"Special Issue on Adaptive Arrays" (IEE, 1983), *IEE Proc. Communications, Radar and Signal Processing,* London, Vol. 130, Pts. F and H, No. 1, 1983, pp. 1–151.
"Special Issue on Adaptive Processing Antenna Systems" (IEEE, 1986), *IEEE Trans. on Antennas and Propagation,* Vol. AP-34, No. 3, March 1986, pp. 273–462.
Starkey, P.G. (1986), "Experimental Performance of High Resolution Direction-Finding Algorithms," *Proc. Military Microwave Conf.,* London, 1986.

Starkey, P.G. (1987), "Direction-Finding with Linear Phased Arrays Using Eigen-Analysis Techniques," *GEC Journal of Research*, Vol. 5, No. 4, 1987, pp. 193–207.

Steinhardt, A.O. (1988), "Householder Transforms in Signal Processing," *IEEE ASSP Magazine*, Vol. 5, No. 3, July 1988, pp. 4–12.

Steyskal, H. (1982), "Synthesis of Antenna Pattern with Prescribed Nulls," *IEEE Trans. on Antennas and Propagation*, Vol. AP-30, No. 2, March 1982, pp. 273–279.

Steyskal, H. (1987), "Digital Beamforming Antennas: An Introduction," *Microwave Journal*, Vol. 30, January 1987, pp. 107–124.

Steyskal, H., and Rose, J.F. (1989), "Digital Beamforming for Radar Systems," *Microwave Journal*, Vol. 32, January 1989, pp. 121–136.

Stoica, P., and Nehorai, A. (1989), "MUSIC, Maximum Likelihood, and Cramer-Rao Bound," *IEEE Trans. on Acoustics, Speech, and Signal Processing*, Vol. ASSP-37, No. 5, 1989, pp. 720–741.

Stoica, P., and Sharman, K.C. (1990), "Maximum Likelihood Methods for Direction-of-Arrival Estimation," *IEEE Trans. on Acoustics, Speech, and Signal Processing*, Vol. 38, No. 7, July 1990, pp. 1132–1143.

Stoica, P., and Nehorai, A. (1990), "MUSIC, Maximum Likelihood, and Cramer-Rao Bound: Further Results and Comparisons," *IEEE Trans. on Acoustics, Speech, and Signal Processing*, Vol. SP-38, No. 12, December 1990, pp. 2140–2150.

Stone, D. (1983), "Open-Loop Adaptivity for Rotating Antennas," *IEE Proc.*, Vol. 130, Pt. F, No. 1, February 1983, pp. 114–117.

Strappaveccia, S. (1987), "Spatial Jammer Suppression by Means of an Automatic Frequency Selection Device" *Proc. of IEE Intl. Radar Conf.*, Radar '87, London, October 19–21, 1987, pp. 582–587.

Strobach, P. (1991), "New Forms of Levinson and Schur Algorithms," *IEEE Signal Processing Magazine*, January 1991, pp. 12–36.

Sung, S., Ham, F.M., and Shelton, W. (1990), "A New Robust Neural Network Method for Coherent Interference Rejection in Adaptive Array Systems," *Proc. of Intl. Joint Conf. on Neural Networks*, San Diego, 1990, pp. II-119–II-124.

Takao, K., Fujita, M., and Nishi, T. (1976), "An Adaptive Antenna Array Under Directional Constraint," *IEEE Trans. Antennas and Propagation*, Vol. AP-24, No. 5, September 1976, pp. 662–669.

Takao, K., and Komiyama, K. (1980), "An Adaptive Antenna for Rejection of Wideband Interference," *IEEE Trans. on Aerospace and Electronic Systems*, Vol. AES-16, No. 4, July 1980, pp. 452–459.

Takao, R., Ito, Y., and Komiyama, K. (1980), "The Behaviour of an Adaptive Array to Wideband Interference," *1980 Intl. Symposium Digest Antennas and Propagation*, Quebec, Canada, 1980, pp. 118–121.

Takao, K., and Uchida, K. (1989), "Beamspace Partially Adaptive Antennas," *IEE Proc.*, Vol. 136, Pt. H, No. 6, December 1989, pp. 439–444.

Tanaka, Y., et al. (1984), "Field Verification of Adaptive Sidelobe Canceler" *Proc. 1984 Intl. Symposium on Noise and Clutter Rejection in Radar and Imaging Sensors*, Tokyo, Japan, 1984, pp. 569–572.

Tang, C.F.T., Liu, K.J.R., and Tretter, S.A. (1991), "On Systolic Arrays for Recursive Complex Householder Transformations with Applications to Array Processing," *Proc. of IEEE Intl. Conf. on Acoustics, Speech, and Signal Processing, ICASSP-91*, Toronto, Canada, May 1991.

Taylor, T.T. (1955), "Design of Line-Source Antennas for Narrow Beamwidth and Low Sidelobes," *IRE Trans. on Antennas and Propagation*, Vol. AP-3, January 1955, pp. 16–28.

Teitelbaum, R. (1991), "A Flexible Processor for a Digital Adaptive Array Radar," *Proc. of the 1991 National Radar Conf.*, Los Angeles, March 12–13, 1991, pp. 103–107.

Theil, A. (1989), "Angle of Arrival Estimation with an Eight-Element Linear Array Operating at 10.5 GHz," *Intl. Conf. on Radar,* April 24–28, 1989, Paris, pp. 535*–540*.
Theil, A. (1990), "On Combining Adaptive Nullsteering with High Resolution Angle Estimation under Main Lobe Interference Conditions," *Proc. IEEE Intl. Radar Conf.,* Washington, DC, May 1990, pp. 295–297.
Tseng, F.I. (1980), "Synthesis of Arrays for Suppressing a Wideband Jammer," *IEEE Trans. on Electromagnetic Compatibility,* Vol. EMC-22, No. 2, May 1980, pp. 107–112.
Turner, R.M. (1977), "Null Placement and Antenna Pattern Synthesis by Control of the Element Steering Phases of a Phased-Array Radar," *Proc. Intl. Radar Conf.,* Radar '77, London, October 25–28, 1977, pp. 222–225.
Turner, R.M., Poirier, A.L., and Wilkins, P.J. (1978), "A Steering-Phase Control Architecture and its Use for Null Steering in a Phased-Array Radar," *Proc. Intl. Conf. on Radar,* Paris, December 4–8, 1978, pp. 441–449.
Valentino, P. (1984), "Digital Beamforming: New Technology for Tomorrow's Radars," *Defense Electronics,* October 1984, pp. 102–107.
Van Brunt, L.B. (1978), *Applied ECM,* Vol. 1, EW Engineering, Inc., Dunn Loring, VA, 1978.
Van Brunt, L.B. (1983), "Radar Antenna System ECCM," *Journal of Electronic Defense,* Vol. 6, No. 8, August 1983, pp. 35–41.
Van Veen, B.D., and Buckley, K.M. (1988), "Beamforming: A Versatile Approach to Spatial Filtering," *IEEE Acoustics, Speech, and Signal Processing Magazine,* Vol. 5, No. 2, April 1988, pp. 4–24.
Van Veen, B.D. (1989), "Systolic Preprocessors for Linearly Constrained Beamforming," *IEEE Trans. on Acoustics, Speech, and Signal Processing,* Vol. ASSP-37, No. 4, April 1989, pp. 600–604.
Viberg, M. (1989), "Sensor Array Processing Using Gated Signals," *IEEE Trans. on Acoustics, Speech and Signal Processing,* Vol. ASSP-37, No. 3, March 1989, pp. 447–450.
Voles, R. (1978), "Simple Null-Steering Antenna for Frequency-Agile Radars," *Proc. IEE,* Vol. 125, July 1978, pp. 623–625.
Vural, A.M. (1979), "Effects of Perturbations on the Performance of Optimum/Adaptive Arrays," *IEEE Trans. on Aerospace and Electronic Systems,* Vol. AES-15, No. 1, January 1979, pp. 76–87.
Vural, P.M., and Stark, M.T. (1980), "A Summary and the Present Status of Adaptive Array Processing Techniques," *Proc. of the 19th IEEE Conf. on Decision and Control,* Albuquerque, NM, December 10–12, 1980, pp. 931–938.
Walach, E. (1984), "On Superresolution Effects in Maximum-Likelihood Adaptive Arrays," IEEE Trans on Antennas and Propagation, Vol. AP-32, No. 3, March 1984, pp. 259–263.
Wang, H., and Kaveh, M. (1984), "Estimation of Angles of Arrival for Wideband Sources," *Proc. of Intl. Conf. on Acoustics, Speech, and Signal Processing, ICASSP '84,* pp. 7.5.1–7.5.4.
Ward, C.R., Robson, A.J., Hargrave, P.J., and McWirther, J.G., (1984), "Application of a Systolic Array to Adaptive Beamforming," *Proc. IEE,* Vol. 131, Pt. F, 1984, pp. 638–645.
Ward, C.R., and Hargrave, P.J. (1985), "A Systolic Array for High Performance Adaptive Beamforming," *AGARD Conf. Preprint No. 380,* Lisbon, Portugal, 1985.
Ward, C.R., Hargrave, P.J., and McWirther, J.G. (1986), "A Novel Algorithm and Architecture for Adaptive Digital Beamforming," *IEEE Trans. on Antennas and Propagation,* Vol. AP-34, No. 3, March 1986, pp. 338–346.
Ward, C.R. and Hargrave, P.J. (1990), "Integrated Digital Adaptive Antennas," *IEE Colloquium on Adaptive Antennas* (Digest No. 098), London June 8, 1990.
Wardrop, B., and Gould, D.M. (1982), "A Digital Sidelobe Canceler for a Linear Phased-Array," *Proc. Military Microwave Conf.,* 1982, London, pp. 575–579.
Wardrop, B. (1983), "Experimental Linear Phased Array with Partial Adaptivity," *IEE Proc.* Vol. 130, Pt. F, No. 1, February 1983, pp. 118–124.
Wardrop, B. (1985), "The Role of Digital Processing in Radar Beamforming," *GEC Journal of Research,* Vol. 3, No. 1, 1985, pp. 34–45.

Watanabe, H., and Nishimoto, S. (1984), "Correlated Interference Rejection Characteristics of Adaptive Antenna Using Spatial Learning," *Proc. of the 1984 Intl. Symposium on Noise and Clutter Rejection in Radars and Imaging Sensors,* Tokyo, Japan, 1984, pp. 585–590.

Webb, J.K. (1983), "Electronic Steering of Antenna Nulls for Interference Reduction," *IEE Proc.,* Vol. 130, Pt. F, No. 5, August 1983, pp. 417–422.

White, W.D. (1978a), "Artificial Noise in Adaptive Arrays," *IEEE Trans. on Aerospace and Electronic Systems,* Vol. AES-14, No. 2, March 1978, pp. 380–384.

White, W.D. (1978b), "Adaptive Cascade Networks for Deep Nulling," *IEEE Trans. on Antennas and Propagation,* Vol. AP-26, No. 3, May 1978, pp. 396–402.

White, W.D. (1978c), "Wideband Cascade Network for Null Steering Antenna," *Proc. Military Microwave Conf.,* London, 1978, pp. 357–365.

White, W.D. (1979a), "Angular Spectra in Radar Applications," *IEEE Trans. on Aerospace and Electronic Systems,* Vol. AES-15, No. 6, November 1979, pp. 895–899.

White, W.D. (1979b), "Author Reply," *IEEE Trans. on Aerospace and Electronic Systems,* Vol. AES-15, No. 6, November 1979, p. 904.

White, W.D. (1983), "Wideband Interference Cancellation in Adaptive Sidelobe Cancelers," *IEEE Trans. on Aerospace and Electronic Systems,* Vol. AES-19, No. 6, November 1983, pp. 915–925.

White, W.D. (1987), "A Multiple-Beam Architecture for Sidelobe Cancelers," *IEEE Trans. on Aerospace and Electronic Systems,* Vol. AES-23, No. 5, September 1987, pp. 612–619.

Widrow, B. (1966), "Adaptive Filters 1: Fundamental," Systems Theory Lab., Stanford University, Stanford, CA, Tech. Rep. 6764-6, December 1966.

Widrow, B., et al. (1967), "Adaptive Antenna Systems, *Proc. IEEE,* Vol. 55, 1967, pp. 2143–2159.

Widrow, B., et al. (1975), "Adaptive Noise Cancelling: Principles and Applications," *Proc. IEEE,* Vol. 63, 1975, pp. 1692–1716.

Widrow, B., et al. (1976), "Stationary and Nonstationary Learning Characteristics of the LMS Adaptive Filter," *Proc. IEEE,* Vol. 64, No. 8, August 1976, pp. 1151–1161.

Widrow, B., Duvall, K.M., Gooch, R.P., and Newman, W.C. (1982), "Signal Cancellation Phenomena in Adaptive Antennas: Causes and Cures," *IEEE Trans. on Antennas and Propagation,* Vol. AP-30, 1982, pp. 469–478.

Widrow, B., and Stearns, S.D. (1985), *Adaptive Signal Processing,* Prentice Hall, Englewood Cliffs, NJ, 1985.

Williams, R.T., Prasad, S., Mahalanabis, A.K., and Sibul, L.H. (1988), "An Improved Spatial Smoothing Technique for Bearing Estimation in a Multipath Environment," *IEEE Trans. on Acoustics, Speech, and Signal Processing,* Vol. ASSP-36, No. 4, April 1988, pp. 425–432.

Wirth, W.D. (1976), "Suboptimal Suppression of Directional Noise by a Sensor Array Before Beam Forming," *IEEE Trans. on Antennas and Propagation,* Vol. AP-24, No 5., September 1976, pp. 741–744.

Wirth, W.D. (1977), "Radar Signal Processing with an Active Receiving Array," *Proc. Intl. Radar Conf.,* London 1977, pp. 218–221.

Wirth, W.D. (1989), "Phased-Array Radar," Course Notes, Selenia, Rome, November 7–8, 1989.

Worms, J. (1985), "On a Single Step Technique for Adaptive Array Processing," Agard Conf., Toulouse (France), Conf. Proc. No. 381, on "Multifunction Radar for Airborne Applications," October 1985, pp. 20-1–20-5.

Worms, J., and Krucker, K. (1985), "On an Improved Gradient Technique for Adaptive Array Processing," Proc. of *IEEE Intl. Radar Conf.,* Washington, DC, May 6–9, 1985, pp. 39–44.

Worms, J. (1987), "Adaptive Jammer Suppression Under Constraints," *Proc. Intl. Radar Conf.,* Radar '87, London, October 19–21, 1987, pp. 269–273.

Worms, J. (1989), "Evaluation of the Filter Coefficient Innovation Rate of a Rotating Partially Adaptive Antenna," *Intl. Conf. on Radar,* April 24–28, 1989, Paris, pp. 279–282.

Yang, J.F., and Kaveh, M. (1990), "Coherent Signal—Subspace Transformation Beamformer," *IEE Proc.,* Vol. 137, Pt. F., N. 4, August 1990, pp. 267–275.

Yeh, C.C. (1986), "Projection Approach to Bearing Estimations," *IEEE Trans. on Acoustics, Speech, and Signal Processing,* Vol. ASSP-34, N. 5, October 1986, pp. 1347–1349.

Yeh, C.C. (1987), "Simple Computation of Projection Matrix for Bearing Estimations," *IEE Proc.,* Vol. 134, Pt. F, No. 2, April 1987, pp. 146–150.

Yubao Zheng, and Peng Yingning (1989), "A New Method of ABF for Large Scale Arrays," *Intl. Conf. on Radar,* April 24–28, 1989, Paris, pp. 298–301.

Yuen, S.M. (1989), "Algorithmic, Architectural and Beam Pattern Issues of Sidelobe Cancellation," *IEEE Trans. on Aerospace and Electronic Systems,* Vol. AES-25, No. 4, July 1989, pp. 459–472.

Yuen, S.M. (1991), "Exaxt Least Squares Adaptive Beamforming Using an Orthogonalization Network," IEEE Trans. on Aerospace and Electronic Systems, Vol. AES-27, No. 2, March 1991, pp. 311–330.

Yuen, S.M., Abend, K., and Berkowitz, R.S. (1988), "A Recursive Least Squares Algorithm for Multiple Inputs and Outputs and a Cylindrical Systolic Implementation," IEEE Trans. on Acoustic, Speech, and Signal Processing, Vol. ASSP-36, No. 12, December 1988, pp. 1917–1923.

Zhang Shouhong, *et al.* (1986), "Some Practical Problems in Adaptive Sidelobe Cancellation," *Intl. Radar Conf.,* Nanjing, China, November 4–7, 1986, pp. 610–617.

Zhu, J.X., and Wang, H. (1989), "A Comparison of Smoothing Approaches for Angle Measurement," *IEEE Trans. on Aerospace and Electronic Systems,* Vol. 25, No. 4, July 1989, pp. 529–535.

Zhu, X. and Wang, N. (1990), "Adaptive Beamforming for Correlated Signal and Interference: A Frequency Domain Smoothing Approach," *IEEE Trans. on Acoustics, Speech, and Signal Processing,* Vol. 38, No. 1, January 1990, pp. 193–195.

Ziskind, I., and Wax, M. (1990), "Resolution Enhancement by Periodicity Constraints," *IEEE Trans. on Acoustics, Speech, and Signal Processing,"* Vol. ASSP-38 No. 4, April 1990, pp. 687–694.

Zohar, S. (1983), "The Steady-State Antenna Patterns of Adaptive Arrays," *IEEE Trans. on Aerospace and Electronic Systems,* Vol. AES-19, No. 4, July 1983, pp. 550–560.

Zoltowski, M.D. (1988), "On the Performance Analysis of the MVDR Beamformer in the Presence of Correlated Interference," *IEEE Trans. on Acoustics, Speech, and Signal Processing,* Vol. ASSP-36, No. 6, June 1988, pp. 945–947.

Zunich, G.T., and Griffiths, L.J. (1991), "A Robust Method in Adaptive Array Processing for Random Phase Errors," *Proc. of IEEE Intl. Conference on Acoustics, Speech, and Signal Processing,* ICASSP-91, Toronto, Canada, May 14–17, 1991, pp. 1357–1360.

Index

Adapted antenna pattern, 151
Adaptive array, 12
Adaptive MTI, 9
Adaptive weight, 96
Airborne warning and control systems (AWACS), 14
Akaike information criteria (AIC), 300
Amplitude error, 28
Amplitude mismatching model, 187
Analog-to-digital conversion (ADC), 34
Analog implementation, 95
Angular resolution limit, 303
Angular superresolution, 293
Antiradiation missiles (ARM), 7
Antenna feed, 24
Antenna gain, 65
Antenna radiating element, 28
Antenna radiation pattern, 28
Aperture blockage, 23
Aperture efficiency, 21
Aperture tapering function, 19
Array manifold, 220
Array radiating element, 25
Autoregressive moving average (ARMA) stochastic sequence, 299
Autoregressive (AR) stochastic sequence, 133
Automatic frequency selection (AFS), 9
Automatic gain control (AGC), 122
Autocorrelation function, 127
Auxiliary antenna (channel), 59
Average sidelobe level, 25

Backlobes, 25
Backsubstitution method, 147
Baseband (BB), 38

Bessel function, 33
Blanking logic, 63
Blanking threshold, 77
Blanking matrix, 150
Burg method, 299
Butler matrix, 39

Calibration, 36
Cancellation ratio (CR), 101
Cassegrain antenna, 24
Chaff, 2
Channel mismatching, 164
Channel transfer function, 193
Cholesky factorization, 131
Classical beamforming, 217
Closed-loop method, 103
Closed-loop adaptivity, 36
Clutter, 13
Coherent limiter, 105
Coherent sources, 217
Constant false alarm rate (CFAR), 6
Control loop noise, 206
Constrained feed array, 26
Conus OTH-B, 59
Co-polarized pattern, 62
Correlation coefficient, 106
Correlated sources, 217
Covariance matrix, 95
Cramer-Rao lower band, 344
Criterion autoregressive transfer (CAT), 300
Crosscorrelation vector, 100
Cross-polarized pattern, 62
Crosstalk of channels, 166

Data-domain algorithm, 138
Data matrix, 325

Data snapshot, 142
Data rate, 139
dBi, 13
Deception, 2
Decorrelation effects, 181
Degree of freedom, 120
Detection loss, 80
Detection threshold, 65
Deterministic spatial filtering, 219
Diagonal loading technique, 131
Dicke-Fix, 9
Difference antenna pattern, 62
Digital beam forming (DBF), 38
Digital implementation, 127
Digital open-loop, 129
Dipole, 96
Direction of arrival (DOA), 217
Direct matrix inversion (DMI) algorithm, 120
Direct solution method, 120
Discrete Fourier transform (DFT), 39
Dolph-Chebyschev tapering function, 21
Doppler cycles averaging (DCA), 333
Doppler filtering, 27
Duty cycle, 59
Dynamic range, 39

Echelon lobes, 220
Eigenanalysis, 95
Effective radiated power (ERP), 2
Eigenvalue, 110
Eigenvalue spread, 120
Eigenvector, 110
Electronic countermeasures (ECM), 1
Electronic counter-countermeasures (ECCM), 1
Electronic support measure (ESM), 2
Electronic warfare (EW), 1
Element failure, 31
Element pattern diversity, 225
ELRA, 57
Error covariance matrix, 134
Estimation noise, 176
Estimation of signal parameters by rotationally invariance technique (ESPRIT), 315

Fast Fourier transform (FFT), 39
Feedback control loop equation, 115
Fiber optics, 56
Fictional interference, 278
Final prediction error (FPE), 300
Finite impulse response (FIR) filter, 48

Forward-backward averaging (FBAV), 300
Frequency agility, 217
Frequency diversity, 8
Frequency spectrum, 8

Gain margin of auxiliary antenna, 73
Gain vector, 135
Gallium arsenide (GaAs) technology, 56
GASP (GEC array signal processor), 57
Gaussian probability density function, 77
Gaussian tapering function, 40
Generalized sidelobe canceler, 150
Givens rotations, 139
Glint errors, 294
Gradient operation, 110
Gram-Schmidt (GS) algorithm, 120
Gram-Schmidt orthogonalization procedure, 122

Hermitian operator, 95
High duty cycle interference, 96
Horn, 96
Householder matrix, 164
Householder reflections, 139
Howells-Applebaum approach, 104
Hybrid analog-digital technique, 120
Hyperbolic householder transformation, 161

Ill-conditioned matrix, 132
Illumination efficiency, 34
Image rejection, 178
Improvement factor (IF), 227
Inner product of vectors, 157
Intermediate frequency, 36
Isotropic antenna gain, 26

Jammer, 1
Jammer direction of arrival (DOA), 95
Jammer-to-noise power ratio (JNR), 77

Kalman filter, 120
Kayser-Bessel tapering function, 22
Karhunen-Loeve expansion, 239
Kumaresan and Tufts method, 315

Lagrange multiplier, 286
Leaky integrator, 116
Least-mean-square (LMS) algorithm, 112
Left and right singular vectors, 325
Levinson-Durbin algorithm, 132
Levinson recursion, 300
Linear array antenna, 28

Linear estimation, 101
Linear prediction theory, 99
Linear subspace, 219
Linear systolic array, 147
Loop noise, 176
Low-noise amplifier, 36
Low probability of intercept (LPI), 8
Low sidelobe antennas, 13

Main antenna, 95
Mainlobe jammer cancellation, 214
Matrix condition number, 140
Maximum eigenvalue, 157
Minimum eigenvalue, 157
Maximum entropy method (MEM), 299
Maximum likelihood estimate (MLE), 344
Microwave absorber, 25
Microwave monolithic integrated circuit (MMIC), 56
Millimetric wave radar, 9
Minimum description length (MDL), 318
Minimum variance (Capon) method, 221
Misadjustment noise, 96
Mismatch of channels, 36
Modified LMS algorithm, 112
Monochromatic jammer, 153
Monopulse antenna system, 214
Monopulse radar, 61
Monte Carlo simulation, 82
Moving target indication (MTI), 9
Multihypotheses testing problem, 77
Multipath, 36
Multiple beam signal processing, 219
Multiple sidelobe canceler (MSLC), 110
Multiple signal classification (MUSIC), 315
Mutual coupling effect, 55

Narrowband jammer, 168
Near-field adaptive nulling, 215
Nonstationary interference, 136
Noise eigenvalues (eigenvectors), 248
Noise subspace, 325
Noise temperature, 14
Norm of a vector, 157
Numerical milling machine, 24

Offset error, 178
Omnidirectional antenna, 193
One-pole low-pass filter, 116
Optimum beamformer, 269

Optimum detection, 225
Open-loop adaptivity, 36

Parallel processing, 138
Parameter target model fitting (PTMF), 336
Partial adaptivity, 220
Path matching, 165
Phased array, 18
Phase detector, 51
Phase error, 28
Phase mismatching model, 187
Phase-only nulling, 220
Phase shifter, 28
Pisarenko method, 345
Plan position indicator (PPI), 10
Planar array antenna, 277
Polarization, 62
Power domain algorithm, 170
Prediction error, 101
Power spectral density, 295
Projection of a vector, 158
Projection matrix, 248
Principal eigenvalues (eigenvectors), 248
Probability density function (pdf), 28
Probability of detection, 62
Probability of false alarm, 63
Probability of false blanking, 80
Probability of jammer blanking, 79
Pulsed interference, 96

QR decomposition, 139
Quadrature phase detector, 179
Quantization, 31
Quantization effects, 116
Quiescent sidelobe level, 96
Quiescent weight error, 152

Radio frequency (RF), 36
Raised cosine amplitude tapering function, 254
Random error, 25
Random search optimization, 291
Ranked eigenvalues, 263
Rayleigh criterion, 292
Receiver, 9
Receiver operating characteristics (ROC), 228
Reflector antenna, 22
Relative sidelobe level (RSL), 29
Retrodirectivity concept, 245
Ricean probability density function, 77

Rotman lens, 40
Round-off error, 138

Sample matrix inversion (SMI), 339
Search radar, 95
Sidelobe blanking (SLB), 7
Sidelobe canceler (SLC), 7
Signal processing, 9
Signal dispersion, 167
Signal-to-noise power ratio (SNR), 36
Signal subspace, 325
Singular values, 325
Singular value decomposition (SVD), 339
Slotted waveguide array, 26
Space feed array, 26
Space-time processor, 213
Spatial frequency, 219
Spatial smoothing, 220
Spatial spectrum, 217
Spill over, 24
Square-law detector, 77
Square-root-free systolic processor, 150
Square-root kalman filter, 138
Stability condition (criterion), 113
Stacked-beam antenna, 27
State estimate, 134
Steady-state cancellation, 128
Subarray, 36
Sum antenna pattern, 62
Suspended stripline feed, 26
Superresolution, 217
Systematic error, 25
Systolic array processor, 137
Swerling target model, 63

Tapped delay line, 187
Target, 67
Taylor tapering function, 22

Thermal noise, 168
Throughput, 37
Time bandwidth product, 182
Time constant, 115
Time-delay steering, 334
Toeplitz matrix, 214
Track braker, 3
Track breaking mode, 262
Tracking radar, 95
Transient time constant, 105
Transition matrix, 134
Transversal digital equalizer, 204
Transmitter, 2
Triangular systolic array, 344
Triangulation of a matrix, 161

Ultra-low sidelobe antenna (ULSA), 13
Unitary matrix, 145
Upper triangular matrix, 148

Very large scale integration (VLSI), 56
Very high performance integrated circuit (VHPIC), 57
Very high speed integrated circuit (VHSIC), 56
Virtual nulling, 263

Wafer scale integration (WSI), 139
Waveform coding, 2
Weight freezing and flushing, 148
Wideband jammer, 168
Wideband radar, 168
Widrow-hoff algorithm, 112
Wiener filter, 120
Word length, 156

Yule-Walker normal equations, 299

Zero-order gain constraint, 308